ANALYSIS

PART I: ELEMENTS

KRZYSZTOF MAURIN

University of Warsaw

ANALYSIS

PART I

Elements

D. REIDEL PUBLISHING COMPANY

DORDRECHT-HOLLAND / BOSTON-U.S.A.

PWN—POLISH SCIENTIFIC PUBLISHERS

WARSAW

Graphic design: *Zygmunt Ziemka*

Translated from the original Polish
Analiza, część I: Elementy, Warszawa 1973
by *Eugene Lepa*

Library of Congress Catalog Card Number 74-80525
ISBN 90 277 0484 8

Co-publication with Państwowe Wydawnictwo Naukowe, Warszawa, Poland

Distributors for Albania, Bulgaria, Chinese People's Republic, Czechoslovakia,
Cuba, German Democratic Republic, Hungary, Korean People's Democratic Republic,
Mongolia, Poland, Rumania, Democratic Republic of Vietnam, the U.S.S.R. and
Yugoslavia
ARS POLONA—RUCH
Krakowskie Przedmieście 7, 00-068 Warszawa 1, Poland

Distributors for the U.S.A., Canada and Mexico
D. REIDEL PUBLISHING COMPANY, INC.
Lincoln Building, 160 Old Derby Street, Hingham, Mass. 02043, U.S.A.

Distributors for all other countries
D. REIDEL PUBLISHING COMPANY
P.O. Box 17, Dordrecht, Holland

1545972

PREFACE

Each word, just as each name, carries with it a specific content of its own, gives rise to various associations depending on personal experience. Hence, the word "analysis" means something different to each mathematician. For some, it encompasses not much more than "differential and integral calculus", whereas for others it is associated with the Riemann–Roch theorem or with harmonic forms.

The position of analysis in mathematics is a special one, quite different from that of algebra, the theory of numbers, or set theory. It is not an independent discipline: it is based on topology and algebra. It came into being at a relatively late date, and at that "to suit the needs" of mechanics and geometry. The problems which have cropped up within it have led to the formation of such vast, "independent" disciplines as set theory, topology, and functional analysis.

The extremely rapid advances made in mathematics since World War II have resulted in analysis becoming an enormous organism spreading in all directions. Gone for good (surely) are the days of the great French "courses of analysis" which embodied the whole of the "analytical" knowledge of the time in three volumes, as the classical work of Camille Jordan, for instance, did. Perhaps this is why present-day textbooks of analysis are disproportionately modest relative to the present state of the art. More: they have "retreated" back to the practice before Jordan and Goursat.

The present three-volume textbook *Analysis* has grown out of the course "Mathematical Analysis for Physicists" which I have been giving at the Warsaw University for the past ten years. The anachronously short time allocated the mathematician by the physicists (four hours a week over four semesters) and a desire to demonstrate to students that analysis is nevertheless alive compelled the author to prepare a course which would be not typical in form (or content). The upshot of this was a mimeographed text *Lectures on Mathematical Analysis*, prepared in four volumes by myself and my assistants, Jerzy Kijowski and Wiktor Szczyrba. The great demand and interest aroused by this publication, beyond the physics community and not only in Warsaw, induced PWN–Polish Scientific Publishers to propose that the mimeographed notes be published in book form. I eagerly took up the proposal "knowing not what I did": much more is expected from a book than from mimeographed lecture notes.

The present book goes far beyond my course for physics students. In order to present all of this material, a course would have to be taught for four hours a week over three entire academic years.

To the best of my knowledge, this is the only textbook which, starting from scratch—to be precise, from rational numbers—goes as far as the theory of distributions (generalized or ideal functions), direct integrals, analysis on complex manifolds, Kähler spaces, the theory of sheaves and vector bundles, etc.

It was my objective to show the young reader the beauty and richness of the unusual world of modern mathematical analysis.

In the case of more advanced topics, such as the theorem on the Atiyah–Singer index, I obviously could not give the complete proofs.

I know, however, that the young mind absorbs beautiful and difficult things, rejoicing in the fact that the world is great and replete with adventure. Wistfully I recall how, during the Nazi occupation, Edward Marczewski introduced the arcana of analysis to me: in those dark days these were for me bright moments, for which I am infinitely grateful to him.

In addition to presenting complete theories, my principal motive in writing this book was to show the horizons to the reader and to encourage him to undertake an independent journey of discovery through the monographs and special texts. Accordingly, the present book is two-fold in character: textbook and monograph, Part I being the textbook (covering two semesters of my course).

The fact that this undertaking is incomplete and lacks some finishing touches is plain to me. However, on the one hand the needs of students (and reminders from the publishers) and, on the other hand, the profound observation of my teacher, Professor Kuratowski, "It is most difficult to stop writing a book", (one could keep on endlessly amending and adding to a book on analysis) induced me to turn the manuscript over to the publishers.

In conclusion, I should like to thank all those persons without whose kind assistance this book could not have come into being; first of all, my thanks go to Jerzy Kijowski and Wiktor Szczyrba, whose elaboration of my lecture notes constituted the backbone of this book. Thanks are also due to Stanisław Woronowicz for writing the section on the introduction of connections by means of jets, Jacek Komorowski for the section on the Lie derivative, and Wiktor Szczyrba for the sections on the Frobenius–Dieudonné theorem. A teacher could scarcely have greater satisfaction than that of being able to learn from his pupils.

This book is dedicated to my students, the students of the Physics Department of the University of Warsaw (previous generations and the present one); I have always been aware that to many of them it was sheer torture to listen to my lectures. This dedication is by way of modest compensation and an expression of gratitude for the lively reaction, freshness and understanding with which they accepted my efforts.

KRZYSZTOF MAURIN

TABLE OF CONTENTS

TABLE OF CONTENTS

TABLE OF CONTENTS

I. SETS
   RELATIONS
   MAPPINGS
   FAMILIES
   REAL NUMBERS

The language of everyday life is inadequate for the needs of science. Each science has its own language. The language of mathematics is perhaps the richest and the purest of the languages of science. Being as it is a living organism, mathematics continuously reproduces its language, continuously enriching and perfecting it. *Some physicists regard mathematics as the language of physics.*

This chapter serves to introduce the notation, but is on no account intended to replace an exposition of set theory. We do not, of course, define the concept of set, the concept of "being an element of a set", etc. Since most concepts of mathematical analysis are definitions of aggregations of objects having a particular property, the notation $\{x: \mathscr{P}\}$ has been introduced, and this is read as "the set of (all) x which possess the property $\mathscr{P}$".

1. LOGICAL NOTATION. DE MORGAN'S LAWS

We shall use the following symbols:

$\wedge$ is read as: and,

$\bigwedge\limits_{x}$ is read as: for all x there follows,

$\vee$ is read as: or (is not exclusive),

$\bigvee\limits_{x}$ is read as: there exists an x such that,

$\neg$ is read as: not,

$\Rightarrow$ is read as: if, ..., then,

$\Leftrightarrow$ is read as: if and only if (implication in both directions),

$:=$ denotes: equal by definition.

Let us write down the *law of contraposition*, the basis of indirect proof (proof by contradiction, *reductio ad absurdum*):

$$(p \Rightarrow q) \Leftrightarrow (\neg q \Rightarrow \neg p).$$

De Morgan's laws, the so-called *rule of negation*, can now be written as

$$\neg(p \wedge q) \Leftrightarrow (\neg p \vee \neg q), \quad \neg(p \vee q) \Leftrightarrow (\neg p \wedge \neg q).$$

Negation thus changes the sign $\wedge$ into $\vee$ and vice versa. It operates in a similar manner on quantifiers:

$$\neg \bigwedge_x B \Leftrightarrow \bigvee_x \neg B, \quad \neg \bigvee_x B \Leftrightarrow \bigwedge_x \neg B,$$

$$\neg \bigvee_x \neg B \Leftrightarrow \bigwedge_x B, \quad \neg \bigwedge_x \neg B \Leftrightarrow \bigvee_x B.$$

2. THE ALGEBRA OF SETS

Now we shall introduce the principal symbols of the algebra of sets and we shall avail ourselves of the opportunity to get some practice in the use of logical notation. As a rule, the convention adopted here is that sets are designated by upper-case letters, and their elements by the corresponding lower-case letters: $a \in A$ is read as "a belongs to (the set) A" or "a is an element of A".

1. Intersection of Sets. The *intersection* or *common part* of sets A and B is denoted by $A \cap B$ and defined as follows: in logical notation

$$(x \in A \cap B) \Leftrightarrow (x \in A) \wedge (x \in B)$$

or in set-theoretical notation

$$A \cap B = \{x : (x \in A) \wedge (x \in B)\}.$$

2. Union of Sets. The *union* of sets A and B is denoted by $A \cup B$. It is a set of the form

$$A \cup B = \{x : (x \in A) \vee (x \in B)\}.$$

The reader will note that the symbols $\cap$, $\wedge$, $\bigwedge$ and $\cup$, $\vee$, $\bigvee$ correspond to each other, hence the choice of symbols of similar shape.

The following simple computational rules hold for unions and intersections:

$$\left. \begin{array}{l} A \cap A = A \\ A \cup A = A \end{array} \right\} \text{ idempotency;}$$

$$\left. \begin{array}{l} (A \cap B) \cap C = A \cap (B \cap C) \\ (A \cup B) \cup C = A \cup (B \cup C) \end{array} \right\} \text{ associativity;}$$

$$\left. \begin{array}{l} A \cap B = B \cap A \\ A \cup B = B \cup A \end{array} \right\} \text{ commutativity;}$$

$$\left. \begin{array}{l} A \cap (A \cup B) = A \\ A \cup (A \cap B) = A \end{array} \right\} \text{ adjunction;}$$

$$\left. \begin{array}{l} A \cap (B \cup C) = (A \cap B) \cup (A \cap C) \\ A \cup (B \cap C) = (A \cup B) \cap (A \cup C) \end{array} \right\} \text{ distributivity.}$$

4

The reader should illustrate these rules with drawings.

Intersections and unions can be determined for any number of sets: let I be an arbitrary set of indices; to each element $i \in I$ we assign a set A_i. The *intersection* $\bigcap_{i \in I} A_i$ of sets A_i, $i \in I$, is taken to mean the set of those objects which belong to all A_i:

$$x \in \bigcap_{i \in I} A_i \Leftrightarrow \bigwedge_{i \in I} x \in A_i;$$

similarly

$$x \in \bigcup_{i \in I} A_i \Leftrightarrow \bigvee_{i \in I} x \in A_i.$$

Thus, in order to operate on sets without any additional reservations it is useful to employ the concept of intersection even when A and B do not have any element in common. To this end, the concept of *null set*, which is denoted by $\emptyset$, is introduced. It could be defined as follows: $\emptyset := \{x: x \neq x\}$, i.e. by means of a condition which cannot be satisfied. There is only one null (empty, void) set; this "profound" statement is justified "philosophically" in that all impossible conditions are equivalent. Sets A and B whose intersection is null are said to be *disjoint*: $A \cap B = \emptyset$.

Now let us go on to a "ticklish" topic:

3. *Partition into Classes.* When the sets A_i are not null for $i \in I$, and are mutually disjoint, $A_i \cap A_k = \emptyset$ for $i \neq k$, and when $B = \bigcup_{i \in I} A_i$, the "family" $(A_i)_{i \in I}$ of all A_i is called the *partition of the set B into classes* A_i, $i \in I$. (In the algebra of sets, classes are always disjoint !) The partition $(C_j)_{j \in J}$ is *finer* than $(A_i)_{i \in I}$ (is a *subpartition*) when $\bigwedge_{j \in J} \bigvee_{i \in I} C_j \subset A_i$.

4. *Inclusion.* The set A is *included in B*, i.e. A is a *subset* of B, which is written as $A \subset B$ or $B \supset A$, when

$$x \in A \Rightarrow x \in B.$$

The following simple relations hold:

$A \subset A$ *reflexivity*,

$(A \subset B) \wedge (B \subset A) \Rightarrow (A = B)$ *identity*,

$(A \subset B) \wedge (B \subset C) \Rightarrow (A \subset C)$ *transitivity*.

5

5. Difference. Complements of Sets. We have:

$$A - B := \{x\colon (x \notin B) \wedge (x \in A)\};$$

it is not assumed that $A \subset B$. The subsets of a particular set ("space" or universe) X are often considered, and then the set $X - A$ is denoted—after Bourbaki—by $\complement A$ and is called the *complement* of the set A. In general, however, we shall use the notation $X - A$. Of course

$$A \cap \complement A = \emptyset, \quad A \cup \complement A = X \quad \textit{complementarity.}$$

Now let us go on to more important matters:

3. THE CARTESIAN PRODUCT. RELATIONS. MAPPINGS. FAMILIES OF SETS

The concept of an *ordered pair* (a, b)—"*a* in first place, *b* in second"—seems to be self-evident, but is not easily defined; it should not be confused with the two-element set $\{a, b\}$. To enliven the exposition let us here present the set-theoretical definition given by Kuratowski and Wiener:

$$(*) \qquad (a, b) := \big\{\{a\}, \{a, b\}\big\};$$

a is called the *first*, and *b* the *second coordinate* of the pair (a, b). Two pairs (a, b), (c, d) are equal,

$$((a, b) = (c, d)) \Leftrightarrow ((a = c) \wedge (b = d)).$$

Definition $(*)$ satisfies this condition.

An ordered *triple* can now be defined by recursion:

$$(a, b, c) := ((a, b), c);$$

similarly, an ordered *n-tuple*

$$(x_1, \ldots, x_n) = ((x_1, \ldots, x_{n-1}), x_n).$$

We can now define the extremely important concept of the *Cartesian product* of sets A and B, which is denoted $A \times B$:

$$A \times B = \{(a, b)\colon a \in A \wedge b \in B\}.$$

The product of n sets $A_1 \times A_2 \times \ldots \times A_n$, which is written as $\underset{i=1}{\overset{n}{\times}} A_i$, is defined in similar fashion.

1. Relations. We can now give a (set-theoretical) definition of the concept of relation.

Relation (or *graph*) $\mathscr{R}$ in a set X is the name given the subset of the Cartesian product $X \times X$, whence

$$\mathscr{R} \subset X \times X.$$

For example, given a relation "$<$" in the set of integers Z; this concerns pairs of integers, e.g. $3 < 5$. Generalizing this concept, we adopt the following definition:

DEFINITION. A subset $\mathscr{R}$ of the product $X \times Y$ is called a *relation*,

$$\mathscr{R} \subset X \times Y;$$

x is said *to be in the relation* $\mathscr{R}$ *with* y, when $(x, y) \in \mathscr{R}$. We denote

$$P_X\mathscr{R} = \{x \in X: \bigvee_y (x, y) \in \mathscr{R}\}.^1$$

Let $\mathscr{R} \subset X \times Y$; then the subset of the product $Y \times X$,

$$\mathscr{R}^{-1} := \{(y, x) \in Y \times Y: (x, y) \in \mathscr{R}\},$$

is called the *inverse relation*.

When in addition to the relation $\mathscr{R}$ we have the relation $\mathscr{S} \subset Y \times Z$, the subset of $X \times Z$

$$\mathscr{R} \circ \mathscr{S} := \{(x, z) \in X \times Z: \bigvee_y (x, y) \in \mathscr{R} \wedge (y, z) \in \mathscr{S}\}$$

is called the *composition* $\mathscr{R} \circ \mathscr{S}$ of relations $\mathscr{R}$ and $\mathscr{S}$. When $X = Y = Z$ we can form the relation $\mathscr{R} \circ \mathscr{R}$ which will further on be designated as $\mathscr{R}^2$.

Examples. Although this entire book may be regarded as examples and properties of relations, at this point we give a number of simple relations.

1. An *empty relation* in the set X: i.e. no two elements are in relation, $\mathscr{R} = \emptyset$.

2. Another extreme example is the *total relation* $\mathscr{R} = X \times X$: any two elements (x, y) are in relation.

3. If $X = Y$, then

$$\mathrm{id}_X := \mathrm{id} := \{(x, y): x = y\}, \quad \text{or} \quad \{(x, x): x \in X\}$$

is called an *identity relation* (*equality*). This relation is known as the *diagonal of the square* $X \times X$.

2. Various Types of Relations. In this section we are concerned only with relations $\mathscr{R}$ in the set X, i.e. with subsets $X \times X$. A relation $\mathscr{R}$ is:

---

[1] In the next we shall write very often $\{x \in X: \mathscr{P}\}$ instead of $\{x: (x \in X) \wedge \mathscr{P}\}$.

I. *reflexive*, when $\bigwedge_{x} (x, x) \in \mathscr{R}$, or $\mathrm{id}_X \subset \mathscr{R}$;

II. *symmetric*, when $\mathscr{R} \subset \mathscr{R}^{-1}$, that is when $\bigwedge_{x} \bigwedge_{y} (x, y) \in \mathscr{R}$ $\Rightarrow (y, x) \in \mathscr{R}$;

III. *transitive*: $\mathscr{R} \circ \mathscr{R} \subset \mathscr{R}$, that is $\bigwedge_{x,y,z} ((x, y) \in \mathscr{R} \wedge (y, z) \in \mathscr{R})$ $\Rightarrow ((x, z) \in \mathscr{R})$;

IV. *identitive*: $\mathscr{R} \cap \mathscr{R}^{-1} \subset \mathrm{id}$, that is $\bigwedge_{x} \bigwedge_{y} ((x, y) \in \mathscr{R} \wedge (y, x) \in \mathscr{R})$ $\Rightarrow (x = y)$;

V. *connected*: $\mathscr{R} \cup \mathscr{R}^{-1} = X \times X$;

VI. *asymmetric*: $\mathscr{R} \subset \complement\mathscr{R}^{-1}$, that is $\bigwedge_{x} \bigwedge_{y} ((x, y) \in \mathscr{R}) \Rightarrow ((y, x)$ $\notin \mathscr{R})$;

VII. *right-unique*: $\mathscr{R}^{-1} \circ \mathscr{R} \subset \mathrm{id}$: $\bigwedge_{x,y,z} [((x, y) \in \mathscr{R} \wedge (x, z) \in \mathscr{R}) \Rightarrow$ $(y = z)]$; [1]

VIII. *left-unique*: $\mathscr{R} \circ \mathscr{R}^{-1} \subset \mathrm{id}$: $\bigwedge_{x,y,z} [((x, y) \in \mathscr{R} \wedge (z, y) \in \mathscr{R}) \Rightarrow$ $(x = z)]$;

IX. *mutually unique*: is right- and left-unique.

In the case of real numbers (they will be defined rigorously somewhat further on) the foregoing types of relations may be visualized geometrically; accordingly, the reflexivity property means that $\mathscr{R}$ contains a diagonal, the symmetry property—that the set $\mathscr{R}$ is symmetric with respect to the diagonal, and the right-uniqueness property—that points lying one above the other do not appear in the graph. The aforementioned terminology will henceforth be used on many occasions. Let us now go on to mappings.

3. Mappings. A *mapping (map)* T of a set X into a set Y is what we call a *right-unique relation* in $X \times Y$, i.e. $T \subset X \times Y$ and

$$\bigwedge_{x,y} ((x, y) \in T \wedge (x, z) \in T) \Rightarrow (y = z),$$

satisfying the additional condition $P_X T = X$. In this case, an element y can be assigned uniquely to each element $x \in X$. Instead of $(x, y) \in T$ we write $y = T(x)$ or $Tx = y$, and read $T: X \to Y$ as: T *maps* X *into* Y.

---

[1] We shall deal with these relations in detail in a moment—these are so-called mappings.

The set X is called the *domain* (*set of arguments*), and $T(X) := \{y \in Y: y = T(x), x \in X\}$[1] is called the *range* (*set of values*) of the map T.

It is seen from the definition itself that two maps $T_1, T_2, T_i \subset X_i \times Y_i$, $i = 1, 2$, are identical when

1) $X_1 = X_2,$

2) $Y_1 = Y_2,$

3) $\bigwedge\limits_{x \in X_1} T_1(x) = T_2(x).$

Remark. Formerly, in many courses, only condition 3) was taken into account, thus causing misunderstanding.

The set $T(A) := \{T(x): x \in A \subset X\}$ is called the *image of a set $A \subset X$*.

For any $B \subset Y$ the set $T^{-1}(B) = \{x: T(x) \in B\}$ is somewhat unfortunately referred to as the *inverse image* of the set B. Plainly the inverse image is a null set when B does not intersect the range of the mapping T, i.e. $B \cap T(X) = \emptyset$.

An *injection* (*embedding*) is the name given one-to-one mapping, i.e. the one-to-one relation:

$$\bigwedge\limits_{x \in X_1} (T(x) = T(x_1)) \Rightarrow (x = x_1).$$

A *surjection* (*surjective map*) or *mapping onto Y* is a mapping T such that we have $T(X) = Y$.

A *bijection* (*bijective map*) is a mapping which is both a surjection and injection; there then exists the inverse mapping $T^{-1}: Y \to X$. Note that $(T^{-1})^{-1} = T$.

Remark. The relation $\mathscr{R}^{-1}$ inverse to the mapping $\mathscr{R}$ is in general only a relation.

A *restriction* of the mapping $T: X \to Y$ is the name applied to any relation $T_1 \subset T$, i.e. this is a map of a subset $X_1 \subset X$; to be more precise: $X_1 = P_X T_1, T_1: P_X T_1 \to Y$, where $\bigwedge\limits_{x_1 \in X_1} T_1(x_1) = T(x_1)$. We then speak of restricting the map T to the (sub)set X_1 and we write $T_1 = T|X_1$. In this case T is also said to be an *extension* of T_1 to the set X. Similarly, when $W \supset Y$, the map $S: X \to W$ such that $S(x) = T(x)$ for every $x \in X$, is a new map; $S \neq T$.

Remark. Previously, the restriction of a map was often identified with the map itself, i.e. the domain and range were not specified exactly.

---

[1] In the next we shall write very often $\{y \in Y: \mathscr{P}, \mathscr{Q}\}$ instead of $\{y \in Y: \mathscr{P} \wedge \mathscr{Q}.$

It may happen that the relation T^{-1} inverse to the map T is right-unique; it is then called the *inverse* of T; of course T^{-1} has the set $T(X) = P_Y T^{-1}$ as its domain. In the case of surjection $T: X \to Y$, $T^{-1}: Y \to X$ as was mentioned above.

The *composition* $S \circ T$ *of maps* is a special case of the composition of relations. Let $T: X \to Y$, $S: Y \to W$; then the composition $S \circ T: X \to W$ is defined by the equality

$$\bigwedge_x (S \circ T)(x) = S(T(x)).$$

Plainly the composition is associative: $(S \circ T) \circ Q = S \circ (T \circ Q)$. When S and T are bijections, $S \circ T$ is a bijection and the inverse map $(S \circ T)^{-1}: W \to X$ may be formed; we then have the equality

$$(S \circ T)^{-1} = T^{-1} \circ S^{-1}.$$

Examples. 1. A *constant mapping* $T: X \to Y$ is one such that $T(X) = \{y_0\}$ is a one-point set: $\bigwedge_x T(x) = y_0$.

2. An *identity* $1 := 1_X := \mathrm{id}_X$ is a mapping $1: X \to X$ such that $\bigwedge_x 1(x) = x$.

3. *Canonical embedding* (*injection*). Let $M \subset X$; the map $j: M \to X$, $j(m): = m$ for all $m \in M$.

It is thus seen that the composition $T \circ j$ is a restriction:

$$T \circ j = T|M;$$

plainly, each restriction is a composition of this kind. Similarly, the range of the mapping $T: X \to Y$, where $W \supset Y$, may be extended by the composition $k \circ T: X \to W$, where $k: Y \to W, k(y) = y, y \in Y$.

4. *Family.* The term "family" of sets $(A_i)_{i \in I}$ was used imprecisely above. This is nothing else than a mapping: each element i of the set I is assigned an object A_i, e.g. a set; hence, this is the mapping $I \ni i \to A_i$. As we see this term is a synonym for "mapping". When to use it? When we are interested more in the range $\{A_i\}$ than in the mapping. This is expressed in a different notation: Instead of $I \ni i \to A(i)$, we write the arguments i as indices, A_i, and the set of arguments (domain) I is called the *set of indices* ("numbers") $A_i = A(i)$; instead of $i \to A(i)$ we accordingly write $(A_i)_{i \in I}$ or, for the convenience of the typesetter, (A_i), $i \in I$.

5. Sequences. Sequences are special families, that is mappings $N \ni n \to a_n$, where $N = I$ denotes the set of natural numbers. When $a_n \in A$, we say (imprecisely) that we have a *sequence of elements of the set* (*space*) A; more precisely, it should be said: *the set of values* $\{a_n : n \in N\}$ *belongs to A*. When $I = N \times N$, we speak of *double sequences*: instead of $x((n,m))$, we usually write x_{nm} and denote the sequence itself by (x_{nm}) or, more precisely, by $(x_{nm})_{n \in N, \, m \in N}$ or (x_{nm}) $(n, m \in N)$.

Remark. A sequence (a_n) should not be confused with its range $\{a_n : n \in N\}$, for these are different concepts; the sequence is a mapping, whereas $\{a_n : n \in N\}$ is simply a set. Failure to distinguish between these concepts frequently leads to misunderstandings and errors and hence our considerable attention to something so "self-evident".

4. EQUIVALENCE RELATIONS. QUOTIENT SPACES AND STRUCTURES

A partition of a set X into classes, $X = \bigcup_{i \in I} A_i$, defines a relation $\mathscr{R} \subset X \times X$, viz. $(x, y) \in \mathscr{R}$, when $\bigvee x, y \in A_i$, i.e. when the elements x, y belong to the same class A_i.[1] This relation is (i) reflexive, (ii) symmetric, (iii) transitive. A relation with these three properties is called an *equivalence* or *equivalence relation*. We shall demonstrate that $\mathscr{R}$ possesses properties (i) to (iii).

(i) — obvious, because $(x, x) \in \mathscr{R}$ means that x belongs to a certain A_i, which is the case since $X = \bigcup_{i \in I} A_i$.

(ii) — also obvious, for if $x, y \in A_i$, then $y, x \in A_i$.

(iii) — somewhat more difficult: $((x, y) \in \mathscr{R}) \wedge (y, z) \in \mathscr{R})$ means that $\bigvee_j \bigvee_i (x, y \in A_i) \wedge (y, z \in A_j)$, whence $y \in A_i \cap A_j \neq \emptyset$; therefore, since "the classes are disjoint", we have $i = j$, that is, $x, y, z \in A_i$, or $(x, z) \in \mathscr{R}$, or $\mathscr{R} \circ \mathscr{R} \subset \mathscr{R}$. $\square$[2]

However, there is the converse

THEOREM I.4.1. Any equivalence relation in X defines a partition of X into classes. Each class is of the form $A_x = \{y \in X : (x, y) \in \mathscr{R}\}$.

---

[1] Usually $(x, y) \in \mathscr{R}$ is written as $x \mathrel{\underset{\sim}{\mathscr{R}}} y$ or $x \sim y$.

[2] The symbol $\square$ indicates that the proof has been completed.

PROOF. If $A_x \cap A_y \neq \emptyset$, there exists a $z \in A_x \cap A_y$, that is $(z \sim x) \wedge \wedge (z \sim y) \Rightarrow x \sim y$, or $A_x = A_y$. Clearly, $x \in A_x$, and hence we have $X = \bigcup_x A_x$. $\square$

The set of classes $\{A_x : x \in X\}$, is called a *quotient space* and is denoted by $X/\mathscr{R}$; the mapping $\varphi : X \to X/\mathscr{R}$, where $\varphi(x) := A_x$, is called a *canonical mapping*. The class A_x will be denoted by $[x]$ or, to be more precise, $[x]_{\mathscr{R}}$.

Associated with the partition into classes is one of the most interesting properties of the human mind: the abstractive faculty, consisting in differences between members of the same class being ignored, and classes being treated as new individuals ("individual differences are abstracted or prescinded"). At first sight it would seem that abstraction impoverishes the world of concepts, but the very opposite is true: most concepts (e.g. mathematical) have arisen, and indeed do now arise, by going over to the quotient space $X/\mathscr{R}$ with respect to the equivalence relation $\mathscr{R}$. The finest partition into classes, that is to say, the finest equivalence relation is 1, identity; it partitions the set X into classes of one element: $((x, y) \in 1) \Leftrightarrow (x = y)$; $A_x = \{x\}$, $x \in X$. Division by this relation does not introduce anything new: each element is a class in itself. At the other extreme, of course, is the total relation $T = X \times X$. The set X is the only class of equivalence; here, we prescind all differences. The examples drawn from "life" (outside mathematics) are legion: on relevant sets, relations such as equally heavy, equally long, of equal age, of the same colour, and equally warm, lead to such "abstract" concepts as weight, length, age, colour, temperature, etc.

Elementary mathematics in this way arrives at such concepts as congruence (e.g. of triangles), similarity, parallelism, equipotence (cardinality), isomorphism; we shall come back to this latter concept repeatedly. Similarly, directions (in a plane) are classes of parallel straight lines; geometric vectors are classes of segments of equal length and identical orientation; rational numbers are classes of fractions $\frac{p}{q}$, where $p, q \in Z$, $q \neq 0$, the pair (p, q) being equivalent to the pair (p_1, q_1) when $pq_1 = qp_1$, that is, in $Z \times Z$ we have a subset $\mathscr{R}$ specified by that definition.

As will be seen in the next section, the set of real numbers, R, is also a quotient space $\mathscr{C}/\sim$ of the set of Cauchy sequences of the rational numbers divided by the appropriate equivalence relation. It is thus evident that the simplest concepts of arithmetic—natural, rational, real numbers, and so on—come into being through various processes of abstraction. Now consider an important example of an equivalence relation (quotient space) which appears time and again in mathematics:

1. Fibration (Stratification). Let $F: X \to Y$; then, depending on the context, $F^{-1}(y_0) = \{x \in X: F(x) = y_0\}$ is called the *stratum, fibre* or *level surface of* y_0 with respect to (the mapping) F. Plainly, the set of fibres $\{F^{-1}(y)\}, y \in F(X)$, is the decomposition of the set X into classes:

$$(*) \qquad X = \bigcup_{y \in F(Y)} F^{-1}(y),$$

for when $y_1 \neq y_2$, then by the definition of the mapping $F^{-1}(y_1) \cap F^{-1}(y_2) = \emptyset$. Since F is defined throughout X, equality $(*)$ holds. What form does the equivalence $\mathscr{R} = \mathscr{R}_F$ defined by F take? The answer is $(x_1, x_2) \in \mathscr{R}$, when $\bigvee_{y_1} F(x_1) = F(x_2) = y_1$. Thus, the mapping F, i.e. the triple (X, Y, F), defines an equivalence relation in X, or the space $X/\mathscr{R}_F$. Sometimes $X/\mathscr{R}_F$ is abbreviated to X/F. Constructs of this type will be considered continually in the course of our exposition, and a chapter will be devoted to so-called fibre bundles in Part III.

Examples. 1. The mapping $F: X \to Y$ is injective; then the relation $\mathscr{R}_F$ is an identity: the classes each have one element.

2. A constant mapping leads to a total relation.

2. Quotient Structures. An ordinary space X, on which an equivalence relation is defined, supports (by the nature of things) some sort of structure, e.g. algebraic: it is a group, vector space, ring, etc. It is especially interesting to go over to a quotient space $X/\mathscr{R}$ when $\mathscr{R}$ is *consistent with the structure given on* X; this structure can then be induced, i.e. transferred to $X/\mathscr{R}$. What does this mean?

Let φ be a canonical projection $\varphi: X \to X/\mathscr{R}$, $\varphi(x) = [x]$, where $[x]$ denotes—as usual—the class to which the element x belongs; $(x, y) \in \mathscr{R} \Leftrightarrow \varphi(x) = \varphi(y)$.

Let $F: X \to X'$ and let an equivalence relation $\mathscr{R}'$ be given on X'; a set $\hat{F}^{-1}[x'] \subset X$ can be assigned to each class $[x'] \in X'/\mathscr{R}'$. In other words, we consider the fibration given by the mapping $\hat{F} := \varphi' \circ F: X \to X'/\mathscr{R}'$, where $\hat{F}^{-1}[x'] = F^{-1} \circ \varphi'^{-1}[x'] = \{x: F(x) \in [x']\}$. In this way we obtain a partition of the space X into classes, and hence an equivalence relation in X. Accordingly, the mapping $F\colon X \to X'$ and the equivalence relation on $\mathscr{R}'$ make it possible to define the equivalence relation on X. Let us denote it by $F^*(\mathscr{R}')$ for the time being and call it the *inverse image of the relation* $\mathscr{R}'$, *with respect to F*. Let $\mathscr{R} := F^*(\mathscr{R}')$. Now we have the mapping $H\colon X/\mathscr{R} \to X'/\mathscr{R}'$ of quotient spaces, defined as

$$H([x]) = [F(x)]_{\mathscr{R}'},$$

that is, if $\varphi\colon X \to X/\mathscr{R}$ is a canonical mapping, then

(*) $\quad H \circ \varphi = \varphi' \circ F,$

which can be visualized by the diagram

$$\text{(D)} \qquad \begin{array}{ccc} X & \overset{F}{\longrightarrow} & X' \\ {\scriptstyle\varphi}\downarrow & & \downarrow{\scriptstyle\varphi'} \\ X/\mathscr{R} & \underset{H}{\longrightarrow} & X'/\mathscr{R}' \end{array}$$

A diagram of this kind, i.e. satisfying (*), is said to be *commutative*.

This situation can be generalized as follows: let an equivalence relation be prescribed on X and let $\varphi\colon X \to X/\mathscr{R}$ be the canonical mapping. Similarly, on X' we are given an equivalence relation $\mathscr{R}'$, and hence also the canonical mapping $\varphi'\colon X' \to X'/\mathscr{R}'$.

DEFINITION. A mapping $F\colon X \to X'$ is said to be *consistent with the relations* (of equivalence) $\mathscr{R}$ and $\mathscr{R}'$ when

$$\big((x, y) \in \mathscr{R}\big) \Rightarrow \big((F(x), F(y)) \in \mathscr{R}'\big),$$

that is

$$(x \overset{\mathscr{R}}{\sim} y) \Rightarrow (F(x) \overset{\mathscr{R}'}{\sim} F(y)).$$

Then there exists a mapping $H\colon X/\mathscr{R} \to X'/\mathscr{R}'$ of quotient spaces such that

$$H \circ \varphi = \varphi' \circ F,$$

i.e. the diagram (D) is commutative. It is sufficient to define H as follows: $H([x]_{\mathscr{R}}) := [F(x)]_{\mathscr{R}'}$.

This definition is valid, for when $x \sim y$, that is, $x, y \in [x]$ then

$$F(x) \sim F(y), \quad \text{or} \quad F(x), F(y) \in [F(x)]_{\mathscr{R}'}.$$

Examples. 1. When $\mathscr{R}'$ is an identity relation, i.e. $\mathscr{R}' = 1_{X'}$, then φ' is an identity mapping and the consistency condition reduces to $x \overset{\mathscr{R}}{\sim} y$ $\Rightarrow F(x) = F(y).$

14

2. Let us consider what it means that F is consistent with 1_X and $\mathscr{R}'$; then φ is an identity: $X = X/\mathscr{R}$, the $\mathscr{R}$-classes are of one element; $(x, y) \in \mathscr{R}$ signifies that $x = y$, whence $(F(x), F(x)) \in \mathscr{R}'$, but any equivalence relation in X' contains a diagonal, $\mathscr{R}' \supset 1_{X'}$. Hence, this case is not interesting in the sense that it always holds and there is no need to speak of consistency.

Let us take another look at the general situation: We have a partition into classes $X = \bigcup_{i \in I} A_i, X' = \bigcup_{j \in J} A'_j$ determined by the relations $\mathscr{R}$ and $\mathscr{R}'$. Denoting $Q_j := F^{-1}(A'_j)$, we obtain a new partition $X = \bigcup_j Q_j = \bigcup_j F^{-1}(A'_j)$, where the sums are taken over all such j that $Q_j \neq \varnothing$; taking a non-empty $Q_j, Q_j \cap Q_k = \varnothing$, when $j \neq k$, we have $A_j \cap A_k = \varnothing$. Thus, we have perhaps a more instructive definition of consistency of the mapping F with the relations (of equivalence) $\mathscr{R}$ and $\mathscr{R}'$: the partition $\{A_i\}$ is finer than (is a *subpartition of*) the partition $\{F^{-1}(A'_j)\}$.

Let us return to the special case: the inverse image of the relation $\mathscr{R}'$ with respect to the mapping $F: X \to X'$.

3. *An Equivalence Relation Induced on a Subset.* Let A' be a subset of the space X', on which an equivalence relation $\mathscr{R}'$ is given. Let $j': A' \to X'$ be a canonical embedding (injection). Then the mapping $\varphi' \circ j': A' \to X'/\mathscr{R}'$ defines on A' an equivalence relation (fibration of the subset A'), the inverse image of the relation $\mathscr{R}'$ with respect to j'. This relation is referred to as the *relation induced by $\mathscr{R}'$* on the subset A' and is denoted by $\mathscr{R}'_{A'}$. When $X' = \bigcup_{k \in K} A'_k$ is the partition corresponding to the relation $\mathscr{R}'$, then $\mathscr{R}'_{A'}$ defines the partition of A' into classes $A'_k \cap A'$.

4. *The Space $\mathscr{F}(X, Y)$.* The set of all mappings $F: X \to Y$ will be denoted by $\mathscr{F}(X, Y)$ or Y^X; the latter notation will be explained further on when general Cartesian products are considered.

When in the set Y we have some operation $\square$ (e.g. Y has an algebraic structure, say that of a vector space), this operation (this structure) can be transferred to $\mathscr{F}(X, Y)$, "raised to $\mathscr{F}(X, Y)$" in a natural manner

$$(T_1 \square T_2)(x) := T_1(x) \square T_2(x);$$

accordingly, when Y is a vector space, for example, $\mathscr{F}(X, Y)$ becomes

15

a vector space if the addition (subtraction) of mappings is defined in a natural manner

$$(T_1 \pm T_2)(x) := T_1(x) \pm T_2(x)$$

and the multiplication of mappings by a number

$$(\lambda T)(x) := \lambda T(x), \quad x \in X.$$

Example. $\mathscr{F}(N, Y)$ is a set of (all) sequences with terms belonging to Y; in particular when $Y = Q$—a set of rational numbers, then $\mathscr{F}(N, Q)$ is a space of sequences of rational numbers. This example is taken up in the next section.

5. *Transferring a Structure (Algebraic) to a Quotient Space.* Usually a space X has a certain algebraic structure, e.g. an operation $\square$ is defined (to focus attention, let us assume that X is a vector space and that $\square$ denotes vector addition). Suppose that in X the equivalence relation $\mathscr{R}$ is consistent with the operation $\square$, i.e. for all $x, x_1, y, y_1 \in X$

(1) $\qquad ((x, x_1) \in \mathscr{R} \wedge (y, y_1) \in \mathscr{R}) \Rightarrow ((x \square y, x_1 \square y_1) \in \mathscr{R}).$

The operation $\square$ can now be transferred to the space $X/\mathscr{R}$ by means of representatives of equivalence classes

(2) $\qquad [x] \square [y] := [x \square y].$

By the implication (1), the definition (2) is valid.

The consistency of every operation in the structure with the relation must, of course, be verified.

In this way we obtain a *quotient structure*, e.g. a quotient group, quotient vector space, quotient algebra, etc.

Example. Suppose that X is a commutative group and H is a subgroup of X. Let us define on X the relation $\mathscr{R}$: $(x, x_1) \in \mathscr{R}$, when $(x-x_1) \in H$; we immediately check that this is an equivalence. The class $[x]$ of the element x will be the subset $x+H := \{x+h: h \in H\}$. We verify condition (1)

$$(x-x_1) \in H \wedge (y-y_1) \in H$$
$$\Rightarrow (x+y)-(x_1+y_1) = ((x-x_1)+(y-y_1)) \in H.$$

Thus, a group operation (addition) can be defined by formula (2) on the quotient space $X/\mathscr{R}$

$$[x]+[y] := [x+y] = (x+y)+H.$$

16

The space $X/\mathcal{R}$ obtained is an Abelian group called a *quotient group* and is denoted by X/H, this being called the *quotient of the group X by the subgroup H.* This procedure will be carried out continually in this exposition, most frequently in the case of vector spaces X and vector subspaces H.

This brings us to the end of our general remarks which are intended to serve as a guide and a glossary to the extensive realm of mathematical analysis.

6. The Product of Mappings. Let $T_i\colon X_i \to Y_i, i \in I$, be a family of mappings; to it we can assign the mapping

$$T\colon \underset{i \in I}{\times} X_i \to \underset{i \in I}{\times} Y_i, \quad T((x_i)_{i \in I}) := (T_i(x_i))_{i \in I};$$

the mapping T is denoted by $\underset{i \in I}{\times} T_i$.

In particular, when $I = \{1, 2\}$, we write $T_1 \times T_2$. Remember that $\underset{i \in I}{\times} X_i$ is the set of all families $(x_i)_{i \in I}, x_i \in X_i$. Since $\prod_{i \in I} X_i$ is frequently written instead of $\underset{i \in I}{\times} X_i$, the notation $\prod_{i \in I} T_i$ is used instead of $\underset{i \in I}{\times} T_i$.

Not infrequently a situation occurs wherein $X_i = X$ for all $i \in I$, i.e. we have a family of mappings

$$S_i\colon X \to Y_i, \quad i \in I.$$

Now we can define the mapping

$$S = (S_i)_{i \in I}\colon X \to \underset{i \in I}{\times} Y, \quad S(x) := (S_i(x))_{i \in I}.$$

If the set of indices I is finite, $I = \{1, 2, ..., n\}$, we write

$$(S_1, ..., S_n).$$

Plainly, (S_1, S_2) *is a completely different mapping than* $S_1 \times S_2$.

5. THE KURATOWSKI–ZORN LEMMA. ORDERING RELATIONS

For propriety's sake we shall say a few words about the axiom of choice and the Kuratowski–Zorn lemma. The reader may easily omit this section on first reading, to return to it when he has been somewhat inured to mathematical abstraction with, *nota bene*, an arsenal of models and examples at his disposal.

1. Order (Ordering) Relations

DEFINITION. A relation $\mathcal{R} \subset X \times X$ which is at one and the same time (i) reflexive, (ii) identitive, and (iii) transitive, is called an *ordering relation*; it is usually denoted by $\leqslant$. Thus, the following formulae hold:

(i) $x \leqslant x$,

(ii) $(x \leqslant y$ and $y \leqslant x) \Rightarrow (x = y)$,

(iii) $(x \leqslant y$ and $y \leqslant z) \Rightarrow (x \leqslant z)$.

A pair $(X, \leqslant)$ is called an *order* (*partial order*) or *ordered space* (*ordered set*). When, moreover, $\mathcal{R}$ is connected, $\mathcal{R} \cup \mathcal{R}^{-1} = X \times X$ (total), i.e.

(iv) any two elements x, y are related by $x \leqslant y$ or $y \leqslant x$, the relation (satisfying (i)–(iv)) is called a *linear* (or *total*, *connected*) *order*, and the pair $(X, \leqslant)$, a *chain*.

When $x \leqslant y$, we say that x is not greater than y, but when along with this $x \neq y$, we write $x < y$ and say that y is *greater than* x.

The *maximal element* M of a subset $X_1 \subset X$, where $(X_1, \leqslant)$ is an order, is what we call a point $M \in X_1$ such that no element greater than M exists in X_1.

Example (general). Let X be a family $(A_i)_{i \in I}$ of subsets of the set A; on writing $A_i \leqslant A_j$ instead of $A_i \subset A_j$, we see that $(\{A_i\}, \subset)$ is an order. $((A_i)_{i \in I}, \subset)$ is a chain when

$$\bigwedge_{i, j \in I} (A_i \subset A_j \text{ or } A_j \subset A_i).$$

Usually, in speaking of a chain, one has in mind this very "model", i.e. a chain of subsets.

We now give three axioms which are equivalent to each other.

I. THE KURATOWSKI–ZORN LEMMA. Let $(X, \leqslant)$ be an ordered set such that for every chain $X_0 \subset X$, (i.e. for every linearly ordered set), there exists an $x_0 \in X$ such that $\bigwedge_{x \in X_0} x \leqslant x_0$; then a maximal element exists in X.

A set $(X, \leqslant)$ which satisfies the assumptions made above is said to be *inductively ordered*. Thus, a maximal element exists in each inductively ordered set.

Kuratowski gave this axiom in set-theoretical language, whereas

18

Zorn (independently, but somewhat later) gave the formulation above; accordingly, the term *Zorn's lemma* is usually seen in the literature.

II. The Axiom of Maximality. Let $\mathscr{E}$ be some family of subsets of a set A. Let $\mathscr{C}$ be a chain contained in $\mathscr{E}$. There then exists a maximal chain $\mathscr{M}$ such that $\mathscr{C} \subset \mathscr{M} \subset \mathscr{E}$.

The reader will note that Axioms I and II are equivalent. Usually the maximality axiom is used in the case of a trivial chain consisting of one "link" C ($\mathscr{C} = \{C\}$). This axiom then guarantees that the subset $C \subset A$, $C \in \mathscr{E}$, may be embedded in a maximal chain $\mathscr{M} \subset \mathscr{E}$. The aforementioned axioms may be shown to be equivalent to:

III. The Axiom of Choice (Zermelo). For any family $(X_i)_{i \in I}$ of non-empty sets there exists a mapping, a so-called *selector*, S: $S(i) \in X_i$, $i \in I$.

This means that the mapping S *selects* one element out of each set X_i.

Example. We shall show by the Kuratowski–Zorn Lemma that

Every vector space X has a basis.

Proof. Let $\mathscr{E} = (A_i)_{i \in I}$ be the family of all subsets $A_i \subset X$ such that every finite aggregate of elements of the set A_i is linearly independent. For every chain $\{A_j\}_{j \in J}$ of elements from $\mathscr{E}$ we have $\bigcup_{j \in J} A_j \in \mathscr{E}$; the hypotheses of Kuratowski–Zorn Lemma are thus satisfied. A maximal element $B = \{x_\alpha\}_{\alpha \in A}$ hence exists in $\mathscr{E}$. Accordingly, when $0 \neq x_0 \notin B$, the set $\{x_0\} \cup \{x_\alpha\}$ is already linearly dependent, i.e. there exist a finite subset $\{x_1, ..., x_n\}$ and a set of numbers $\{a_0, ..., a_n\}$ such that

$$a_0 x_0 + a_1 x_1 + ... + a_n x_n = 0, \quad \text{where} \sum_{i=0}^{n} |a_i| > 0;$$

since $x_1, ..., x_n$ are linearly independent, therefore $a_0 \neq 0$. Thus

$$x_0 = \frac{a_1}{a_0} x_1 + ... + \frac{a_n}{a_0} x_n.$$

Corollary. Every set $\{y_\beta\}$ of linearly independent elements of the vector space X can be extended (i.e. complemented) to the basis X.

For the proof it is sufficient to take the family $\mathscr{E}_1 \subset \mathscr{E}$ of such $A_i \in \mathscr{E}$ that $A_i \supset \{y_\beta\}$ and apply Axiom I or II to $\mathscr{E}_1$.

Remark. The axiom of choice is often used unconsciously, e.g. in the definition of the product $\underset{i \in I}{\times} X_i$ the elements of this set are mappings $I \ni i \to x_i \in X_i$, and hence we make use of the possibility of choosing one element x_i from each set X_i.

6. THE CANTOR THEORY OF REAL NUMBERS

Mathematical analysis is based on real numbers; it may even be said that mathematics is scarcely imaginable without the real-number concept: e.g. the (n-dimensional) Euclidean space R^n is the product of n replicas of the space R of real numbers. The whole of differential geometry is based on the space R^n.

Several theories of real numbers exist. The axiomatic introduction of the set R would seem most convenient. Usually, however, a price must be paid for convenience, and the price in this case is the lack of a construct and, more importantly, the axiomatic approach is not very instructive. It is useful and sometimes downright indispensable when some theory is already rich in content and is becoming obscure. On our part, we do not assume any knowledge of real numbers. The theory of Cantor (1845–1918), the brilliant founder of the theory of sets, takes for the starting point the rational number set (field) Q and considers sequences of rational numbers. Among them he distinguishes the space $\mathscr{C}$ of Cauchy sequences and introduces the equivalence relation $\sim$ in it. The set R is, according to Cantor, a quotient space: $R = \mathscr{C}/\sim$. Algebraic operations are immediately carried over from the rational number set Q to the space R. Thus, Cantor's theory may be regarded as an example for Section 4. The Cantor method has the additional advantage that it leads immediately (as Hausdorff noted, cf. Chapter II) to a general procedure for complementing the space $\mathscr{P}$, i.e. embedding the set $\mathscr{P}$ in the set $\tilde{\mathscr{P}}$. Very often entirely new concepts in mathematics are arrived at by complementation. As we shall see, $R = \tilde{Q}$, i.e. the space R is the complement of the set of rational numbers.

These introductory remarks, which are undoubtedly still incomprehensible to the reader, are intended to advertise the method of Cantor and to encourage the reader to make a careful study of this section and the next: for this is the first theory which we present. The Cantor method will be applied on many occasions to objects more complicated than rational numbers.

6. THE CANTOR THEORY OF REAL NUMBERS

Sequences of rational numbers are the starting point. The reader is assumed here to be familiar with algebraic operations on rational numbers and how to operate with the relation < (greater than). The set of rational numbers will be denoted by Q. We recall here only the concept of absolute value and its properties.

DEFINITION. The *absolute value* of a number $a \in Q$ is a rational number

$$|a| := \begin{cases} a, & \text{when } a \geqslant 0, \\ -a, & \text{when } a < 0. \end{cases}$$

The so-called *triangle inequality*

$$|a+b| \leqslant |a|+|b|$$

holds, and implied directly from it are the inequalities

$$|a-c| \leqslant |a-b|+|b-c| \quad \text{(also called the } triangle\ inequality\text{)},$$

$$||a|-|b|| \leqslant |a-b|, \quad a, b, c \in Q.$$

Cauchy Sequences. In the set of sequences of rational numbers, i.e. in the set of mappings $\mathscr{F}(N, Q)$, we distinguish Cauchy sequences.

DEFINITION. A sequence (of rational numbers) $(a_n) \in \mathscr{F}(N, Q)$ is called a *Cauchy sequence* when

$$\bigwedge_{Q \ni \varepsilon > 0} \bigvee_{\mathscr{N} \in N} \bigwedge_{N \ni n,m > \mathscr{N}} |a_n - a_m| < \varepsilon.$$

The set of Cauchy sequences is denoted by the letter $\mathscr{C}$ and the equivalence relation $\sim$ is introduced into the set $\mathscr{C}$:

DEFINITION. $(a_n) \sim (b_n)$, $a_n, b_n \in Q$, when

$$\bigwedge_{Q \ni \varepsilon > 0} \bigvee_{\mathscr{N} \in N} \bigwedge_{n > \mathscr{N}} |a_n - b_n| < \varepsilon.$$

To put it descriptively: "remote terms differ (arbitrarily) little."

Let us verify that $\sim$ is indeed an equivalence:

(i) Reflexivity: $(a_n) \sim (a_n)$ is obvious.

(ii) Symmetry: $((a_n) \sim (b_n)) \Rightarrow ((b_n) \sim (a_n))$ is immediate since $|a_n - b_n| = |b_n - a_n|$.

(iii) Transitivity: $((a_n) \sim (b_n)) \wedge ((b_n) \sim (c_n)) \Rightarrow ((a_n) \sim (c_n))$.

Indeed, the triangle inequality yields

$$|a_n - c_n| \leqslant |a_n - b_n| + |b_n - c_n|.$$

When $|a_n - b_n| < \frac{1}{2}\varepsilon$, $|b_n - c_n| < \frac{1}{2}\varepsilon$ for $n > \mathscr{N}$, then $|a_n - c_n| < \varepsilon$ for $n > \mathscr{N}$. $\square$

Now we can give the following definition:

DEFINITION. The *set of real numbers* R is a quotient space $\mathscr{C}/\sim$. Q is identified with the set of classes of constant sequences. Thus $Q \subset R$ (more precisely, Q has been embedded in R).

Before we go on to define the algebraic and ordering structures in the real-number set, we give

Examples. 1. The sequence (a_n), $a_n := 1/n$, $n \in N$, is a Cauchy sequence since

$$\left| \frac{1}{n} - \frac{1}{m} \right| \leqslant \frac{1}{n} + \frac{1}{m} \leqslant \frac{2}{m} \quad \text{for } m \leqslant n.$$

It is thus sufficient to set $\mathscr{N} > 2/\varepsilon$.

2. Let us take the decimal expansion of $\sqrt{2}$ ("with deficiency and with excess")

$$\begin{aligned}
a_1 &= 1, & b_1 &= 2, \\
a_2 &= 1.4, & b_2 &= 1.5, \\
a_3 &= 1.41, & b_3 &= 1.42, \\
a_4 &= 1.414, & b_4 &= 1.415, \text{ etc.}
\end{aligned}$$

Thus we have $|a_n - b_n| \leqslant 2/10^{n-1}$. Both sequences (a_n), $(b_n) \in \mathscr{C}$; moreover, $(a_n) \sim (b_n)$, that is $[(a_n)] = [(b_n)]$, and hence they represent the same real number, $\sqrt{2}$ of course. Accordingly, $[(a_n)] := \sqrt{2}$.

3. On the other hand, the sequence (a_n), $a_n = n$, $n \in N$, is not a Cauchy sequence, since $|a_n - a_m| = |n - m| \geqslant 1$ for $n \neq m$. The latter sequence is an example of an unbounded sequence and, as is shown by the next lemma, Cauchy sequences are bounded.

LEMMA. I.6.1. If $(a_n) \in \mathscr{C}$, then

$$(*) \qquad \bigvee_{0 < M \in Q} \bigwedge_{n \in N} |a_n| < M.$$

Inequality $(*)$ states that the sequence (a_n) is *bounded*. Lemma I.6.1 can thus be formulated as:

Every Cauchy sequence is bounded.

As will be seen in Chapter II, this lemma holds (*mutatis mutandis*) for Cauchy sequences in any metric space.

PROOF OF LEMMA I.6.1. Let $\varepsilon > 0$ and $\mathscr{N} \in N$ be taken from the definition of Cauchy sequences; then for

$$(**) \qquad M := \max(|a_1|, \ldots, |a_{\mathscr{N}+1}| + \varepsilon)$$

the inequality $(*)$ holds. $\square$

7. OPERATIONS ON REAL NUMBERS. THE LIMIT
OF A SEQUENCE OF REAL NUMBERS

As we know from our general considerations, algebraic operations in the rational-number set Q carry over in the usual manner to sequences of rational numbers, i.e.

$$(a_n)+(b_n) := (a_n+b_n).$$

Thus, in particular, we have certain algebraic operations in the set $\mathscr{C}$ (of Cauchy sequences). Hence, the idea which suggests itself is to introduce algebraic operations through representatives into R, as a quotient space $\mathscr{C}/\sim$; more precisely:

DEFINITION. If $(a_n), (b_n) \in \mathscr{C}$, then

$$[(a_n)] \pm [(b_n)] := [(a_n \pm b_n)], \qquad [(a_n)] \cdot [(b_n)] := [(a_n \cdot b_n)];$$

if $[(b_n)] \neq 0$, then $b_n \neq 0$ for remote n; for them we form a_n/b_n,

$$\frac{[(a_n)]}{[(b_n)]} := \left[\left(\frac{a_n}{b_n}\right)\right].$$

Although this definition is "natural", its validity must be verified; in our case we must check:

(i) Whether the sequence $(a_n \pm b_n)$ is a Cauchy sequence (similarly for the product and the quotient)?

(ii) Whether the class $[(a_n+b_n)]$ does not depend on the representatives, i.e. when $(a'_n) \sim (a_n)$ and $(b'_n) \sim (b_n)$, whether

$$[(a'_n+b'_n)] = [(a_n+b_n)],$$

i.e. whether

$$(a'_n+b'_n) \sim (a_n+b_n)?$$

Ad (i): $|(a_n+b_n)-(a_m+b_m)| \leqslant |a_n-a_m|+|b_n-b_m| < 2\varepsilon$ for $n, m > \mathscr{N}$, since then $|a_n-a_m|, |b_n-b_m| < \varepsilon$.

Ad (ii): $|(a'_n+b'_n)-(a_m+b_m)| \leqslant |a'_n-a_m|+|b'_n-b_m| < 2\varepsilon$, since

$$((a'_n) \sim (a_n)) \wedge ((b'_n) \sim (b_n))$$
$$\Rightarrow ((|a'_n-a_m| < \varepsilon) \wedge (|b'_n-b_m| < \varepsilon) \text{ for } n, m > \mathscr{N}).$$

The reader can prove (i) and (ii) for the product and the quotient by making use of the fact that Cauchy sequences are bounded (i.e. by using Lemma I.6.1).

It is somewhat more difficult to define the inequality of real numbers: a "natural" definition in terms of representatives is not good; as we

23

know, $(a_n) > (b_n)$ means that $a_n > b_n$ for $n = 1, 2, \ldots$ It would be incorrect to conclude from this that $[(a_n)] > [(b_n)]$ since, on setting $a_n = b_n + 1/n$, we have $a_n > b_n$, but $[(a_n)] = [(b_n)]$. Accordingly, we adopt another definition:

DEFINITION. Let $(a_n), (b_n) \in \mathscr{C}$. We say that $[(a_n)] > [(b_n)]$ when

$$\bigvee_{0 < r \in Q} \bigvee_{\mathscr{N} \in N} \bigwedge_{n > \mathscr{N}} b_n < a_n - r.$$

This definition does not suffer from the aforementioned defect and does not depend on the choice of representatives. Let $(b_n') \sim (b_n)$ and $(a_n') \sim (a_n)$. We have

$$r < a_n - b_n \leqslant |a_n - a_n'| + a_n' - b_n' + |b_n' - b_n|,$$

whence $r - 2\varepsilon < a_n' - b_n'$ for $n > \mathscr{N}$. Let $\varepsilon > 0$ be such that $r - 2\varepsilon = r' > 0$; then $b_n' < a_n' - r'$ for $n > \mathscr{N}'$.

COROLLARY I.7.1. $\bigwedge_{R \ni \varepsilon > 0} \bigvee_{Q \ni r > 0} 0 < \frac{1}{2}r < \varepsilon.$

PROOF. Let us assume that $\varepsilon = [(a_n)] > 0$, thus there exists $0 < r \in Q$ such that $0 < a_n - r$ for remote n. Hence $0 < \frac{1}{2}r < a_n - \frac{1}{2}r$ what is nothing but $\frac{1}{2}r < \varepsilon$. $\square$

LEMMA I.7.2. Let $(a_n), (c_n) \in \mathscr{C}$; then

$$\left(\bigvee_{\mathscr{N} \in N} \bigwedge_{n > \mathscr{N}} a_n \leqslant c_n\right) \Rightarrow \left([(a_n)] \leqslant [(c_n)]\right).$$

INDIRECT PROOF. Suppose that $[(a_n)] > [(c_n)]$; thus $\bigvee_{q > 0} \bigvee_{\mathscr{N} \in N} \bigwedge_{n > \mathscr{N}} a_n - q > c_n$, hence a contradiction. $\square$

LEMMA I.7.3. The triangle inequality holds for real numbers (the absolute value of a real number is defined in the same way as for rational numbers)

$$|a - c| \leqslant |a - b| + |b - c|; \quad a, b, c \in R.$$

PROOF. Let $a = [(a_n)], b = [(b_n)], c = [(c_n)], a_n, b_n, c_n \in Q$. From the triangle inequality in Q we have

$$|a_n - c_n| \leqslant |a_n - b_n| + |b_n - c_n|$$

for any n, whence by Lemma I.7.2 and by the definition we have

$$[(|a_n - c_n|)] \leqslant [(|a_n - b_n| + |b_n - c_n|)] = [(|a_n - b_n|)] + [(|b_n - c_n|)]$$
$$= |[(a_n - b_n)]| + |[(b_n - c_n)]|.$$

Thus, using the simple fact that $|[(a_n)]| = [(|a_n|)]$, we get $|[(a_n)] - [(c_n)]|$
$\leqslant |[(a_n - b_n)]| + |[(b_n - c_n)]|$, that is

$$|a - c| \leqslant |a - b| + |b - c|. \quad \square$$

We have embedded the set Q in the space R. It will now be seen that the set Q is dense in R or, in other words, that any real number can be arbitrarily approximated by rational numbers. This is stated by

LEMMA I.7.4. $\bigwedge\limits_{x \in R} \bigwedge\limits_{\varepsilon > 0} \bigvee\limits_{q \in Q} |x - q| < \varepsilon$.

PROOF. Let $x = [(x_n)]$, $x_n \in Q$. Since $(x_n) \in \mathscr{C}$, there thus exists an $\mathscr{N} \in N$ such that $|x_n - x_m| < \frac{1}{2}\varepsilon$ for $n, m > \mathscr{N}$. Let us take the constant sequence $a_n = x_{\mathscr{N}+1}, n = 1, 2, \ldots$; it defines the rational number $q = (a_n)$. By definition and by Lemma I.7.1, however, we have

$$|x - q| = |[(x_n)] - [(a_n)]| = [(|x_n - a_n|)]$$
$$= [(|x_{\mathscr{N}+1} - x_n|)] \leqslant \frac{1}{2}\varepsilon < \varepsilon. \quad \square$$

In the next chapter we shall begin to use the following denotations: $[a, b]$, $]a, b[$, $[a, b[$, $]a, b]$, $]-\infty, a]$, $]-\infty, a[$, $[a, \infty[$, $]a, \infty[$, where $a, b \in R, a < b$. We define only two of them, namely

$$[a, b[:= \{x \in R: a \leqslant x < b\},$$
$$]-\infty, a] := \{x \in R: x \leqslant a\}.$$

In the light of these, the form of the other definitions is clear.

1. The Limit of a Sequence of Real Numbers. At this point we give the central concept of mathematical analysis and topology, the concept of the limit of a sequence. This will be gradually generalized, so that we arrive at the notion of the limit of a generalized sequence and that of a filter (cf. Chapter XIV).

DEFINITION. Let (x_n) be a sequence of real numbers. This sequence is said to have a *limit* x or to *converge to* x and we write $\lim\limits_{n \to \infty} x_n = x$ or $x_n \to x, n \to \infty$, when

$$\bigwedge\limits_{\varepsilon > 0} \bigvee\limits_{\mathscr{N} \in N} \bigwedge\limits_{n \geqslant \mathscr{N}} |x_n - x| < \varepsilon.$$

Clearly, ε may be taken to be rational without restricting the generality of the considerations, as is shown by Corollary I.7.1.

As demonstrated by the example of the decimal expansion of $\sqrt{2}$, a Cauchy sequence of rational numbers may not converge to any rational number but, as is evident from the following lemma, a Cauchy sequence in Q converges to a real number, of which it is a representative.

LEMMA I.7.5. $((x_n) \in \mathscr{C}) \Rightarrow (\lim_{n \to \infty} x_n = [(x_n)])$.

PROOF. Let $x := [(x_n)]$. Then for $\varepsilon > 0$ there exists an $\mathscr{N}$ such that
$$|x - x_n| \leqslant |x - x_{\mathscr{N}+1}| + |x_{\mathscr{N}+1} - x_n| < \tfrac{1}{2}\varepsilon + \tfrac{1}{2}\varepsilon = \varepsilon$$
for $n > \mathscr{N}$

(cf. the proof of Lemma I.7.4). $\square$

2. *The Completeness of the Space* R. Note that a sequence convergent in R is a Cauchy sequence. Indeed, we have
$$|x_n - x_m| \leqslant |x_m - x| + |x - x_n| < 2\varepsilon \quad \text{for } m, n > \mathscr{N}.$$

The converse question comes to mind: is every Cauchy sequence in R convergent? An affirmative answer is provided by the famous

THEOREM I.7.6 (Cauchy). Every Cauchy sequence of real numbers is convergent:
$$((x_n)\text{—Cauchy sequence, } x_n \in R) \Rightarrow (\bigvee_{x \in R} x = \lim_{n \to \infty} x_n).$$

This fact will later be called the *completeness* of the (metric) space R.

PROOF. As we know from Lemma I.7.4, every term x_n can be approximated arbitrarily by rational numbers, i.e.
$$\bigwedge_{x_n} \bigvee_{a_n \in Q} |x_n - a_n| < \frac{1}{n}.$$

First of all, we shall demonstrate that (a_n) is a Cauchy sequence and then that $x := [(a_n)] = \lim_{n \to \infty} x_n$:
$$|a_n - a_m| \leqslant |a_n - x_n| + |x_n - x_m| + |x_m - a_m| \leqslant \frac{1}{n} + \varepsilon + \frac{1}{m}$$

for $n, m > \mathscr{N}$, because (x_n) is a Cauchy sequence. Let us now verify that $x = \lim_{n \to \infty} x_n$: As we know from Lemma I.7.5, $a_n \to x$, whereby
$$|x - x_n| \leqslant |x - a_n| + |a_n - x_n| < \varepsilon_1 + \frac{1}{n}. \quad \square$$

Cauchy's theorem plays an important role in many proofs: to demonstrate that a sequence is convergent, it is sufficient to show the sequence to be a Cauchy sequence, and this is in general much easier to do.

8. THEOREMS ON THE LIMITS OF SEQUENCES

We shall prove that algebraic operations in the space R are continuous; to put it briefly, R will be shown to be a topological field. More precisely, there holds the

THEOREM I.8.1 (On the Continuity of Algebraic Operations).

(i) $((x_n \to x) \wedge (y_n \to y)) \Rightarrow (x_n \pm y_n \to x \pm y),$

(ii) $((x_n \to x) \wedge (y_n \to y)) \Rightarrow (x_n \cdot y_n \to x \cdot y),$

(iii) $((x_n \to x) \wedge (y_n \to y) \wedge (y \neq 0)) \Rightarrow \left(\lim_{n \to \infty}(x_n/y_n) = x/y\right).$

Remark. Owing to the assumption that $y \neq 0$, in (iii) the distant terms $y_n \neq 0$; only for such terms do we form the quotients x_n/y_n.

PROOF. Since all the points are proved in a similar manner, only (ii) will be demonstrated here:

$$|x_n y_n - xy| = |x_n(y_n - y) + y(x_n - x)|$$
$$\leqslant |x_n||y_n - y| + |y||x_n - x| < M(|y_n - y| + |x_n - x|) < \varepsilon$$

for large n, because (x_n), as a Cauchy sequence, is bounded (cf. Lemma I.6.1). $\square$

Before we proceed to further theorems on limits, note that the "behaviour" of a finite number of terms in the sequence has no effect on the limit.

THEOREM I.8.2. (On the Preservation of Inequalities).

$$\left(\lim_{n \to \infty} x_n = x, \ \lim_{n \to \infty} y_n = y; \ \bigvee_{\mathcal{N}} \bigwedge_{n > \mathcal{N}} x_n \leqslant y_n\right) \Rightarrow (x \leqslant y).$$

INDIRECT PROOF. Let $x = y + 2\varepsilon$, $\varepsilon > 0$ and let $\mathcal{N}_1 > \mathcal{N}$ be such that $x_n > x - \varepsilon$ and $y_n < y + \varepsilon$ for $n > \mathcal{N}_1$; thus

$$y_n - x_n < y - x + 2\varepsilon = 0, \quad \text{a contradiction.} \ \square$$

THEOREM I.8.3 (On Three Sequences). Let (x_n), (y_n), (z_n) be sequences; if

(i) $\bigvee_{\mathcal{N}} \bigwedge_{n > \mathcal{N}} x_n \leqslant z_n \leqslant y_n;$

(ii) $\lim x_n = \lim y_n = x$

then the sequence (z_n) converges and $\lim z_n = x$.

PROOF. Let $\varepsilon > 0$. Since $x_n \to x$, there exists a number $\mathcal{N}_1 > 0$ such that $x - \varepsilon < x_n < x + \varepsilon$ for $n > \mathcal{N}_1$.

Since $y_n \to x$, there thus is a number $\mathcal{N}_2$ such that $y_n < x+\varepsilon$ for $n > \mathcal{N}_2$. By hypothesis, however, we have $x_m \leqslant z_m \leqslant y_m$ for $m > \mathcal{N}$, whereby for $n > \max(\mathcal{N}, \mathcal{N}_1, \mathcal{N}_2)$ we at the same time have

$$x - \varepsilon < x_n \leqslant z_n \leqslant y_n < x + \varepsilon.$$

Accordingly

$$-x - \varepsilon < -z_n < -x + \varepsilon,$$

or

$$-\varepsilon < x - z_n < \varepsilon, \text{ i.e. } |x - z_n| < \varepsilon$$

for $n > \max(\mathcal{N}, \mathcal{N}_1, \mathcal{N}_2)$, and this precisely means that (z_n) converges to x. $\square$

Besides the Three-Sequence Theorem, use is often made of the following criterion:

A sequence (x_n) is said to be *increasing* (*non-decreasing*) when $x_{n+1} > x_n$ ($x_{n+1} \geqslant x_n$) for all $n \in N$. Similarly, a sequence (y_n) for which $y_{n+1} < y_n$ ($y_{n+1} \leqslant y_n$) is called *decreasing* (*non-increasing*). All four of these types of sequences are called *monotonic* or (recently) *isotonic*. Consequently, we have

THEOREM I.8.4. A monotonic, bounded sequence is convergent.

PROOF. Plainly it would suffice to assume that the sequence is bounded and, from a certain N onwards, monotonic. Suppose, for example, that (x_n) is non-decreasing (and bounded). It is sufficient to show that it is a Cauchy sequence. This will be done by *reductio ad absurdum*. Thus, suppose that (x_n) is not a Cauchy sequence; then

$$\bigvee_{\varepsilon > 0} \bigwedge_{\mathcal{N}_k} \bigvee_{n_k, m_k > \mathcal{N}_k} |x_{n_k} - x_{m_k}| \geqslant \varepsilon.$$

Take any $\mathcal{N}_1$ and the corresponding n_1, m_1, and let $n_1 < m_1$. Since (x_n) is non-decreasing, then $|x_{m_1} - x_{n_1}| = x_{m_1} - x_{n_1} \geqslant \varepsilon$. Now, take $\mathcal{N}_2 > m_1$ and the corresponding $n_2 < m_2$, such that $n_2, m_2 > \mathcal{N}_2 > m_1$. Thus we have, respectively,

$$x_{n_2} \geqslant x_{m_1}, \quad x_{m_2} \geqslant x_{n_2} + \varepsilon, \quad x_{m_2} \geqslant x_{m_1} + \varepsilon \geqslant x_{n_1} + 2\varepsilon.$$

Continuing this procedure, after k steps we arrive at

$$x_{m_k} \geqslant x_{n_1} + k\varepsilon.$$

Now $\varepsilon > 0$, and thus $k\varepsilon$ is greater than any number for a sufficiently large $k \in N$, that is, the sequence (x_n) is not bounded. A contradiction. $\square$

Remark. The concept of bounded sequence has been used on several occasions thus far. Let us generalize this concept to any function.

DEFINITION. The set $A \subset R$ is *bounded* when

$$\bigvee_{M>0} \bigwedge_{a \in A} |a| < M.$$

A mapping (function) $f: X \to R$ is said to be *bounded* when its range $f(X) = \{f(x): x \in X\}$ is bounded.

On setting $X = N$, we obtain the concept of bounded sequence. The concept of bounded mapping $T: X \to Y$ could have been defined in similar fashion, provided that in the space Y we had a distinct family of bounded sets: "T is bounded when its range $T(X)$ is bounded."

It will be seen shortly that this can be done for a broad class of spaces called metric spaces.

II. METRIC SPACES
CONTINUOUS MAPPINGS

The metric spaces constitute an extensive class of spaces. Since Euclidean space is metric, and every subspace of a metric space is in a natural fashion a metric space, in speaking of space we usually have metric space in mind. These spaces, for which Hausdorff and Fréchet formulated an axiomatics, were investigated thoroughly by Polish mathematicians in the period between the two World Wars. Hence the name *Polish spaces* which Bourbaki has given to complete (separable) metric spaces.

1. THE CONCEPT OF DISTANCE AND METRIC SPACE

DEFINITION. Let X be a set. A function

$$d: X \times X \to R_+ := \{r \in R: r \geqslant 0\}$$

satisfying the conditions

(i) $d(x, y) = d(y, x)$ (*symmetry*),
(ii) $d(x, y) + d(y, z) \geqslant d(x, z)$ (*triangle inequality*),
(iii) $\big(d(x, y) = 0\big) \Leftrightarrow (x = y)$,

is called a *distance*. When only points (i) and (ii) are satisfied, d is called a *semi-distance*.

A pair (X, d), where d is a (semi-) distance, is called a (*semi-*) *metric space*.

Remark. Distance is sometimes called a *metric*, hence the name of metric space.

Examples. 1. Any subset Z of a metric space (X, d) is in a natural manner a metric space $(Z, d|Z \times Z)$.

2. $X = R$, and $d(x, y) := |x - y|$.

3. $X = R^2$, and $d(x, y) := \sqrt{(x_1 - y_1)^2 + (x_2 - y_2)^2}$.

4. More generally, $X = R^n$ is in a natural manner a metric space when the distance is given by the formula $d(x, y) = (\sum_{i=1}^{n} (x_i - y_i)^2)^{1/2}$; this is the so-called *Pythagorean distance*.

Example 4 suggests the following generalization.

2. PRODUCTS OF METRIC SPACES

Suppose that $(X_i, d_i), i = 1, 2, \ldots, n$ are metric spaces; then their product

$$\underset{i=1}{\overset{n}{\times}} X_i = \{(x_1, \ldots, x_n): x_i \in X_i, i = 1, \ldots, n\}$$

is a metric space, when it is assumed that

$$(*) \qquad d(x, y) := \sum_{i=1}^{n} d_i(x_i, y_i)$$

or again

$$\bar{d}(x, y) := \Big(\sum_{i=1}^{n} d_i(x_i, y_i)^2\Big)^{1/2}.$$

The proof is carried out for $n = 2$ in the case of the distance $(*)$. The reader can supplement the proof inductively for any n by noting that $X_1 \times \ldots \times X_n = (X_1 \times \ldots \times X_{n-1}) \times X_n$.

(i) Symmetry: $d(x, y) = d((x_1, x_2), (y_1, y_2)) = d_1(x_1, y_1) + d_2(x_2, y_2) = d_1(y_1, x_1) + d_2(y_2, x_2) = d(y, x)$. We have made use of the symmetry of d_1, d_2.

(ii) A triangle inequality is obtained by using the triangle inequality for d_1, d_2:

$$d(x, z) = d((x_1, x_2), (z_1, z_2)) = d_1(x_1, z_1) + d_2(x_2, z_2)$$
$$\leqslant d_1(x_1, y_1) + d_1(y_1, z_1) + d_2(x_2, y_2) + d_2(y_2, z_2)$$
$$\leqslant d((x_1, x_2), (y_1, y_2)) + d((y_1, y_2), (z_1, z_2))$$
$$= d(x, y) + d(y, z).$$

It has already been shown that d is a semi-distance.

(iii) Let $0 = d(x, y) = d((x_1, x_2), (y_1, y_2)) = d_1(x_1, y_1) + d_2(x_2, y_2)$. But the sum of two non-negative numbers is 0 only when both vanish, and hence $d_i(x_i, y_i) = 0$, that is, $x_i = y_i, i = 1, 2$. Thus $x = y$. $\square$

Every set X can be metrized, i.e. distance can be introduced in it, e.g.

$$d(x, y) = \begin{cases} 0, & \text{when } x = y, \\ 1, & \text{when } x \neq y. \end{cases}$$

Distance so defined is called *discrete distance*, and (X, d) *discrete* (metric) *space*.

3. BOUNDS OF SET

DEFINITION. If (X_i, d_i), $i = 1, 2$, are metric, an injection $T: X_1 \to X_2$ preserving distance, i.e. satisfying the condition $d_2(T(x_1), T(y_1)) = d_1(x_1, y_1)$ for any $x_1, y_1 \in X_1$, is called an *isometry*.

DEFINITION. The set

$$K(x_0, r) := \{x: d(x, x_0) < r\}$$

is called a *ball* of radius r and centre x_0.

DEFINITION. A set A of a metric space is *bounded* when it is a subset of a ball.

The latter condition is equivalent to the condition "contained in a ball of given centre x_0", for when $A \subset K(x_1, r)$, then $A \subset K(x_0, r + d(x_0, x_1))$.

Examples of Bounded Sets. 1. Any ball.

2. A subset of a bounded set.

3. The sum of a finite number of bounded sets, for if $A_i \subset K(x_0, r_i)$, $i = 1, 2, ..., p$, then $\bigcup_{i=1}^{p} A_i \subset K(x_0, \max(r_1, ..., r_p))$.

4. A finite set.

3. THE BOUNDS OF A SET

DEFINITION. Let A be a subset of the real axis: $A \subset R^1$.

The set A has a *least upper bound* (*supremum*) $M \in R^1$, which fact we write as $M = \sup A$, when

(i) $\bigwedge_{x \in A} x \leqslant M$, $\qquad$ *for all x there follows ,*

(ii) $\bigwedge_{\varepsilon > 0} \bigvee_{x_1 \in A} x_1 > M - \varepsilon$. $\qquad$ *there exists an x such that*

The set A has a *greatest lower bound* (*infimum*) $m \in R^1$, which fact we write as $m = \inf A$, when

(i) $\bigwedge_{x \in A} x \geqslant m$,

(ii) $\bigwedge_{\varepsilon > 0} \bigvee_{x_1 \in A} x_1 < m + \varepsilon$.

It is easily seen that if a set has both supremum and infimum, it is bounded. We note, moreover, that a bound need not be an element of the set, e.g. $\inf \{1/n: n \in N\} = 0$, but zero is not of the form $1/n$.

The definition of bound and the completeness of the space R imply

PROPOSITION II.3.1. *A bounded set $A \subset R$ has both supremum and infimum.*

The proof will be given for the least upper bound M. Two cases are possible:

a) The set A has a greatest element a, and then plainly $a = \sup A$;

b) The set A does not have the greatest element. It may then be assumed that there is an $M \in R$ such that $\bigwedge_{x \in A} x < M$ and that the interval $[M-1, M]$ contains an infinite number of elements from A. Let us take one of them, a_0, and divide the interval into two parts: $\left[M-1, M-\frac{1}{2}\right]$, $\left]M-\frac{1}{2}, M\right]$. If the "right" interval contains an element from A, it contains an infinite number of such elements and $a_1 \in A, a_1 > a_0$, may be chosen from them. If the "right" interval does not contain elements from A, we take $a_1 > a_0$ contained in the "left" interval. Now, let us divide the interval from which a_1 was chosen, into two parts in similar fashion and choose an element $a_2 > a_1$ from the "right" interval (provided it contains elements from A) or from the "left" one (if the right-hand part contains no elements from A). Proceeding further in this way, we construct an increasing sequence (a_n) which, by Theorem I.8.4, has a limit $a = \lim_{n \to \infty} a_n$. It will be demonstrated that $a = \sup A$. Show that

(i) $\bigwedge_{x \in A} x \leqslant a,$

(ii) $\bigwedge_{\varepsilon > 0} \bigvee_{x_1 \in A} x_1 > a - \varepsilon.$

Note that case (ii) follows from the fact that $a = \lim_{n \to \infty} a_n$. Suppose therefore, that case (i) does not hold. Let there exist $x_0 \in A$, $x_0 > a$. Divide the interval $[M-1, M]$ into two parts as before. Note that if $a \in \left[M-1, M-\frac{1}{2}\right]$, $x_0 \in \left]M-\frac{1}{2}, M\right]$, we would have a contradiction with the construction of the sequence (a_n). Hence, a and x_0 lie in the same part of the interval $[M-1, M]$, whereby $|x_0 - a| < \frac{1}{2}$. On subdividing each of these parts into two and reasoning in analogous fashion, we see that $|x_0 - a| < 1/2^2$, and after n steps we have $|x_0 - a| < 1/2^n$, whence $x_0 = a$, which contradicts the assumption that $x_0 > a$. It has thus been shown that the set A has for its bound either its greatest (least) element, or the limit of a sequence of elements $a_n \in A$ (point of accumulation). $\square$

4. OPEN SETS. THE TOPOLOGY OF A SPACE

In the present section we single out open sets, the family of sets of metric space that is fundamental to our further considerations.

DEFINITION. A subset $\mathscr{O}$ of a metric space (X, d) is an *open set in X* when it is empty or when

$$\bigwedge_{x \in \mathscr{O}} \bigvee_{r > 0} K(x, r) \subset \mathscr{O}.$$

The family $\mathscr{T}$ of all open sets is called the *topology* of the space X.

Example. The ball $K(x_0, r_0)$ is an open set, for if $x \in K(x_0, r_0)$, then $K(x, r - d(x, x_0)) \subset K(x_0, r_0)$. This follows immediately from the triangle inequality.

The following theorem characterizes the topology of a metric space.

THEOREM II.4.1. (i) The union (of any cardinality) of open sets is an open set

$$(\mathscr{O}_i \in \mathscr{T}, i \in I) \Rightarrow \left(\bigcup_{i \in I} \mathscr{O}_i \in \mathscr{T} \right).$$

(ii) The common part (intersection) of a finite number of open sets is an open set

$$(\mathscr{O}_k \in \mathscr{T}, k = 1, ..., p) \Rightarrow \left(\bigcap_{k=1}^{p} \mathscr{O}_k \in \mathscr{T} \right).$$

PROOF. (i) Let $x \in \bigcup_{i \in I} \mathscr{O}_i$, that is, let x belong to an $\mathscr{O}_{i_0}$, but since it is open $\mathscr{O}_{i_0}$ contains a ball $K(x, r)$, hence *a fortiori* $K(x, r) \subset \bigcup_{i \in I} \mathscr{O}_i$.

(ii) It is sufficient to give the proof for $p = 2$. Let $x \in \mathscr{O}_1 \cap \mathscr{O}_2$. Since the sets $\mathscr{O}_k$ are open, there exist the balls $K(x, r_k) \subset \mathscr{O}_k$, $k = 1, 2$. Setting $r = \min(r_1, r_2)$, we have $K(x, r) \subset \mathscr{O}_1 \cap \mathscr{O}_2$. $\square$

Remark. Topological space is the name given a set X with the specified family $\mathscr{T} = \{\mathscr{O}_i, i \in I\}$ of subsets, which satisfies conditions (i) and (ii) and

(iii) The empty set and X belong to $\mathscr{T}$.

Theorem II.4.1 now leads itself to concise formulation:

Every metric space is a topological space.

However, (X, d) satisfies one more condition:

PROPOSITION II.4.2. Let $x_1 \neq x_2$; then there exist open sets $\mathscr{O}_k$, $k = 1, 2$, such that $x_k \in \mathscr{O}_k$ and $\mathscr{O}_1 \cap \mathscr{O}_2 = \varnothing$.

PROOF. It is sufficient to take $\mathscr{O}_k = K(x_k, \frac{1}{2} d(x_1, x_2))$. $\square$

$\theta_1 = K(x_1, \frac{1}{2} d(x_1, x_2))$

$\theta_2 = K(x_2, \frac{1}{2} d(x_1, x_2))$

37

Remark. A topological space for which Proposition II.4.2 is true, is called a *Hausdorff space*. Thus we arrive at

COROLLARY II.4.3. (X, d) is a Hausdorff space.

Hausdorff spaces will be taken up in Chapter XIV.

Examples of Open and Non-open Sets. 1. $X = R$, $d(x, y) := |x-y|$. $]a, b[= \{x \in R: a < x < b\}$; an interval without end-points is an open set in R.

2. The same interval, treated as a subset of the plane R^2 is not open since it does not contain any ball.

These examples demonstrate eloquently that the very same sets have different topological properties, depending on the space in which they are considered.

3. Let (X, d) be a discrete metric space; then every one-point set $\{x_0\}$ is an open set because it contains (is identical with) the ball $K\left(x_0, \frac{1}{2}\right)$. Accordingly, every subset A as the union $\bigcup_{a \in A} \{a\}$ of open sets is itself an open set.

4. A point $x_0 \in R^2$, where R^2 has the Pythagorean distance, is not an open set because it does not contain any ball. When R^2 is provided with the discrete topology, however, then x_0 becomes an open set. This example shows forcibly that one and the same subset may be open under one topology and not open under another.

The *interior* int$A$ (or $\mathring{A}$) *of a set* A *in* (X, d) is (*ex definitione*) the greatest open set contained in A, i.e. int$A = \bigcup \mathcal{O}_i$, where $\mathcal{O}_i$ runs over all open sets of (X, d), contained in A.

Plainly, the interior of a set may be empty, e.g. int$\{x_0\}$ when $x_0 \in R$.

Exercise. A point $a \in$ int$A$ is called an *interior point* of the set A. Show that a is an interior point if and only if

$$\bigvee_{r > 0} K(a, r) \subset A.$$

DEFINITION. A *neighbourhood* $\mathcal{O}(x)$ of a point x is a set which contains an open set $\mathcal{O} \ni x$, e.g. a ball $K(x, r)$.

5. CLOSED SETS. CLOSURE OF A SET

DEFINITION. A set $A \subseteq X$ is *closed in* (X, d) when its complement $X - A$ is an open set in (X, d).

5. CLOSED SETS. CLOSURE OF SET

On going over to complements in Theorem II.4.1, we have

THEOREM II.5.1. Closed sets of the space (X, d) have the following properties:

(i) The empty set and the entire space X are closed.

(ii) The common part $\bigcap_{i \in I} A_i$ of the closed sets A_i is closed.

(iii) The union of two closed sets, hence of a finite number of such sets, is itself a closed set.

It is thus seen that, given the closed sets of a given space, we know its topology.

Examples of Closed Sets. 1. A *closed ball* with centre x_0 and radius r, i.e. the set $\bar{K}(x_0, r) := \{x: d(x_0, r) \leqslant x\}$, is a closed set. To prove this, it must be shown that $\mathcal{O} := \mathbf{C}\bar{K}(x_0, r) = \{x' \in X: d(x', x_0) > r\}$ is an open set. Indeed, let $d(x', x_0) =: r_1 > r$; then $K(x', r_1 - r) \subset \mathcal{O}$. This follows immediately from the triangle inequality.

2. Since a point (i.e. a one-point set) is a closed set, point (iii) of Theorem II.5.1 implies that each finite set is closed.

3. In a discrete space (X, d) each set $\mathcal{O}$ is closed because $\mathcal{O} = \mathbf{C}(X - \mathcal{O})$, and $X - \mathcal{O}$ is open. Thus, in a discrete space each set is both closed and open at the same time.

In the space (X, d) one may speak of the limit of a sequence of points.

DEFINITION. It is said that x is the *limit of a sequence* (x_n) or that the *sequence* (x_n) *converges* (*is convergent*) *to* x when

$$\bigwedge_{\varepsilon > 0} \bigvee_{\mathcal{N} \in N} \bigwedge_{n > \mathcal{N}} d(x, x_n) < \varepsilon,$$

that is, when the remote terms of the sequence lie close to x.

This is written as: $x = \lim x_n$ or, more briefly, $x = \lim_{n \to \infty} x_n$ or again $x_n \to x$.

The following theorem characterizes closed sets in terms of limits of sequences.

THEOREM II.5.2. $(A$ is closed in $(X, d)) \Leftrightarrow (\bigwedge_{A \ni a_n \to a \in X} a \in A)$.

PROOF. $\Rightarrow$ *Indirect Proof*: Let $a_n \in A$, $a_n \to a$, but $a \notin A$. Since $\mathbf{C}A$ is open, there exists a $K(a, r) \subset \mathbf{C}A$, which contradicts the assumption $a_n \to a$.

39

⇐ *Indirect Proof*: Suppose that $X - A$ is not open; thus, there exist an $a_0 \notin A$ and a sequence $a_n \in K(a_0, 1/n)$ such that $a_n \notin X - A$, that is, $a_n \in A$ and $a_n \to a_0$, whence by the assumption $a_0 \in A$, we arrive at a contradiction. □

DEFINITION. The *closure of a set* $A \subset X$ *in* (X, d) is the set $\bar{A}$ which is the least closed set containing A. In other words

$$\bar{A} = \bigcap_{i \in I} Z_i,$$

where Z_i ranges over all closed sets containing A: $Z_i \supset A$, $i \in I$.

Plainly, when A is closed, $A = \bar{A}$.

Exercise. Show that the closure operation satisfies the following conditions:

(a) $\bar{\bar{A}} = \bar{A}$,

(b) $\bar{\emptyset} = \emptyset$,

(c) $\overline{A \cup B} = \bar{A} \cup \bar{B}$,

(d) $\bar{A} \supset A$.

Prove this for any topological space. K. Kuratowski demonstrated that topological space could be defined in terms of operations which satisfy axioms (a)–(d). To be more precise: if in a space X an operation $A \to \bar{A}$ satisfying (a)–(d) is given, then it defines the topology consisting of all sets $\mathcal{O} \subset X$ such that $X - \mathcal{O} = \overline{X - \mathcal{O}}$. This is the unique topology in X, for which the closure operation coincides with the initial one.

DEFINITION. Let A be a subset of the metric space X. It is said that $x_0 \in X$ is a *point of accumulation of the set* A if

$$\bigwedge_{r > 0} \bigvee_{x_0 \neq a \in A} d(a, x_0) < r.$$

In other words, the point x_0 is the limit of the sequence of elements from A which are different from x_0.

We can give a new characterization of closed sets:

(A set A of the metric space X is closed in X) ⇔ (A contains all of its points of accumulation).

DEFINITION. A set $A \subset X$ is *dense in* (X, d) if $\bar{A} = X$, i.e.

$$\bigwedge_{x \in X} \bigvee_{a_n \in A} a_n \xrightarrow[n \to \infty]{} x.$$

Example. Q is dense in R, cf. Lemma I.7.4.

6. CAUCHY SEQUENCES. COMPLETENESS OF METRIC SPACES

Just as we obtained the definition of a convergent sequence in (X, d) replacing the absolute value sign $|x-y|$ by $d(x, y)$, so now we get the definition of a Cauchy sequence.

DEFINITION. A sequence $N \ni n \to x_n \in X$ in a metric space (X, d) is a *Cauchy sequence* when

$$\bigwedge_{\varepsilon > 0} \bigvee_{\mathscr{N}} \bigwedge_{n, m > \mathscr{N}} d(x_n, x_m) < \varepsilon.$$

DEFINITION. If in (X, d) every Cauchy sequence has a limit (belonging to X), the space (X, d) is *complete*.

Accordingly, Cauchy's theorem may be formulated briefly:

THEOREM II.6.1. The space R is complete under the natural distance $d(x, y) = |x-y|$.

The space of rational numbers $(Q, | \ |)$ was not complete (the sequence of decimal expansions of $\sqrt{2}$ was a Cauchy sequence but did not converge in Q since $\sqrt{2}$ is not a rational number !). Therefore, it had to be completed, i.e. embedded isometrically in a complete space—the space R—so that Q be dense in it. Figuratively speaking, we have adjoined the limits of nonconvergent Cauchy sequences. Hausdorff noticed that any space (X, d) could be complemented by analogous procedure.

We now prove

PROPOSITION II.6.2. Suppose that the spaces (X_j, d_j), $j = 1, 2, ..., p$, are complete. Their product $X = \underset{j=1}{\overset{p}{\times}} X_j$ then is a complete space.

PROOF. It is easily seen that the convergence in $\underset{j=1}{\overset{p}{\times}} X_j$ is the convergence of every coordinate, i.e.

$$\overset{n}{x} = (\overset{n}{x_1}, ..., \overset{n}{x_p}) \to x = (x_1, ..., x_p),$$

$$\text{when } \overset{n}{x_j} \to x_j, \ j = 1, 2, ..., p.$$

Let $(\overset{n}{x})$ be a Cauchy sequence in X; then every j-th coordinate $(\overset{n}{x_j})$ of this sequence is a Cauchy sequence in X_j. But (X_j, d_j) is complete, whereby $\overset{n}{x_j} \to x_j \in X$ as $n \to \infty$. Thus $\lim \overset{n}{x} = x = (x_1, ..., x_p)$. $\square$

PROPOSITION II.6.3. A closed subset of a complete space is a complete space.

The proof is an immediate conclusion from Theorem II.5.2.

41

7. CONTINUOUS MAPPINGS

When we have a mapping $T: X_1 \rightarrow X_2$ of metric spaces (topological, in general), we may speak of its continuity. First of all, we give a so-called neighbourhood definition, which operates only with open sets, hence allowing itself to be carried over unchanged into general topological spaces. Later we shall give a sequential definition. In the case of metric spaces, the only kind we are considering here, these two definitions are equivalent. As will be seen later, the sequential definition may be so modified (generalized sequences) that it also allows itself to be carried over to topological spaces.

DEFINITION 1 (Neighbourhood definition). (a) Let (X_i, d_i), with topology $\mathcal{T}_i$, $i = 1, 2$, be metric spaces. The mapping $T: X_1 \rightarrow X_2$ is *continuous* when

$$\bigwedge_{\mathcal{O}_2 \in \mathcal{T}_2} T^{-1}(\mathcal{O}_2) \in \mathcal{T}_1,$$

i.e. the inverse image of any open set is an open set.

(b) T is *continuous* at a point x_0 if for every neighbourhood $\mathcal{O}_2$ of the point $T(x_0)$ the set $T^{-1}(\mathcal{O}_2)$ is a neighbourhood of the point x_0.

COROLLARY II.7.1. (The mapping T is continuous) $\Leftrightarrow$ (T is continuous at every point).

It is left to the reader to prove this corollary. $\Rightarrow$ is obvious.

DEFINITION 2 (Sequential definition). $T: X_1 \rightarrow X_2$ is *continuous* at $x_0 \in X_1$ if for every sequence $x_n \rightarrow x_0$ the sequence $T(x_n) \rightarrow T(x_0)$. T is *continuous* if it is continuous at every point. If for every sequence (x_n) convergent to x_0 the sequence $T(x_n) \rightarrow y_0$, then y_0 is called the *limit* of T in x_0 and is denoted by $\lim_{x \rightarrow x_0} T(x)$. Thus, T is continuous in x_0 when $\lim_{x \rightarrow x_0} T(x) = T(x_0)$.

Before we proceed to prove the equivalence of the two definitions consider some

Examples. 1. A constant mapping $T(X_1) = \{y_0\}$ is continuous, for if $\mathcal{O}_2 \not\ni y_0$, then $T^{-1}(\mathcal{O}_2) = \emptyset$, and hence is open; if, on the other hand, $\mathcal{O}_2 \ni y_0$, then $T^{-1}(\mathcal{O}_2) = X_1$, and hence is open.

Exercise. Carry out the proof by means of the sequential definition.

2. $X_2 = X_1$. $T = \mathrm{id}_{X_1}$ (identity mapping) is clearly continuous for $T^{-1}(\mathcal{O}_2) = \mathcal{O}_2$.

3. $X_1 = X \times X$, $X_2 = R$, where (X, d) is a metric space. It will be shown that the distance $d: X \times X \to R$ is a continuous function. This follows from the triangle inequality: let $x, y, x', y' \in X$; then

$$d(x, y) \leqslant d(x, x') + d(x', y') + d(y', y),$$

thus, by symmetry,

$$|d(x, y) - d(x', y')| \leqslant d(x, x') + d(y, y'),$$

whence, if $(x_n, y_n) \to (x, y)$, or $d(x_n, x)$, $d(y_n, y) \to 0$, $n \to \infty$, then $d(x_n, y_n) \to d(x, y)$, as $n \to \infty$.

Exercise. Give the proof with the aid of the neighbourhood definition. Note that it is sufficient to show that for every $r > 0$, $d^{-1}([0, r[)$ is an open set in $X \times X$.

4. Let $X_1 \times X_2$ be the (Cartesian) product of the spaces (X_1, d_1), (X_2, d_2). By p_i denote the projection onto the i-th coordinate: $p_i(x_1, x_2) = x_i$. The projections will be shown to be continuous. To this end it is sufficient to examine the inverse image of an arbitrary ball $K(x_i, r_i)$ in X_i. However

$$p_1^{-1}\big(K(x_1, r_1)\big) = K(x_1, r_1) \times X_2,$$

and this is an open set in $X_1 \times X_2$.

Remark. As we know, every open set in a metric space is the union of balls (contained in it). Thus, in order to demonstrate the continuity of $T: X_1 \to X_2$, it is sufficient to show that for every ball $K(x_2, r_2)$ in X_2, $T^{-1}\big(K(x_2, r_2)\big)$ contains a ball.

THEOREM II.7.2. The two definitions of continuity are equivalent, that is

$$\text{(Neighbourhood definition)} \Leftrightarrow \text{(Sequential definition)}.$$

PROOF. $\Rightarrow$: Let $x_n \to x_0$, that is

$$(*) \qquad \bigwedge_{\delta > 0} \bigvee_{\mathcal{N} \in N} \bigwedge_{n > \mathcal{N}} d_1(x_n, x_0) < \delta, \quad \text{or} \quad x_n \in K(x_0, \delta).$$

However, by the neighbourhood definition (cf. remark above), we have

$$\bigwedge_{\varepsilon > 0} \bigvee_{\delta > 0} T(K(x_0, \delta)) \subset K(T(x_0), \varepsilon),$$

whence

$$(**) \qquad T(x_n) \in K(T(x_0), \varepsilon), \quad \text{that is} \quad T(x_n) \to T(x_0).$$

Since x_0 was arbitrary, the mapping T is sequence-wise continuous.

⇐ (a.a.):[1] Thus, let (∗) and (∗∗) hold, and let T be a mapping which is neighbourhood-wise non-continuous at x_0, i.e. let

$$(∗∗∗) \qquad \bigvee_{\varepsilon>0} \bigwedge_{\delta>0} \bigvee_{x_\delta \in X_1} d_1(x_\delta, x_0) < \delta \quad \text{and} \quad d_2\big(T(x_\delta), T(x_0)\big) \geqslant \varepsilon.$$

Take $\delta = 1/n$ and the corresponding points $x_\delta = x_n$ such that (∗∗∗) holds. Hence, $x_n \to x_0$, whereas $T(x_n) \not\to T(x_0)$. A contradiction.[2] □

THEOREM II.7.3. A composition of continuous mappings is a continuous mapping.

PROOF. Let $T_1: X \to Y$, $T_2: Y \to Z$; that is $T_2 \circ T_1: X \to Z$. But $(T_2 \circ T_1)^{-1}(A) = T_1^{-1}\big(T_2^{-1}(A)\big)$. However, when A is open, $T_2^{-1}(A)$ is open in Y, whence $T_1^{-1}\big(T_2^{-1}(A)\big)$ is open in X. □

In Chapter I we became acquainted with two kinds of products of mappings, $T_1 \times T_2$ and (T_1, T_2). It will now be shown that both are continuous, i.e. that there holds

PROPOSITION II.7.4. (a) Let $T_1: X_1 \to Y_1$, $T_2: X_2 \to Y_2$ be continuous; then $T_1 \times T_2: X_1 \times X_2 \to Y_1 \times Y_2$ is continuous.

(b) $(S_i: X \to Y_i, i = 1, 2,$ are continuous$) \Rightarrow \big((S_1, S_2): X \to Y_1 \times Y_2$ is continuous$)$.

PROOF. (a) $(T_1 \times T_2)(x_1, x_2) := \big(T_1(x_1), T_2(x_2)\big)$. Let $x_i \underset{n\to\infty}{\overset{n}{\to}} x_i$, $i = 1, 2$. Because of continuity, $T_i(\overset{n}{x_i}) \to T_i(x_i)$, that is $\big(T_1(\overset{n}{x_1}), T_2(\overset{n}{x_2})\big) \to \big(T_1(x_1), T_2(x_2)\big)$, for convergence in the product is the convergence of both coordinates.

(b) Let $x_n \to x$, $n \to \infty$; hence

$$(S_1, S_2)(x_n) := \big(S_1(x_n), S_2(x_n)\big) \underset{n\to\infty}{\to} \big(S_1(x), S_2(x)\big),$$

for the same reason as just above. □

As a corollary, we arrive at the important

THEOREM II.7.5. Let the mappings $S_i: X \to Y$ be continuous for $i = 1$, 2, and let a continuous operation □, i.e. the mapping □: $Y \times Y \to Y$, be given in Y; then $S_1 \square S_2: X \to Y$ is continuous.

---

[1] The abbreviation a.a. (*ad absurdum*) signifies the indirect proof.

[2] The reader will note that the axiom of choice has been used in proving this theorem.

PROOF. $(S_1 \square S_2)(x) := S_1(x) \square S_2(x)$, that is $S_1 \square S_2$ is continuous as the composition of continuous mappings (S_1, S_2) and $\square$. $\square$

On associating this theorem with Theorem I.8.1, we immediately arrive at

COROLLARY II.7.6. *Let $f, g: X \to R$ be continuous functions on X. Then*

(i) $f+g$,

(ii) $f-g$,

(iii) fg,

(iv) f/g

are continuous.

Remark. Point (iv) should be taken to mean: the quotient of continuous functions is continuous at those points where it is determined, i.e. wherever g does not vanish.

THEOREM II.7.7. *Given a mapping $T: X \to Y$.*

(T is continuous on X) $\Leftrightarrow$ (The inverse image of every closed set in Y is a closed set in X).

PROOF. $\Leftarrow$: Let $A \subset Y$. We make use of the relation $X - T^{-1}(A) = T^{-1}(Y-A)$. If the set A is closed in Y, and $T^{-1}(A)$ is closed in X, then $Y-A$ is open in Y, and $T^{-1}(Y-A)$ is open in X, that is, T is continuous.

$\Rightarrow$: Conversely, if T is continuous and A closed in Y, then $Y-A$ is open in Y, whereby $X - T^{-1}(A)$ is open in X, that is, $T^{-1}(A)$ is closed in X. $\square$

If $T: X \overset{\text{onto}}{\to} Y$ and is one-to-one, the inverse mapping $T^{-1}: T(X) \overset{\text{onto}}{\to} X$ exists.

If the mappings T and T^{-1} are continuous, T is called a *topological mapping* or a *homeomorphism*. Such a mapping takes open sets over into open sets, and closed sets over into closed ones. Theorems II.7.5 and II.7.7 imply

COROLLARY II.7.8. *If the mapping $T: X \to Y$ is continuous, then*

(i) $\bigwedge\limits_{y_0 \in Y}$ *the set $\{x \in X: T(x) = y_0\} = T^{-1}(y_0)$ is closed,*

(ii) *the set $T^{-1}(Y - \{y_0\}) = \{x \in X: T(x) \neq y_0\}$ is open.*

COROLLARY II.7.9. *(The function $f: X \to R$ is continuous) $\Leftrightarrow$ ($\bigwedge\limits_{c \in R}$ the set $\{x \in X: f(x) > c\}$ is open in X and $\bigwedge\limits_{c \in R}$ the set $\{x \in X: f(x) < c\}$ is open in X).*

PROOF. $\Rightarrow$ follows from Definition 1 (p. 42).

$\Leftarrow$: Let $c_1 = f(x_0) - \varepsilon$, $c_2 = f(x_0) + \varepsilon$; $\varepsilon > 0$.

The set $Z_1 = \{x \in X: f(x) > c_1\}$ is open and $x_0 \in Z_1$, and hence there is a $K(x_0, r_1) \subset Z_1$.

The set $Z_2 = \{x \in X: f(x) < c_2\}$ is open and $x_0 \in Z_2$, and hence there is a $K(x_0, r_2) \subset Z_2$.

The set $K(x_0, r_1) \cap K(x_0, r_2)$ is non-empty and open as the intersection of two open sets. Thus, there exists a ball $K(x_0, r) \subset K(x_0, r_1) \cap \cap K(x_0, r_2)$; if $x \in K(x_0, r)$, then

$$\left. \begin{array}{l} f(x) > f(x_0) - \varepsilon \\ f(x) < f(x_0) + \varepsilon \end{array} \right\} \Rightarrow |f(x) - f(x_0)| < \varepsilon. \ \square$$

DEFINITION (Uniform Continuity of Mappings). The mapping T: $X \to Y$ is said to be *uniformly continuous on X* if

$$\bigwedge_{\varepsilon > 0} \bigvee_{\delta > 0} \bigwedge_{x, x' \in X} \left(d(x, x') < \delta\right) \Rightarrow \left(d(T(x), T(x')) < \varepsilon\right).$$

Clearly, a uniformly continuous mapping of the set X is continuous on X.

An example of a continuous, but not uniformly continuous, function: $(x) = x^2$ is continuous on R, but $(x + \alpha)^2 - x^2 = 2\alpha x + \alpha^2$ may assume any value for a particular, arbitrarily small α; hence, this function is not uniformly continuous on R.

8. COMPACTNESS

DEFINITION. A *subsequence* of the sequence (x_n) is what we call the sequence $N \ni k \to x_{n_k}$, where $k \to n_k$ is an increasing sequence of natural numbers.

LEMMA II.8.1. If the sequence (x_n) converges to the point x, each of its subsequences also converges to x.

PROOF. This follows from the definition of subsequence and the definition of convergence of a sequence (x_n). $\square$

DEFINITION. A set K which is a subset of the metric space X is said to be *compact* if out of each sequence (x_n) of elements of the set K we can extract a subsequence (x_{n_k}) which converges to a limit contained within the set K.

THEOREM II.8.2. (Bolzano, Weierstrass). An interval $[a, b] \subset \mathbf{R}^1$ is compact.

PROOF (by induction). Given a sequence (x_k), $x_k \in [a, b] \subset \mathbf{R}^1$.

Divide the interval into two equal parts. Take that part which contains an infinite number of points x_k (if both parts contain an infinite number of points, take either part). Choose x_k contained in this part and denote it by $x_{k_1} = x_k$. Divide this part of the interval in two and once again select x_{k_2} from the part containing an infinite number of points. Proceeding further in this way, we obtain the subsequence (x_{k_n}) of the sequence (x_k). This subsequence is a Cauchy sequence, because

$$|x_{k_l} - x_{km}| \leqslant \frac{(a-b)}{2^s}, \quad s = \min(l, m).$$

Since the space $\mathbf{R}^1$ is complete, there exists an $x = \lim_{k \to \infty} x_k$, but $a \leqslant x_{k_n} \leqslant b$, whence $a \leqslant x \leqslant b$, that is, $x \in [a, b]$. $\square$

THEOREM II.8.3. $(K_1 \subset X_1, K_2 \subset X_2,$ the sets K_1, K_2 are compact) $\Leftrightarrow (K_1 \times K_2$ is compact).

PROOF. $\Rightarrow$: Take a sequence of points $(x_n, y_n) \in (K_1 \times K_2) \subset X_1 \times X_2$; K_1 is compact, and hence a subsequence $x_{n_k} \to x \in K_1$ can be selected from (x_n). Take a subsequence $(x_{n_k}, y_{n_k}) \in (K_1 \times K_2)$; K_2 is compact, and hence a subsequence $y_{n_{k_l}} \to y \in K_2$ can be extracted from (y_{n_k}). The subsequence $x_{n_{k_l}} \to x$ as a subsequence of a convergent sequence. Thus the subsequence $(x_{n_{k_l}}, y_{n_{k_l}})$ of the sequence (x_n, y_n) converges to $(x, y) \in K_1 \times K_2$.

$\Leftarrow$: We prove that K_1 is compact. Let us take (x_n) in K_1 and $y \in K_2$. There exists a convergent subsequence (x_{n_k}, y) of the sequence (x_n, y) in $K_1 \times K_2$; $(x_{n_k}, y) \to (x, y) \in K_1 \times K_2$. Thus $x_{n_k} \to x \in K_1$. $\square$

THEOREM II.8.4. A compact set in a metric space X is bounded and closed in X.

PROOF. *Closedness* (a.a.): Suppose that the set K is not closed. Accordingly, there exists a point of accumulation x_0 of K, which does not belong to K. Take the sequence $K \ni x_n \to x_0$. Each of its subsequences also converges to x_0, and thus the set is not compact. A contradiction.

Boundedness (a.a.): Suppose that the set K is not bounded, then

$$\bigwedge_{A \subset K} (A\text{---finite}) \Rightarrow \left(\bigvee_{x \in K} \bigwedge_{a \in A} d(x, a) > 1 \right).$$

47

Thus we can construct, by induction, a sequence (x_n) in K such that $d(x_n, x_m) > 1$ for $n \neq m$.

So, no subsequence of this sequence is a Cauchy sequence, and hence is not convergent. A contradiction. $\square$

The converse theorem is not in general true. Take, for example, a discrete space X with an infinite number of elements. Since $d(x, y) \leqslant 1 < 2$, the set X is bounded. Closedness: cf. Example 3, p. 39. On the other hand, the set X is not compact, for if we take a sequence such that elements are not repeated in it, a constant subsequence cannot be selected from such a sequence, and hence a convergent subsequence cannot be extracted.

The converse theorem is, however, valid in R^n.

Note at this point that a set of a finite number of elements is always compact.

THEOREM II.8.5. A set K which is bounded and closed in R^n is compact.

PROOF. We use the Bolzano–Weierstrass method; cf. Theorem II.8.2. The set is subdivided into "cells" by hyperplanes. Such a division is possible since the set is bounded. The diameter of the cells tends to zero, whereby the subsequence obtained is a Cauchy sequence. By virtue of the completeness of R^n, the sequence has a limit and it follows from the closedness that the limit is contained in K. $\square$

The definitions of compactness and completeness imply that a compact space is complete. The converse theorem is not in general true since, for example, R^1 is complete but is not compact (because it is not bounded). As a closed set, a compact set on R^1 contains its bounds since only a point of accumulation or an element of a set can be a bound.

DEFINITION. If $\{A_i\}_{i \in I}$ is such a family of subsets of a set X that $X = \bigcup_{i \in I} A_i$, then it is called a *covering* of the set X. In the case all A_i are open (closed) we speak about an *open* (*closed*) *covering* of the set X.

THEOREM 2.8.6 (Borel, Lebesgue). Let X be a metric space, then X is compact if and only if for every open covering $\{A_i\}_{i \in I}$ of X there exists a finite set $\{i_1, \ldots, i_k\}$, $k \in N$, such that $\{A_{i_n}\}_{n=1}^k$ is a covering of X.

PROOF. $\Rightarrow$: Let $\{A_i\}_{i \in I}$ be an open covering of X. Firstly, we shall prove that there exists a $\lambda > 0$ such that

$$\bigwedge_{x \in X} \bigvee_{i \in I} K(x, \lambda) \subset A_i.$$

Let us assume that this is not true, i.e.

$$(*) \qquad \bigwedge_{n \in N} \bigvee_{x_n \in X} \bigwedge_{i \in I} K\left(x_n, \frac{1}{n}\right) \not\subset A_i.$$

But from the sequence (x_n) we can take a subsequence (x_{n_k}) convergent to $x_0 \in X$. Since $\{A_i\}_{i \in I}$ is an open covering, there exist $r_0 > 0$ and $i_0 \in I$ such that $K(x_0, r_0) \subset A_{i_0}$. By the convergence of (x_{n_k})

$$\bigvee_{k \in N} \bigwedge_{k > k_0} |x_{n_k} - x_0| < \frac{1}{2} r_0.$$

Thus, taking k so great that $k > k_0$ and $n_k > \dfrac{2}{r_0}$, we obtain $K\left(x_{n_k}, \dfrac{1}{n_k}\right)$

$\subset K\left(x_{n_k}, \dfrac{1}{2} r_0\right) \subset K(x_0, r_0) \subset A_{i_0}$; a contradiction with $(*)$.

Having $\lambda > 0$ as above we proceed as follows:
we choose $y_1 \in X$,
then $y_2 \in X - K(y_1, \lambda)$,
then $y_3 \in X - \big(K(y_1, \lambda) \cup K(y_2, \lambda)\big)$,
and so on.

If the infinite number of such choices were possible, we would get a sequence (y_n) such that $|y_n - y_m| \geqslant \lambda$ for every $n, m \in N$. But such sequence does not have a convergent subsequence; a contradiction.

Thus the above construction stops after k steps (for some $k \in N$), i.e. there exist $y_1, \ldots, y_k$ such that $\bigcup\limits_{n=1}^{k} K(y_n, \lambda) = X$. But each $K(y_n, \lambda)$ is contained in some A_{i_n}, hence $\bigcup\limits_{n=1}^{k} A_{i_n} = X$.

$\Leftarrow$ (a.a): Let be given a sequence (x_n) containing no convergent subsequences. Thus its range $A := \{x_n : n \in N\}$ is a closed set. Since $X - A$ is open, $\bigwedge\limits_{x \in X - A} \bigvee\limits_{r_x > 0} K(x, r_x) \subset X - A$. On the other hand $\bigwedge\limits_{x_n} \bigvee\limits_{\varepsilon_n}$ $K(x_n, \varepsilon_n) \cap A$ is a finite set. Obviously $\{K(x, r_x)\}_{x \in X - A} \cup \{K(x_n, \varepsilon_n)\}_{n \in N}$ is an open covering of X. But, if we take any finite number of elements of this covering, then their sum will contain only finite number

49

of balls $K(x_n, \varepsilon_n)$ and, hence, only finite number of points of the (infinite) set A. A contradiction. $\square$

Remark. The above property of compact metric spaces is taken as a definition of compactness in the case of more general topological spaces (Hausdorff spaces); cf. Chapter XII.

9. CONTINUOUS FUNCTIONS AND MAPPINGS ON COMPACT SETS

DEFINITION. A mapping $T\colon X \to Y$ is said to be *bounded* if the set of values $T(X) \subset Y$ is a bounded set.

THEOREM II.9.1. Let X be a compact set and $T\colon X \to Y$ a continuous mapping on X; then $T(X) \subset Y$ is compact.

PROOF. Let $y_n \in T(X) \subset Y$; for any y_n there is an $x_n \in X$ such that $T(x_n) = y_n$. From the sequence (x_n) we select a subsequence which converges to $x \in X$,

$$x_{n_k} \to x \in X.$$

Since T is continuous on X, therefore $y_{n_k} = T(x_{n_k}) \to T(x) \in T(X)$. $\square$

COROLLARY II.9.2. A continuous mapping on a compact set is bounded.

COROLLARY II.9.3. A continuous function on a compact set K attains its bounds, i.e.

$$\bigvee_{x_0 \in K} f(x_0) = \sup f(K), \qquad \bigvee_{x_0' \in K} f(x_0') = \inf f(K).$$

THEOREM II.9.4. (X is a compact space; the mapping $T\colon X \to Y$ is continuous on X) $\Rightarrow$ (T is uniformly continuous on X).

PROOF (a.a.). Suppose that T is not uniformly continuous on X, i.e.

$$\bigvee_{\varepsilon > 0} \bigwedge_{0 < \delta = 1/n} \bigvee_{x_n} \bigvee_{x_n'} d_1(x_n, x_n') < \frac{1}{n} \text{ and } d_2\big(T(x_n), T(x_n')\big) \geqslant \varepsilon.$$

Thus we have two sequences, (x_n) and (x_n'). Since the space X is compact, subsequences $x_{n_k} \to x_0 \in X$, $x_{n_k}' \to x_0' \in X$ could be selected; however, since $d_1(x_n, x_n') < 1/n$, these subsequences may be chosen so that $x_0 = x_0'$; by the continuity of T it follows that

$$\bigwedge_{\varepsilon > 0} \bigvee_{\mathcal{N}} \bigwedge_{k > \mathcal{N}} d_2\big(T(x_{n_k}), T(x_{n_k}')\big) < \varepsilon,$$

which contradicts the assumption. $\square$

THEOREM II.9.5. (The set X is compact; the mapping $T: X \to Y$ is a bijection and is continuous on X) $\Rightarrow$ (T^{-1} is continuous on Y).

PROOF. Let $y_n \to y_0 = T(x_0)$; since T is a bijection, $y_n = T(x_n)$ and $T^{-1}(T(x_n)) = x_n$. Suppose that $x_n \not\to x_0$; thus either $x_n \to x_0' \neq x_0$, or the sequence (x_n) is not convergent. The first alternative is incompatible with the continuity of T, for $T(x_n) \to T(x_0) \neq T(x_0')$. If the second alternative holds, the compactness of X implies that from the sequence (x_n) we can extract a subsequence convergent to $x \in X$, where $x \neq x_0$, which is also incompatible with the continuity of the mapping. $\square$

10. CONNECTED SPACES

DEFINITION. A space (X, d) is *not connected* (*non-connected*) if there exist $X_1 \subset X$, $X_2 \subset X$ such that
(i) $X_1 \cup X_2 = X$,
(ii) $X_1 \cap X_2 = \emptyset$,
(iii) $X_1, X_2 \neq \emptyset$,
(iv) X_1, X_2 are open in X.

Otherwise, we say the space is *connected*.

A subset $A \subset X$ is connected (not connected) if it is connected (not connected) when treated as a metric space with the metric induced from X.

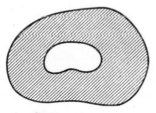 

Connected set   Non-connected set
Fig. 1

Note that if the space X is not connected, that is, if there are two sets X_1, X_2 which are non-empty and open in X, and also $X_1 \cup X_2 = X$, $X_1 \cap X_2 = \emptyset$; then

$$\complement X_1 = X_2, \quad \complement X_2 = X_1.$$

These sets are thus simultaneously open and closed. Accordingly, in the definition we can speak either of closed sets or open sets.

THEOREM II.10.1. (X is a connected space, $f: X \to R^1$ is a continuous function, $f(x_1) < f(x_2)$) $\Rightarrow$ ($\bigwedge\limits_{c \in R^1} f(x_1) < c < f(x_2) \bigvee\limits_{x_0 \in X} f(x_0) = c$).

In other words, a continuous function on a connected set assumes all intermediate values.

PROOF (a.a.). Let $\bigwedge\limits_{x \in X} f(x) \neq c$; then for every x, either $f(x) > c$, or $f(x) < c$. The sets

$$X_1 = \{x \in X: f(x) < c\}, \quad X_2 = \{x \in X: f(x) > c\}$$

are not empty, for $x_1 \in X_1$, $x_2 \in X_2$, and moreover do not intersect, are open sets, and their union is the entire space X. Thus, X is not connected. A contradiction. $\square$

THEOREM II.10.2. (X is a connected space, $T: X \to Y$ is a continuous mapping) $\Rightarrow$ ($T(X)$ is connected).

PROOF (a.a.). Suppose that $T(X)$ is not connected; thus, there exist such sets Y_1, Y_2 that

$$Y_1, Y_2 \text{—open sets in } T(X),$$

$$Y_1 \cup Y_2 = T(X), \quad Y_1 \cap Y_2 = \varnothing, \quad Y_1, Y_2 \neq \varnothing;$$

however, since T is continuous, therefore

$$T^{-1}(Y_1) \text{ open in } X,$$

$$T^{-1}(Y_2) \text{ open in } X,$$

$$T^{-1}(Y_1), T^{-1}(Y_2) \neq \varnothing,$$

$$T^{-1}(Y_1) \cup T^{-1}(Y_2) = X,$$

$$T^{-1}(Y_1) \cap T^{-1}(Y_2) = \varnothing,$$

and, consequently, the space X is not connected. A contradiction. $\square$

DEFINITION. A set $W \subset R^1$ is said to be *convex* if for all $a, b \subset W$ the interval $[a, b] \subset W$.

THEOREM II.10.3. $X \subset R^1$, (X is a convex set) $\Leftrightarrow$ (X is connected).

PROOF. $\Rightarrow$ (a.a.): Suppose that X is not connected; it thus decomposes into a union $X_1 \cup X_2$ of non-empty and non-intersecting open sets in X. Let us take $a \in X_1$, $b \in X_2$, and let $a < b$. Let

$$X_1': = \{x \in X_1: x < b\}.$$

This set has the least upper bound $c = \sup X_1'$, $a \leqslant c \leqslant b$.

If $c \in X_1$, then $c < b$; in that event since every x satisfying $a < x < b$ belongs to X and since X_1 is open in X, there exists an interval $[c, c+\varepsilon[$ contained in X_1 and in $[a, b]$. This, however, contradicts the definition of least upper bound.

If $c \in X_2$, then $a < c$; accordingly, since every $x \in [a, b]$ belongs to X and since X_2 is open in X, there exists an interval $]c-\varepsilon, c] \subset X_2$ and $\subset [a, b]$, and this is also incompatible with the definition of bound.

Thus $c \notin X_1$ and $c \notin X_2$, whence $c \notin X$, and consequently X is not convex. A contradiction.

$\Leftarrow$: Suppose that the set X is not convex; then there exist points $a, b \in X$ and $a < c < b$ such that $c \notin X$. Let us take the identity mapping $f(x) = x$, which, as is known, is continuous. As we recall, f takes on all intermediate values, including c. A contradiction. $\square$

COROLLARY II.10.4. The only connected sets in R^1 are:

  (i) intervals $[a, b]$, $]a, b[$, $[a, b[$, $]a, b]$,

  (ii) semi-lines,

  (iii) the entire line.

PROOF. It is not difficult to show that every convex set in R^1 is one of the above list. $\square$

COROLLARY II.10.5. It follows from Theorems II.10.1 and II.10.3 that a continuous function on a convex set in R^1, assumes all intermediate values.

III. DIFFERENTIATION AND INTEGRATION
OF FUNCTIONS OF ONE VARIABLE

In the present chapter we shall concern ourselves with what is called "differential and integral calculus" and what until not long ago was a synonym for "mathematical analysis". The concepts of differential and derivative, as well as integral, of function of one variable, are presented in a form such that, without any major changes, they can be carried over to the general case of mapping of Banach space (cf. Chapter VII). The derivative was introduced by the founders of "analysis", Newton and Leibniz, the former arriving at this concept in the course of his inquiries into the foundations of dynamics: "velocity is the derivative of path with respect to time". What that phrase means will be seen in a moment. The concept of the derivative as introduced by these great men did not, of course, possess the requisite precision since the contemporary concept of function did not then exist.

The precise concept of the integral is due to Riemann. Bernhard Riemann (September 17, 1826–July 20, 1866) is one of the greatest figures in the history of mathematics. His ideas to this day remain one of the principal impulsions in modern analysis, geometry, and topology; this is an amazing phenomenon in mathematics in this day when theories are thrown onto the scrap heap after ten years.

The concept of the differential, as will be seen in subsequent chapters, runs throughout the whole of mathematical analysis, being generalized as it is in very diverse directions. These generalizations are born out of both the needs of other areas of mathematics and the endeavours to create a sufficiently rich language for the needs of the physicists. Indeed, from the point of view of physics, "mathematics is the language of physics" (E. Wigner).

1. THE DERIVATIVE AND THE DIFFERENTIAL

DEFINITION. Let $\mathcal{O} \subset R^1$ be an open subset of the real axis, and $f: \mathcal{O} \to R^1$ a function on $\mathcal{O}$. If for a given point $x_0 \in \mathcal{O}$ there exists a linear mapping $l: R^1 \to R^1$ such that the function $r(x_0, \cdot): \mathcal{O} \to R^1$

$$r(x_0, x) := f(x) - f(x_0) - l(x - x_0)$$

has the following property

$$\frac{|r(x_0, x)|}{|x - x_0|} \to 0, \quad \text{as } 0 \neq |x - x_0| \to 0,$$

then we say that f is *differentiable at* x_0. The mapping l, being linear, is the multiplication by a real number, say l; the number l is called the *derivative of the function f at the point* x_0 and is denoted by: $f'(x_0)$, $f^{(1)}(x_0)$, $\dfrac{df(x_0)}{dx}$, $\dfrac{df}{dx}(x_0)$, $\dfrac{d}{dx}f(x_0)$. The linear part of the increment $f(x)-f(x_0)$, that is, $l(x-x_0)$, is called the *differential of the function f at the point x_0, corresponding to the increment $(x-x_0)$ of the argument* and is designated by $df(x_0, x-x_0)$. With these notations we can write the defining equation

$$(f(x)-f(x_0))-df(x_0, x-x_0) = r(x, x_0).$$

Remark. The linear part of the mapping f is at times called "tangential mapping" since its graph is a straight line which at the point $(x_0, f(x_0))$ is "tangent" to the graph of the function f. We emphasize that only with the definition of derivative is the intuitive concept of "tangency" defined precisely: Thus the derivative cannot be defined as tangent mapping without previously stipulating what "to be tangent" means. We could first of all define that two mappings $f_i: \mathcal{O} \to R^1$, $i = 1, 2$, are tangent at the point $(x_0, f_1(x_0))$, when $f_1(x_0) = f_2(x_0)$ and $f_1(x) - f_2(x) = r(x_0, x)$, where $|r(x_0, x)|/|x-x_0| \to 0$ for $|x-x_0| \to 0$. Now we could recast the definition as follows:

DEFINITION. $f: \mathcal{O} \to R^1$ is *differentiable at* x_0 when f has at $(x_0, f(x_0))$, a tangent mapping of the form $\psi(x) = f(x_0)+l(x-x_0)$, where l is a linear mapping.

The reader is advised to sketch the graphs.

A question arises immediately: can a function have several derivatives? The answer is provided by

LEMMA III.1.1. A function can have at most one derivative at x_0.

PROOF. Let us take any "two" decompositions mentioned in the definition:

$$f(x)- f(x_0) = l_1(x-x_0)+r_1(x_0, x) = l_2(x-x_0)+r_2(x_0, x).$$

Subtracting, we get

$$0 = (l_1-l_2)(x-x_0)+r_1(x_0, x)-r_2(x_0, x).$$

Dividing by $|x-x_0|$, passing to the limit $x \to x_0$, and bearing in mind that $|r_i(x_0, x)/(x-x_0)| \to 0$, we find that $l_1-l_2 = 0$. $\square$

2. PROPERTIES OF DERIVATIVES

COROLLARY III.1.2. (A function f is differentiable at the point x_0) $\Rightarrow$ (f is continuous at x_0).

The inverse theorem is not in general true.

Examples. 1. Constant function: $f(x) = \text{const}$;

$$f(x)-f(x_0) = \text{const}-\text{const} = 0, \quad \text{that is } f'(x) \underset{x}{\equiv} 0.$$

2. Linear function: $f(x) = ax$;

$$f(x)-f(x_0) = a(x-x_0), \quad f'(x) \underset{x}{\equiv} a.$$

3. Continuous non-differentiable function: $f(x) = |x|$;

$$f(x)-f(0) = |x| = \begin{cases} +x & \text{for } x \geqslant 0, \\ -x & \text{for } x < 0; \end{cases}$$

the decomposition for positive increments is different from that for negative ones. This function is not differentiable at zero.

A notation frequently used is

$$x-x_0 = h, \quad f(x+h)-f(x) = f'(x)h+r(x,h);$$

on dividing both sides by h, we have

$$\frac{f(x+h)-f(x)}{h} = f'(x) + \frac{r(x,h)}{h}.$$

If $h \to 0$, the right-hand side tends to $f'(x)$ since $r(x,h)/h \to 0$. Thus, we have the formula

$$f'(x) = \lim_{h \to 0} \frac{f(x+h)-f(x)}{h}.$$

2. THE PROPERTIES OF DERIVATIVES

THEOREM III.2.1. If f, g are differentiable at x_0, then

(i) $(f\pm g)'(x_0) = f'(x_0)\pm g'(x_0)$,

(ii) $(fg)'(x_0) = f'(x_0)g(x_0)+f(x_0)g'(x_0)$,

(iii) $g(x_0) \neq 0, \left(\dfrac{f}{g}\right)'(x_0) = \dfrac{f'(x_0)g(x_0)-g'(x_0)f(x_0)}{[g(x_0)]^2}$.

PROOF.

(i) $\dfrac{(f\pm g)(x_0+h)-(f\pm g)(x_0)}{h} = \dfrac{f(x_0+h)-f(x_0)}{h} \pm$

$\pm \dfrac{g(x_0+h)-g(x_0)}{h} \to f'(x_0)\pm g'(x_0)$.

(ii)
$$\frac{(fg)(x_0+h)-(fg)(x_0)}{h}=\frac{f(x_0+h)g(x_0+h)-f(x_0)g(x_0)}{h}$$

$$=\frac{(f(x_0+h)-f(x_0))g(x_0+h)}{h}+\frac{(g(x_0+h)-g(x_0))f(x_0)}{h}$$

$$\to f'(x_0)g(x_0)+g'(x_0)f(x_0);$$

in passing to the limit we made use of the fact that

$$f(x_0+h)\underset{h\to0}{\to}f(x_0),$$

$$g(x_0+h)\underset{h\to0}{\to}g(x_0),$$

which follows from the continuity of f and g at x_0.

(iii)
$$\left[\frac{f}{g}(x_0+h)-\frac{f}{g}(x_0)\right]\frac{1}{h}=\frac{f(x_0+h)g(x_0)-g(x_0+h)f(x_0)}{hg(x_0+h)g(x_0)}$$

$$=\frac{(f(x_0+h)-f(x_0))g(x_0)-(g(x_0+h)-g(x_0))f(x_0)}{hg(x_0+h)g(x_0)}$$

$$\underset{h\to0}{\to}\frac{f'(x_0)g(x_0)-g'(x_0)f(x_0)}{[g(x_0)]^2}.\quad\square$$

THEOREM III.2.2 (The Rolle Theorem).

Hypothesis: The function f is

(i) continuous on $[a, b]$,

(ii) differentiable on $]a, b[$,

(iii) $f(a) = f(b)$.

Thesis: There exists a $c \in]a, b[$ such that $f'(c) = 0$.

PROOF. We make use of the fact that a function continuous on $[a, b]$ attains its bounds. Thus, there is a $c \in [a, b]$ such that $f(c) = \sup f[a, b]$ and there is a point $d \in [a, b]$ such that $f(d) = \inf f[a, b]$. Of course, $f(c) \geqslant f(d)$; if $f(c) = f(d)$, then the function is constant and its derivative is identically equal to zero. If $f(c) > f(d)$, then either c or d is contained within the interval $]a, b[$. Let c be contained in the interior of $[a, b]$; then for small h

$$\frac{f(c+h)-f(c)}{h}\begin{cases}\geqslant 0 & \text{if } h < 0,\\ \leqslant 0 & \text{if } h > 0,\end{cases}$$

whence $0 \leqslant f'(c) \leqslant 0$, and thus $f'(c) = 0$. $\square$

2. PROPERTIES OF DERIVATIVES

THEOREM III.2.3 (The Lagrange Mean-Value Theorem of Differential Calculus).

Hypothesis: The function f is
  (i) continuous on $[a, b]$,
  (ii) differentiable on $]a, b[$.

Thesis: There exists a $c \in]a, b[$ such that

$$f'(c) = \frac{f(b) - f(a)}{b - a}.$$

PROOF. Let

$$F(x) := f(x) - \frac{f(b) - f(a)}{b - a}(x - a).$$

It is easily seen that this function satisfies the hypothesis of the Rolle theorem; thus there is a $c \in]a, b[$ such that $F'(c) = 0$. But

$$F'(x) = f'(x) - \frac{f(b) - f(a)}{b - a},$$

$$0 = F'(c) = f'(c) - \frac{f(b) - f(a)}{b - a},$$

whereby $f'(c) = \frac{f(b) - f(a)}{b - a}$. $\square$

This theorem can be rewritten: For a function continuous on $[a, b]$ and differentiable in the interior of $[a, b]$ there exists a $\theta, 0 < \theta < 1$, such that

$$f(b) - f(a) = f'(a + \theta(b - a)) \cdot (b - a).$$

COROLLARY III.2.4.

Hypothesis:

(i) f is differentiable on an open connected set $\mathcal{O} \subset R^1$; cf. Corollary II.10.4,

(ii) $f'(x) \underset{x \in \mathcal{O}}{\equiv} 0$.

Thesis: $f = \text{const}$ on $\mathcal{O}$.

PROOF. Since $\mathcal{O}$ is connected, the interval $[a, b] \subset \mathcal{O}$ for all $a, b \in \mathcal{O}$ such that $a < b$; by Theorem III.2.3, there exists a $c \in]a, b[$ such that

$$0 = f'(c) = \frac{f(b) - f(a)}{b - a}, \quad \text{whence } f(b) - f(a) = 0. \; \square$$

COROLLARY III.2.5. If f is a differentiable function on $]a, b[$, then $(f\text{—non-decreasing (increasing)}) \Leftrightarrow (f' \geqslant 0 \; (f' > 0) \text{ on }]a, b[)$.

PROOF. $\Rightarrow$: Let $x \in]a, b[$, then

$$\frac{f(x+h)-f(x)}{h} \geqslant 0.$$

But a non-negative, convergent sequence has the non-negative limit.

$\Leftarrow$ (a.a.): Let us assume that f is not non-decreasing, i.e. there exist $a < x_1 < x_2 < b$ such that $f(x_2) < f(x_1)$. But then there exists $c \in]x_1, x_2[$ for which

$$f'(c) = \frac{f(x_2)-f(x_1)}{x_2-x_1} < 0.$$

A contradication. $\square$

Let us now consider the differentiability of the inverse function, if such does exist.

THEOREM III.2.6.

Hypothesis:

(i) f is differentiable on $\mathcal{O} \subset R^1$, where $\mathcal{O}$ is an open connected set (i.e. an open interval, open half-line, or R^1),

(ii) $\bigwedge\limits_{x \in \mathcal{O}} f'(x) > 0$ (or < 0).

Thesis:

(i) f^{-1} exists,

(ii) $(f^{-1})'(f(x_0)) = \dfrac{1}{f'(x_0)}$.

PROOF. (i) (a.a.): If f^{-1} does not exist, i.e. there are $a, b \in \mathcal{O}$ such that $a \neq b, f(a) = f(b)$, then (by the Rolle theorem) there exists $c \in]a, b[$ such that $f'(c) = 0$. This is in contradiction with item (ii) of the hypothesis.

(ii) Since f is invertible, for every $\Delta y \neq 0$ such that $f(x_0)+\Delta y \in f(\mathcal{O})$ there exists a Δx such that $f(x_0+\Delta x) = f(x_0)+\Delta y$; we thus have

$$\frac{f^{-1}(f(x_0)+\Delta y)-f^{-1}(f(x_0))}{\Delta y} = \frac{f^{-1}(f(x_0+\Delta x))-f^{-1}(f(x_0))}{f(x_0+\Delta x)-f(x_0)}$$

$$= \frac{x_0+\Delta x-x_0}{f(x_0+\Delta x)-f(x_0)} = \frac{1}{\dfrac{f(x_0+\Delta x)-f(x_0)}{\Delta x}} \xrightarrow{\Delta x \to 0} \frac{1}{f'(x_0)}.$$

Now showing $\lim\limits_{\Delta y \to 0} \Delta x = 0$ will do. Suppose this is not true. Thus a sequence (x_n) exists such that $f(x_n) \to f(x_0)$ and $\bigwedge\limits_{n} x_n \notin K(x_0, \varepsilon)$. But by Corollary III.2.5 f is monotonic, so

$$\bigwedge_n |f(x_n) - f(x_0)|$$

$$\geq \min\{|f(x_0 + \varepsilon) - f(x_0)|, |f(x_0 - \varepsilon) - f(x_0)|\} > 0.$$

A contradiction. $\square$

THEOREM III.2.7.

Hypothesis: Function g is differentiable in x and f is differentiable in $g(x)$.

Thesis:

(i) $f \circ g$ is differentiable in x,

(ii) $(f \circ g)'(x) = f'(g(x))g'(x)$.

PROOF. We have

$$(f \circ g)(x+h) - (f \circ g)(x)$$
$$= f(g(x+h)) - f(g(x))$$
$$= f(g(x) + g'(x)h + r_1(x, h)) - f(g(x))$$
$$= f(g(x)) + f'(g(x))(g'(x)h + r_1(x, h))$$
$$\quad + r_2(g(x)g'(x)h + r_1(x, h)) - f(g(x))$$
$$= f'(g(x))(g'(x)h + r_1(x, h)) + r_2(g(x)g'(x)h + r_1(x, h))$$
$$= f'(g(x))g'(x)h + f'(g(x))r_1(x, h) + r_2(g(x)g'(x)h + r_1(x, h)).$$

Now it is sufficient to prove that

$$\frac{1}{h}[f'(g(x))r_1(x, h) + r_2(g(x), g'(x)h + r_1(x, h))] \xrightarrow[h \to 0]{} 0.$$

Let us stress that our x is fixed and we are interested in h's $\neq 0$. Of course, $\frac{1}{h}f'(g(x))r_1(x, h) \xrightarrow[h \to 0]{} 0$. We take $0 \neq h_n \to 0$; then $h'_n :=$

$$:= g'(x)h_n + r_1(x, h_n) \to 0. \text{ But}$$

$$\frac{1}{h_n}r_2(g(x), h'_n)$$

$$= \begin{cases} r_2(g(x), 0)/h_n = 0, & \text{when } h'_n = 0, \\ \dfrac{r_2(g(x), h'_n)}{h'_n} \cdot \dfrac{h'_n}{h_n} = \dfrac{r_2(g(x), h'_n)}{h'_n}\left[g'(x) + \dfrac{r_1(x, h_n)}{h_n}\right], \\ \text{when } h'_n \neq 0. \end{cases}$$

Hence, we see that $\frac{1}{h_n}r_2(g(x), h'_n) \to 0$ as $n \to \infty$.

Thus, differentiability has been proved and so has been the relation

$$(f \circ g)'(x) = f'(g(x))g'(x). \quad \square$$

Given a differentiable function $f: \mathcal{O} \to R^1$, where $\mathcal{O}$—open subset of R^1. If the function

$$\mathcal{O} \ni x \to F(x) := f'(x) \in R^1$$

is differentiable at $x_0 \in \mathcal{O}$, then we say that f is *twice* (2 *times*) *differentiable at* x_0; the derivative $F'(x_0)$ is called the *second derivative* of the function f at x_0, and it is denoted by:

$$f''(x_0), \ f^{(2)}(x_0), \ \frac{d^2 f(x_0)}{dx^2}, \ \frac{d^2 f}{dx^2}(x_0), \ \frac{d^2}{dx^2} f(x_0).$$

We define analogously the property of being *p times differentiable* and the corresponding *p-th derivative* (*derivative of the p-th order*) is denoted by: $f^{(p)}(x_0)$, $\dfrac{d^p f(x_0)}{dx^p}$, and so on. By $C^p(]a, b[)$ we shall denote the real vector space of all functions on $]a, b[$ which are p times differentiable on $]a, b[$, while $C^p([a, b])$ will denote the real vector space of all continuous functions on $[a, b]$ which are p times differentiable on $]a, b[$ and whose all derivatives up to the p-th order allow the (unique) continuous extensions to the entire interval $[a, b]$.

3. DIRECTED SETS. NETS
(THE GENERAL THEORY OF LIMITS)

DEFINITION. A relation $\prec$ in a set Π is called the *directing relation* (*in the set Π*) when it is reflexive, transitive, and

$$(*) \qquad \bigwedge_{\pi_1, \pi_2 \in \Pi} \ \bigvee_{\pi_3 \in \Pi} (\pi_1 \prec \pi_3) \wedge (\pi_2 \prec \pi_3).$$

The pair $(\Pi, \prec)$ is called a *directed set*. A subset Π' of a directed set Π is said to be *cofinal* with Π when

$$\bigwedge_{\pi \in \Pi} \ \bigvee_{\pi' \in \Pi'} \pi \prec \pi'.$$

When $\pi_1 \prec \pi_2$ we say that π_2 is *later* than π_1, or that π_1 is *earlier* than π_2.

Remark. Usually the further assumption

$$(\pi_1 \prec \pi_2) \wedge (\pi_2 \prec \pi_1) \Rightarrow (\pi_1 = \pi_2)$$

is made about the relation $\prec$.

Examples of Directed Sets. 1. The set N of natural numbers with the relation $\leqslant$.

2. Let Π be the set $\mathscr{R}(X)$ of all subsets of a set X. If $A_1, A_2 \subset X$, we write $A_1 \prec A_2$, when $A_1 \subset A_2$, that is, $\mathscr{R}(X)$ is *directed by inclusion*. (We would also be able to define $A_1 \prec A_2$ when $A_2 \subset A_1$.)

3. Let $(\mathscr{O}_i(x_0))$ be the family of all neighbourhoods of a point x_0 of a metric space (X, d) or, more generally, topological space. We write $\mathscr{O}_i(x_0) \prec \mathscr{O}_j(x_0)$ when $\mathscr{O}_j(x_0) \subset \mathscr{O}_i(x_0)$. Let us verify the condition (∗): For $\mathscr{O}_1(x_0), \mathscr{O}_2(x_0)$ let us take $\mathscr{O}_3(x_0) = \mathscr{O}_1(x_0) \cap \mathscr{O}_2(x_0)$, which is also a neighbourhood of the point x_0. The reader will note that we operated with this kind of direction in the neighbourhood definition of continuity.

4. Let us associate the set $N - A$ with each finite subset $A \subset N$. Let $\Pi = \{N - A : A$—finite sets$\}$ and let Π be directed by the inclusion in Example 2. On furnishing the set N with the discrete topology, we may regard Π as the set of neighbourhoods of infinity.

DEFINITION. The mapping $(\Pi, \prec) \to X$ is called a *net* (*or generalized sequence*) and is written $(x_\pi)_{\pi \in \Pi}$ or, briefly, (x_π).

When X is a metric (or, more generally, a topological) space, one may speak of the limit of the net. To obtain the definition, the set of indices $N = (N, \leqslant)$ in the definition of the sequence, is replaced by the directed set $(\Pi, \prec)$.

DEFINITION. A net (x_π) in a metric (topological) space *converges* to x_0 when

$$\bigwedge_{K(x_0, r)} \bigvee_{\pi_0} \bigwedge_{\pi \succ \pi_0} x_\pi \in K(x_0, r).$$

(The ball $K(x_0, r)$ may, of course, be replaced by an arbitrary neighbourhood $\mathscr{O}(x_0)$ of the point x_0.) We write $x_0 = \lim x_\pi$, $x_0 = \lim_{\pi \in \Pi} x_\pi$ or, at last, $x_\pi \to x_0$.

We see why we can operate with (ordinary) sequences in metric space, for there holds

PROPOSITION III.3.1. The family of balls $K(x, 1/n)$, $n \in N$, is cofinal with the net of neighbourhoods $\{\mathscr{O}(x)\}$ of the point x.

IMMEDIATE PROOF: Every neighbourhood $\mathscr{O}(x)$ contains a ball $K(x, r)$; let us take $n > 1/r$. □

In general topological spaces there is no sequence of neighbourhoods which is cofinal with the set of all neighbourhoods of an arbitrary point x; hence the need to operate with nets. Chapter XII of this book is devoted to these topics.

Many properties of numerical sequences are carried over to numerical nets. The most important will be collected together in the following theorem:

THEOREM III.3.2. Let $(\Pi, \prec)$ be a directed set, and (a_π), (b_π), $\pi \in \Pi$ convergent nets in R^1. Then

  (i) $\lim(a_\pi \pm b_\pi) = \lim a_\pi \pm \lim b_\pi$,

  (ii) $\lim(a_\pi b_\pi) = (\lim a_\pi)(\lim b_\pi)$,

  (iii) $(a_\pi \leqslant b_\pi, \pi \in \Pi) \Rightarrow (\lim a_\pi \leqslant \lim b_\pi)$—preservation of inequality,

  (iv) The theorem on three nets—analogous to Theorem I.8.3—holds.

DEFINITION. A net (a_π) is said to be *non-increasing* (*non-decreasing*) when, for every $\pi_1 \succ \pi_2$, $a_{\pi_1} \leqslant a_{\pi_2}$ $(a_{\pi_1} \geqslant a_{\pi_2})$. Such nets are called *monotonic*.

  (v) A non-increasing (non-decreasing), bounded net (a_π) in R^1 converges to $m = \inf\{a_\pi : \pi \in \Pi\}$ $(M = \sup\{a_\pi : \pi \in \Pi\})$.

The proofs are identical with those for sequences. As an example, we give the proof of (v).

$m = \inf\{a_\pi\}$ means that $\bigwedge_{\varepsilon > 0} \bigvee_{\pi_\varepsilon} m \leqslant a_{\pi_\varepsilon} < m + \varepsilon$ and $\bigwedge_\pi m \leqslant a_\pi$.

Since this net is non-increasing, for $\pi \succ \pi_\varepsilon$ we have $a_\pi \leqslant a_{\pi_\varepsilon}$ and, hence, $m \leqslant a_\pi \leqslant a_{\pi_\varepsilon} < m + \varepsilon$, that is, $|m - a_\pi| < \varepsilon$ for $\pi \succ \pi_\varepsilon$. $\square$

It is frequently convenient to assign the limit ∞ $(-\infty)$ to a non-decreasing (non-increasing) and unbounded net. Property (v) can then be formulated as follows:

  (v) Every monotonic net is convergent (*in the broader sense*).

The Upper (Lower) Limit of a Net. Let $f : X \to R^1$ and let $(\Pi, \prec)$ be a *filtration family* in the set X, i.e. Π is a set of subsets of X, $(\pi_1 \prec \pi_2) \Leftrightarrow (\pi_1 \subset \pi_2)$ for π_1, $\pi_2 \in \Pi$. Let us form a net $a_\pi := \sup f(\pi)$ $\big(a_\pi := \inf f(\pi)\big)$, $\pi \in \Pi$. The net (a_π) is non-decreasing (non-increasing), and hence is convergent in the broader sense. The limit of this net is denoted by $\lim_{\pi \in \Pi} \sup f(\pi)$, and is read *limes superior* (similarly, $\lim_{\pi \in \Pi} \inf f(\pi)$ is read as *limes inferior*) and is called the *upper (lower) limit of the function f, with respect to the filtration family Π.*

3. DIRECTED SETS. NETS

The following examples are particularly important.

Examples. 1. $X = N$, $\Pi = \{N - A: A\text{—finite sets}\}$. Thus, we are dealing with a numerical sequence $(f(n))$. In this case the usual notation is $\liminf_{n \to \infty} f(n)$ and $\limsup_{n \to \infty} f(n)$.

2. $\Pi = \{\mathcal{O}(x_0)\}$ is the net of all neighbourhoods of a point x_0 (directed by inclusion). Then $\limsup_{\{\mathcal{O}(x_0)\}} f(x)$ is written as $\limsup_{x \to x_0} f(x)$ and analogously for limes inferior.

3. If $X = R^1$, and Π is the set of directed neighbourhoods of $+\infty$, i.e. complements of the half-lines $]-\infty, x]$, $x \in R^1$, then we write $\limsup_{x \to +\infty} f(x)$ and $\liminf_{x \to +\infty} f(x)$, respectively.

4. X is a locally compact space, i.e. a space in which every point has a compact neighbourhood. When $\{K\}$ denotes all compact sets of X, then $\Pi = \{\complement K\} = \{X - K: K \in \{K\}\}$ is called the *set of directed neighbourhoods of the point* x_∞—point at infinity; this phraseology will be justified in Chapter XV. Thus, one may again speak of

$$\limsup_{\pi \in \Pi} f(\pi) =: \limsup_{x \to x_\infty} f(x).$$

The concept of upper limit is, as we see, quite a difficult one. It will be used only on several occasions.

For a function continuous at x_0 the equality $\limsup_{x \to x_0} = \liminf_{x \to x_0} = \lim_{x \to x_0}$ holds. However, let us take the function

$$f(x) = \begin{cases} 0 & \text{for } x \geqslant 0, \\ 1 & \text{for } x < 0. \end{cases}$$

Then

$$\limsup_{x \to 0} f(x) = 1, \quad \liminf_{x \to 0} f(x) = 0,$$

and the limit $\lim_{x \to 0} f(x)$ does not exist at all. It turns out that we have

LEMMA III.3.3. Let $f: X \to R^1$. Then

(i) $\liminf_{x \to x_0} f(x) \leqslant \limsup_{x \to x_0} f(x)$,

(ii) $\left(\limsup_{x \to x_0} f(x) = \liminf_{x \to x_0} f(x) = g\right) \Leftrightarrow$ (there exists $\lim_{x \to x_0} f(x) = g$),

(iii) (there exists $g = \lim\limits_{x \to x_0} h(x)$ and $g > 0$)

$$\Rightarrow \left(\lim\limits_{x \to x_0} \sup \left(f(x) h(x) \right) = [\lim\limits_{x \to x_0} \sup f(x)] \, [\lim\limits_{x \to x_0} h(x)] \right).$$

4. THE RIEMANN INTEGRAL

Given an interval $[a, b]$ on the real axis. Let us take a set of intervals $(P_k)_{k=1,2,\dots,s}$ which are contained in the interval $[a, b]$, and which are such that

(i) $\bigcup\limits_{k=1}^{s} P_k = [a, b]$,

(ii) $P_i \cap P_j = \emptyset$ for $i \neq j$.

The set $(P_k)_{k=1,2,\dots,s}$ is a partition of the interval $[a, b]$ and is denoted by $\pi = \{P_k\}$; cf. 3 on p. 5.

A partition $\pi' = \{P'_s\}$ of the interval $[a, b]$ is a *subpartition* of the partition $\pi = \{P_k\}$, when $\bigwedge\limits_{P'_i} \bigvee\limits_{P_j} P'_i \subset P_j$, and we write $\pi \prec \pi'$.

The set Π of all partitions of the interval $[a, b]$ is a set directed by the relation "to be a subpartition", for it is easily seen that all the conditions of the definition are satisfied:

(i) reflexivity is obvious,

(ii) transitivity follows from the transitivity of the inclusion relation,

(iii) the condition (∗): for given $\pi_1 = \{P_i^1\}$, $\pi_2 = \{P_j^2\}$ one should take $\pi_3 := \{P_{ij}\}$, where $P_{ij} := P_i^1 \cap P_j^2$, provided that P_{ij} is not an empty set.

The positive number

$$\delta(\pi) := \max_{i=1, 2, \dots, s} |P_i|,$$

where $|P_i|$ is the length of the interval P_i, is called the *diameter of a partition* $\pi = \{P_i\}_{i=1,2,\dots,s}$.

Let us take an interval $[a, b] \subset R^1$ and a bounded function f on $[a, b]$ and let us form two nets

$$\Pi \ni \pi \to \bar{S}(f, \pi) := \sum_k \sup f(P_k) \cdot |P_k|,$$

$$\Pi \ni \pi \to \underline{S}(f, \pi) := \sum_k \inf f(P_k) \cdot |P_k|.$$

The number $\overline{S}(f, \pi)$ is called the *upper sum* corresponding to the function f and the partition π, and the number $\underline{S}(f, \pi)$ is the *lower sum* corresponding to the function f and the partition π.

Let us now define the *upper integral*

$$\overline{\int_{[a, b]}} f := \inf_{\pi \in \Pi} \overline{S}(f, \pi) \quad \text{when the infimum exists,}$$

and, correspondingly, the *lower integral*

$$\underline{\int_{[a, b]}} f := \sup_{\pi \in \Pi} \underline{S}(f, \pi) \quad \text{when the supremum exists.}$$

DEFINITION. A function bounded on $[a, b]$ is said to be *Riemann-integrable* when upper and lower integrals exist and

$$\overline{\int_{[a, b]}} f = \underline{\int_{[a, b]}} f.$$

We then write

$$\overline{\int_{[a, b]}} f = \underline{\int_{[a, b]}} f =: \int_{[a, b]} f.$$

LEMMA III.4.1. The net $\overline{S}(f, \pi)$ is non-increasing and the net $\underline{S}(f, \pi)$ is non-decreasing.

PROOF. Let us prove the first assertion, for instance: let $\pi_2 \succ \pi_1$, that is,

$$\bigwedge_{P_k^2} \bigvee_{P_i^1} P_k^2 \subset P_i^1 \quad (k = 1, \ldots, s; i = 1, \ldots, l, l \leqslant s).$$

This and the properties of the partition imply that

(1) $\qquad P_i^1 = \bigcup_{\lambda=1}^{m} P_{k\lambda}^2, \quad m \leqslant s, \quad \text{and} \quad |P_i^1| = \sum_{\lambda=1}^{m} |P_{k\lambda}^2|.$

From the properties of the least upper bound it follows that

(2) $\qquad \sup f(P_i^1) \geqslant \sup f(P_k^2).$

By (1) and (2) we have $\overline{S}(f, \pi_1) \geqslant \overline{S}(f, \pi_2)$ for $\pi_2 \succ \pi_1$. $\square$

COROLLARY III.4.2. If upper (lower) integral exists, then

$$\lim_{\pi \in \Pi} \overline{S}(f, \pi) = \overline{\int_{[a, b]}} f \quad \left(\lim_{\pi \in \Pi} \underline{S}(f, \pi) = \underline{\int_{[a, b]}} f\right).$$

THEOREM III.4.3 (Riemann). A function f continuous on $[a, b]$ is integrable.

PROOF. Since a function continuous on $[a, b]$ is uniformly continuous, therefore

$$\bigwedge_{\varepsilon > 0} \bigvee_{\eta > 0} \bigwedge_{x_1, x_2} (|x_1 - x_2| < \eta) \Rightarrow (|f(x_1) - f(x_2)| < \varepsilon).$$

Let us take π_ε such that $\delta(\pi_\varepsilon) < \eta$. For $\pi \succ \pi_\varepsilon$, $\delta(\pi) \leqslant \delta(\pi_\varepsilon)$,

$$\bar{S}(f, \pi) - \underline{S}(f, \pi) = \sum_k \sup f(P_k) \cdot |P_k| - \sum_k \inf f(P_k) \cdot |P_k|$$

$$= \sum_k (\sup f(P_k) - \inf f(P_k)) \cdot |P_k| \leqslant \varepsilon \sum_k |P_k| \leqslant \varepsilon |b - a|,$$

but by Lemma III.4.1 and Corollary III.4.2

$$0 \leqslant \overline{\int} f - \underline{\int} f \leqslant \bar{S}(f, \pi) - \underline{S}(f, \pi),$$

therefore

$$0 \leqslant \overline{\int} f - \underline{\int} f \leqslant \varepsilon |b - a|.$$

Now, ε is arbitrary, whence

$$\overline{\int}_{[a, b]} f = \underline{\int}_{[a, b]} f. \quad \square$$

LEMMA III.4.4.

Hypothesis: Functions f_1, f_2 are bounded on $X = [a, b]$.
Thesis:

(i) $\inf f_1(X) + \inf f_2(X) \leqslant \sup(f_1 + f_2)(X) \leqslant \sup f_1(X) + \sup f_2(X)$,
$\inf f_1(X) + \inf f_2(X) \leqslant \inf(f_1 + f_2)(X) \leqslant \sup f_1(X) + \sup f_2(X)$,

(ii) $\sup(\lambda f_1)(X) = \begin{cases} \lambda \sup f_1(X) & \text{for } \lambda > 0, \\ \lambda \inf f_1(X) & \text{for } \lambda < 0. \end{cases}$

PROOF. (i) By the definition of the bound for every $x \in X$ we have

$$\inf f_1(X) \leqslant f_1(x), \quad f_1(x) \leqslant \sup f_1(X),$$

$$\inf f_2(X) \leqslant f_2(x), \quad f_2(x) \leqslant \sup f_2(X).$$

Adding the inequalities, we obtain

$$\inf f_1(X) + \inf f_2(X) \leqslant f_1(x) + f_2(x) \leqslant \sup f_1(X) + \sup f_2(X).$$

Since this holds for any x, therefore

$$\inf f_1(X) + \inf f_2(X) \leqslant \sup(f_1 + f_2)(X) \leqslant \sup f_1(X) + \sup f_2(X)$$

and

$$\inf f_1(X) + \inf f_2(X) \leqslant \inf(f_1 + f_2)(X) \leqslant \sup f_1(X) + \sup f_2(X).$$

(ii) Self-evident. $\square$

THEOREM III.4.5.

Hypothesis: Functions f, f_1, f_2 are integrable on $[a, b]$.

Thesis: Functions λf, $f_1 + f_2$ and $|f|$ are integrable on $[a, b]$ and:

(i) $\displaystyle\int_{[a,b]} (f_1 + f_2) = \int_{[a,b]} f_1 + \int_{[a,b]} f_2,$

$\left.\rule{0pt}{40pt}\right\}$ *linearity of integral,*

(ii) $\displaystyle\int_{[a,b]} \lambda f = \lambda \int_{[a,b]} f,$

(iii) $(f \geqslant 0) \Rightarrow \left(\displaystyle\int_{[a,b]} f \geqslant 0\right)$, *positivity of integral.*

(iv) $\left| \displaystyle\int_{[a, b]} f \right| \leqslant \int_{[a, b]} |f|.$

PROOF. (i) By Lemma III.4.4 we have

$$\bar{S}(f_1 + f_2, \pi) = \sum_k \sup(f_1 + f_2)(P_k)|P_k| \leqslant \sum_k \sup f_1(P_k)|P_k| +$$

$$+ \sum_k \sup f_2(P_k)|P_k|,$$

$$\underline{S}(f_1 + f_2, \pi) = \sum_k \inf(f_1 + f_2)(P_k)|P_k| \geqslant \sum_k \inf f_1(P_k)|P_k| +$$

$$+ \sum_k \inf f_2(P_k)|P_k|.$$

Thus, we arrive at

$$\underline{S}(f_1, \pi) + \underline{S}(f_2, \pi) \leqslant \underline{S}(f_1 + f_2, \pi) \leqslant \bar{S}(f_1 + f_2, \pi)$$
$$\leqslant \bar{S}(f_1, \pi) + \bar{S}(f_2, \pi).$$

The right and left sides tend to $\displaystyle\int_{[a,b]} f_1 + \int_{[a,b]} f_2$ and therefore by Theorem III.3.2 (iv), we have

$$\lim_{\pi \in \Pi} \underline{S}(f_1 + f_2, \pi) = \lim_{\pi \in \Pi} \bar{S}(f_1 + f_2, \pi) = \int_{[a,b]} f_1 + f_2$$

$$= \int_{[a,b]} f_1 + \int_{[a,b]} f_2.$$

(ii) Let us consider two cases:

a) $\lambda \geqslant 0$; $\bar{S}(\lambda f, \pi) = \lambda \bar{S}(f, \pi)$, $\underline{S}(\lambda f, \pi) = \lambda \underline{S}(f, \pi)$;

b) $\lambda < 0$; $\bar{S}(\lambda f, \pi) = \lambda \underline{S}(f, \pi)$, $\underline{S}(\lambda f, \pi) = \lambda \bar{S}(f, \pi)$.

It is thus evident that for any λ

$$\int_{[a,b]} \lambda f = \lambda \int_{[a,b]} f.$$

(iii) $f \geqslant 0$, and thus $\sup f(P_k) \geqslant 0$ which implies $\bar{S}(f, \pi) \geqslant 0$ and therefore $\int_{[a,b]} f \geqslant 0$.

(iv) Let us define functions f^+, f^-

$$f^+(x) := \max\{f(x), 0\}, \quad f^-(x) := \max\{-f(x), 0\}$$

so that

$$f = f^+ - f^-, \quad |f| = f^+ + f^- = 2f^+ - f.$$

We now show that f^+ is integrable whenever f is, and thus by repeated application of (i) and the above formula integrability of $|f|$ will follow.

For every interval $I \subset [a, b]$ we have

$$0 \leqslant \sup f^+(I) - \inf f^+(I) \leqslant \sup f(I) - \inf f(I).$$

Now

$$0 \leqslant \bar{S}(f^+, \pi) - \underline{S}(f^+, \pi) = \sum_k \left(\sup f^+(P_k) - \inf f^+(P_k) \right) |P_k|$$

$$\leqslant \sum_k \left(\sup f(P_k) - \inf f(P_k) \right) |P_k| = \bar{S}(f, \pi) - \underline{S}(f, \pi).$$

Since the right-hand side in this inequality tends to 0 we obtain

$$\int_{[a,b]} f^+ = \lim_{\pi \in \Pi} \bar{S}(f^+, \pi) = \lim_{\pi \in \Pi} \underline{S}(f^+, \pi) = \int_{[a,b]} f^+$$

as claimed.

Now by linearity of the integral and (iii)

$$\int_{[a,b]} |f| = \int_{[a,b]} f^+ + \int_{[a,b]} f^- \leqslant \left| \int_{[a,b]} f^+ - \int_{[a,b]} f^- \right| = \left| \int_{[a,b]} f \right|. \quad \square$$

COROLLARY III.4.6. Let $C([a, b])$ denote the set of all continuous functions on $[a, b]$; then the function (*functional*)

$$C([a, b]) \ni f \to \int_{[a,b]} f \in R^1 \text{ is positive and linear.}$$

Remark. We denote

$$\int\limits_a^b f := \int\limits_a^b f(x)\,dx := \begin{cases} \displaystyle\int\limits_{[a,b]} f, & \text{when } b > a, \\[2mm] -\displaystyle\int\limits_{[a,b]} f, & \text{when } b < a. \end{cases}$$

THEOREM III.4.7.

Hypothesis: $a < b < c$, a function f is integrable on $[a, b]$ and $[b, c]$.
Thesis: f is integrable on $[a, c]$ and

$$\int\limits_{[a,\,c]} f = \int\limits_{[a,\,b]} f + \int\limits_{[b,\,c]} f.$$

PROOF. Let us take a partition π' of the interval $[a, c]$ such that one of the intervals P_k has an end-point at b. Then $\pi' = \pi_1' \cup \pi_2'$, where π_1' is the partition of the interval $[a, b]$, and π_2' the partition of the interval $[b, c]$, and for every $\pi \succ \pi'$ we have an analogous decomposition $\pi = \pi_1 \cup \pi_2$. Thus

$$\left.\begin{aligned} \overline{S}(f, \pi) &= \overline{S}(f, \pi_1) + \overline{S}(f, \pi_2) \\ \underline{S}(f, \pi) &= \underline{S}(f, \pi_1) + \underline{S}(f, \pi_2) \end{aligned}\right\} \quad \text{for } \pi \succ \pi'.$$

Passage to the limit yields

$$\lim_{\pi \in \Pi} \overline{S}(f, \pi) = \lim_{\pi \in \Pi} \underline{S}(f, \pi) = \int\limits_a^c f = \int\limits_a^b f + \int\limits_b^c f. \quad \square$$

THEOREM III.4.8 (On the Mean-Value in Integral Calculus).

Hypothesis: $f \in C([a, b])$.
Thesis:

$$\bigvee_{c \in [a,\,b]} f(c) = \frac{1}{b-a} \int\limits_a^b f(x)\,dx.$$

PROOF. The inequalities

$$\inf f([a, b]) \leqslant \frac{1}{|b-a|} \int\limits_a^b f \leqslant \sup(f[a, b])$$

hold. Being continuous on $[a, b]$, the function f attains its bounds on $[a, b]$ and all intermediate values; thus there exists a point $c \in [a, b]$ such that

$$f(c) = \frac{1}{b-a} \int\limits_a^b f. \quad \square$$

Remark. The value $f(c) = \dfrac{1}{b-a} \int\limits_a^b f$ is called the *mean value of the function* on the interval $[a, b]$. A different way of writing the Mean-Value Theorem is

$$\bigvee_{0 \leqslant \theta \leqslant 1} f(a + \theta(b-a)) = \frac{1}{b-a} \int\limits_a^b f(x)\,dx.$$

We shall now prove an extremely important theorem linking together differentiation and integration.

THEOREM III.4.9 (Fundamental Theorem of Differential and Integral Calculus).

Hypothesis: $f \in C([a, b])$ and $F(x) := \int\limits_a^x f(t)\,dt$ for $x \in [a, b]$.

Thesis: F is differentiable on $]a, b[$ and $F'(x) = f(x)$ for $x \in]a, b[$.

PROOF. We have

$$\frac{F(x+h) - F(x)}{h} = \frac{\int\limits_a^{x+h} f(t)\,dt - \int\limits_a^x f(t)\,dt}{h}$$

$$= \frac{\int\limits_a^x f(t)\,dt + \int\limits_a^{x+h} f(t)\,dt - \int\limits_a^x f(t)\,dt}{h} = \frac{\int\limits_x^{x+h} f(t)\,dt}{h};$$

by the Mean-Value Theorem, this is seen to be equal to

$$\frac{hf(x+\theta h)}{h} = f(x+\theta h), \quad 0 \leqslant \theta \leqslant 1.$$

If $h \to 0$, then $f(x+\theta h) \to f(x)$, for f is continuous, whereby $F'(x) = f(x)$. $\square$

DEFINITION. F is a *primitive function* of a given function f if $F'(x) = f(x)$.

Note that if a given function f on an interval, has two primitive functions, they differ by a constant. For, if

$$F_1'(x) = f(x), \quad F_2'(x) = f(x),$$

then

$$(F_1(x) - F_2(x))' = 0,$$

whence

$$F_1(x) - F_2(x) = \text{const.}$$

A primitive function exists for every continuous function; it is the integral

$$F(x) = \int_a^x f(t)\,dt.$$

It is seen that

$$\int_a^b f(t)\,dt = F(b) + \text{const},$$

where F is an arbitrary primitive function of the given function f. To determine the constant, we set $a = b$:

$$0 = \int_a^a f = F(a) + \text{const} \Rightarrow \text{const} = -F(a),$$

and we thus have the formula

$$\int_a^b f(t)\,dt = F(b) - F(a) =: [F]_a^b.$$

THEOREM III.4.10 (On Integration by Parts).

Hypothesis: U, V are functions differentiable on $]a, b[$ and continuous on $[a, b]$.

Thesis: $\int_a^b U'(x)V(x)dx = [UV]_a^b - \int_a^b U(x)V'(x)\,dx.$

PROOF. By the rules of differentiation we know that $U'V + V'U = (UV)'$. Therefore

$$\int_a^b U'V + \int_a^b UV' = [UV]_a^b.$$

On taking one of the integrals to the other side, we obtain the thesis. □

THEOREM III.4.11 (On Change of Variables in the Riemann Integral).

Hypothesis: $[a, b] \ni t \to \varphi(t) \in [\varphi(a), \varphi(b)]$, $\varphi' \neq 0$, φ'—continuous, and $f \in C([\varphi(a), \varphi(b)])$.

Thesis: $\displaystyle\int_{\varphi(a)}^{\varphi(b)} f(x)\,dx = \int_a^b f(\varphi(t))\,\varphi'(t)\,dt.$

PROOF. Let

$$F(z) := \int_{\varphi(a)}^{z} f(x)\,dx, \quad G(s) := \int_a^s f(\varphi(t))\varphi'(t)\,dt.$$

Note that

$$F'(z) = f(z), \quad F'(\varphi(s)) = f(\varphi(s)),$$

$$\frac{d}{ds}F(\varphi(s)) = F'(\varphi(s))\varphi'(s) = f(\varphi(s))\varphi'(s),$$

$$G'(s) = f(\varphi(s))\varphi'(s).$$

Thus, $G'(s) = (F \circ \varphi)'(s)$, which implies that these functions differ by a constant. But $G(a) = 0$ and $F(\varphi(a)) = 0$, whence $G(b) = F(\varphi(b))$, that is

$$\int_a^b f(\varphi(t))\varphi'(t)dt = \int_{\varphi(a)}^{\varphi(b)} f(x)dx.$$

This formula can be recast in a slightly different form:

$$\int_a^b f(x)dx = \int_{\varphi^{-1}(a)}^{\varphi^{-1}(b)} f(\varphi(t))\varphi'(t)dt. \quad \square$$

5. THE LOGARITHM AND THE EXPONENTIAL FUNCTION

As the reader is not assumed to be familiar with elementary functions, we take this opportunity to give the "historical" definition of the logarithm in terms of the integral. This is how F. Klein taught the logarithm should be derived. This definition has many advantages: continuity, differentiability, invertibility, etc., are immediately evident.

DEFINITION. $]0, \infty[\ni x \to \log x := \int_1^x \frac{1}{t} dt$ is the function *logarithm*.

The properties of the logarithm:

(i) log is defined on $]0, \infty[$;

(ii) $\log 1 = 0$;

(iii) $(\log x)' = \dfrac{1}{x} > 0$, that is, the function log is increasing;

(iv) $\log(xy) = \log x + \log y$.

PROOF. We have

$$\frac{d}{dx}\log(xy) = \frac{1}{xy}\frac{d}{dx}(xy) = \frac{y}{xy} = \frac{1}{x},$$

whence $\log(xy) = \log x + \text{const}$; for $x = 1$ we have $\log y = 0 + \text{const}$, $\log y = \text{const}$. $\square$

From the preceding we have

(v) $\log(a^n) = n \log a$, $\log(\prod\limits_{i=1}^{k} a_i) = \sum\limits_{i=1}^{k} \log a_i$.

(vi) $\log \dfrac{1}{x} = -\log x$.

PROOF.

$$0 = \log 1 = \log \frac{1}{x} x = \log \frac{1}{x} + \log x, \quad \log x = -\log \frac{1}{x}. \ \square$$

(vii) $\log(]0, \infty [) = R^1$.

PROOF. It follows from (iii), (v) and (vi). $\square$

Since $\dfrac{d}{dx}(\log x) > 0$, there exists the inverse function

$$\exp x := \log^{-1} x;$$

this function is called the *exponential function*.

The properties of the exponential function:

(i) The exponential function maps R^1 onto the half-axis $]0, \infty[$;

(ii) $\dfrac{d}{dx} \exp x = \exp x$.

PROOF. Let $\log y = x$; then

$$\frac{d}{dx} \exp x = \left. \frac{1}{\left(\dfrac{d}{dy} \log y \right)} \right|_{y = \exp x} = y|_{y - \exp x} = \exp x. \ \square$$

(iii) $\exp 0 = 1$ (since $\log 1 = 0$).

(iv) $\exp(x+y) = \exp x \cdot \exp y$.

PROOF. Let $x = \log \xi, y = \log \eta$; then

$$\exp(x+y) = \exp(\log \xi + \log \eta) = \exp(\log(\xi \cdot \eta)) = \xi \cdot \eta$$
$$= \exp x \cdot \exp y. \ \square$$

By property (iv)

(v) $\exp(\sum\limits_{i=1}^{n} a_i) = \prod\limits_{i=1}^{n} \exp a_i$, $\exp nu = (\exp a)^n$.

(vi) $\exp(-a) = \dfrac{1}{\exp a}$.

PROOF. We have

$$1 = \exp(a-a) = \exp a \cdot \exp(-a),$$

whence $\exp(-a) = \dfrac{1}{\exp a}$. $\square$

We introduce the *number e* and the *symbol e^x*:

$$e := \exp(1), \quad e^x := \exp x.$$

Note that for natural n,

$$(e)^n = (\exp 1)^n = e^n$$

and

$$\frac{1}{(e)^n} = (e)^{-n} = (\exp 1)^{-n} = e^{-n}.$$

We now introduce the *function a^x* where $a > 0$:

$$R^1 \ni x \to a^x := \exp(x \log a).$$

This is the so-called *general power*.

The properties of the general power:

(i) $a^0 = \exp(0 \log a) = \exp 0 = 1,$

(ii) $(a^x)' = (\exp(x \log a))' = \exp(x \log a)(x \log a)'$
$= \exp(x \log a) \log a = a^x \log a.$

It is seen that $(a_x)' > 0$ for $a > 1$; $(a^x)' < 0$ for $0 < a < 1$. An inverse function can be derived for $0 < a \neq 1$: if $y = a^x$, then $x =: \log_a y$. We have

$$(\log_a x)' = \frac{1}{\left(\dfrac{d}{dy} a^y\right)}\Bigg|_{y=\log_a x} = \frac{1}{a^y \log a}\Bigg|_{y=\log_a x} = \frac{1}{\log a}\frac{1}{x}.$$

We can write

$$\log x = \log_e x \quad \text{or} \quad \log_e x = \ln x.$$

Let us consider the (power) function $]0, \infty[\ni x \to x^\alpha := e^{\alpha \log x}$, $\alpha \in R^1$. We have

(i) $x^{\alpha+\beta} = e^{(\alpha+\beta)\log x} = e^{\alpha \log x} e^{\beta \log x} = x^\alpha x^\beta,$

(ii) $x^0 = 1,$

(iii) $x^{-\alpha} = \dfrac{1}{x^\alpha},$

(iv) $(x^\alpha)' = (e^{\alpha \log x})' = e^{\alpha \log x}(\alpha \log x)' = x^\alpha \dfrac{\alpha}{x} = \alpha x^{\alpha-1}.$

6. FUNCTION exp AND LOGARITHM AS LIMITS

Note that $(e^x)' = e^x$. The question is whether this is the only function whose derivative is equal to the function itself. The answer is provided by the following theorem:

THEOREM III.5.1. $(f'(x) = f(x), f(0) = 1) \Rightarrow (f(x) = e^x)$.

PROOF. Suppose that there is a function ψ such that $\psi'(x) = \psi(x)$ and $\psi(0) = 1$; let us form the quotient $f(x) = \psi(x)/e^x$. Then we have

$$f'(x) = \frac{\psi'(x)e^x - e^x\psi(x)}{(e^x)^2} = \frac{(\psi'(x) - \psi(x))e^x}{(e^x)^2} = 0,$$

whereby $f(x) = \text{const} = \psi(x)/e^x$; but $\psi(0) = 1$ and $\text{const} = \psi(0)/e^0 = 1$, and consequently $\psi(x) = e^x$. $\square$

6. THE FUNCTION exp AND THE LOGARITHM AS LIMITS

$$\frac{1}{x} = (\log x)' = \lim_{h \to 0} \frac{\log(x+h) - \log x}{h} = \lim_{h \to 0} \frac{\log\left(\dfrac{x+h}{x}\right)}{h}.$$

Let $t = 1/x$;

$$t = \lim_{h \to 0} \frac{\log(1+th)}{h} = \lim_{h \to 0} \log(1+th)^{1/h}.$$

Since exp is a continuous function

$$e^t = \lim_{h \to 0} (1+th)^{1/h},$$

and, in particular,

$$e = \lim_{h \to 0} (1+h)^{1/h} \quad \text{or} \quad e = \lim_{n \to \infty} (1+1/n)^n.$$

Since $(a^x)' = a^x \log a$, we have

$$\lim_{h \to 0} \frac{a^{x+h} - a^x}{h} = a^x \log a,$$

that is,

$$a^x \lim_{h \to 0} \frac{a^h - 1}{h} = a^x \log a, \quad \text{whence} \quad \log a = \lim_{h \to 0} \frac{a^h - 1}{h}.$$

Example of Application of Exponential Function. We consider radio-active decay. It is known that the number of atoms decaying per unit time is proportional to the number of atoms, $-\dfrac{dN}{dt} = \lambda N$. Hence, $N(t) = \text{const} \cdot \exp(-\lambda t)$, when $t = 0$, $N(t) = N(0) = \text{const}$, and therefore

$$N(t) = N(0)e^{-\lambda t}.$$

The time needed for half of the atoms to decay is called the half-life:

$$\frac{1}{2}N(0) = N(0)e^{-\lambda T}, \qquad T = \frac{1}{\lambda}\log 2.$$

7. EXTENSION OF CONTINUOUS MAPPINGS

Now, let us consider the following problem: Given continuous mappings T_1, T_2 of a metric space (X, d_1) into a metric space (Y, d_2). Suppose that $T_1 = T_2$ on $A \subset X$. Does $T_1 = T_2$ on X? Plainly this is not true for every A, but it is true for subsets dense in X.

PROPOSITION III.7.1.

Hypothesis: T_k—mappings, $T_k : X \to Y$, $k = 1, 2$;
  (i) T_k continuous on X;
  (ii) $T_1 = T_2$ on $A \subset X$;
  (iii) A dense in X.

Thesis: $T_1 = T_2$ on X.

PROOF. Let $x \in X$; there then exists a sequence (x_n) in A such that $x_n \to x$. We have

$$T_1(x_n) = T_2(x_n),$$

but $T_1(x_n) \to T_1(x)$, $T_2(x_n) \to T_2(x)$, because T_1, T_2 are continuous. $\square$

In elementary courses of algebra, as we recall, functions x^α, a^x are defined for rational numbers. Our definitions coincide with them for rational numbers. The set of rational numbers is dense in the set of real numbers, and it thus follows from Theorem III.5.1 that this is the proper extension of these functions to R^1, i.e. the definition given by elementary algebra would lead to the same continuous (!) functions.

Extension by Closure. We shall now prove the important

THEOREM III.7.2.

Hypothesis: Given metric spaces (X, d_1), (Y, d_2); the set $X \supset A$ is dense; $T: A \to Y$.

Thesis: (There exists a continuous mapping $\overline{T}: X \to Y$, $T \subset \overline{T}$) $\Leftrightarrow (\bigwedge_{x \in X}$ there exists the limit $\lim_{A \ni y \to x} T(y))$.

PROOF. As follows from Proposition III.7.1, there is only one mapping $\overline{T}$.

$\Rightarrow$: Obvious, for since $T \subset \overline{T}$, $\overline{T}$ being continuous, therefore

$$\overline{T}(x) = \lim_{A \ni y \to x} \overline{T}(y) = \lim_{A \ni y \to x} T(y).$$

$\Leftarrow$: We define $\overline{T}(x) := \lim_{A \ni y \to x} T(y)$. It is sufficient to show the mapping $\overline{T}$ so defined is continuous at every point $x \in X$. Let $\mathcal{V}$ be a closed neighbourhood of the point $\overline{T}(x)$. From the hypothesis we know that for the point x there exists an open neighbourhood $\mathcal{U}$ such that

$$T(\mathcal{U} \cap A) \subset \mathcal{V} \quad \text{(because } T \text{ is continuous on } A).$$

But $\mathcal{U}$ is a neighbourhood of each of its points. Therefore, for every $u \in \mathcal{U}$ we have

$$\overline{T}(u) = \lim_{\mathcal{U} \cap A \ni y \to u} T(y),$$

and consequently

$$\overline{T}(u) \in \overline{T(\mathcal{U} \cap A)} \subset \mathcal{V}, \quad \text{because} \quad \mathcal{V} = \overline{\mathcal{V}}.$$

Since a closed neighbourhood is contained in every open neighbourhood, the theorem is proved. $\square$

Remark. The proof, without any changes, is also valid for the more general situation when X is a Hausdorff space and Y is a so-called regular space (cf. Chapter XII).

8. HYPERBOLIC FUNCTIONS

Let us define the *hyperbolic sine* and the *hyperbolic cosine*:

$$\sinh x := \frac{e^x - e^{-x}}{2}, \quad \cosh x := \frac{e^x + e^{-x}}{2}.$$

The properties of the two functions:

(i) $\cosh^2 x - \sinh^2 x = \frac{1}{4}(e^{2x} + e^{-2x} + 2 - e^{2x} - e^{-2x} + 2) = 1,$

(ii) $\cosh 0 = 1, \sinh 0 = 0,$

(iii) $\cosh(-x) = \cosh x, \sinh(-x) = -\sinh x,$

(iv) $(\cosh x)' = \frac{1}{2}(e^x + e^{-x})' = \frac{1}{2}(e^x - e^{-x}) = \sinh x,$

(v) $(\sinh x)' = \frac{1}{2}(e^x - e^{-x})' = \frac{1}{2}(e^x + e^{-x}) = \cosh x.$

Since $(\sinh x)' = \cosh x > 0$, the hyperbolic sine is an invertible function. If $y = \sinh x$, then $x =: \operatorname{arsinh} y$ (*area sine*). Let $e^x = t$; then $y = \frac{1}{2}(t - 1/t)$, whence.

$$t^2 - 2yt - 1 = 0, \quad t = y \pm \sqrt{y^2 + 1}.$$

To choose the proper sign, note that $t = e^x > 0$, which implies that the minus sign should be discarded. Thus we have

$$e^x = y + \sqrt{y^2 + 1}, \quad x = \operatorname{arsinh} y = \ln(y + \sqrt{1 + y^2}).$$

Therefore

$$(\operatorname{arsinh} y)' = \left(\ln(y + \sqrt{1 + y^2})\right)' = \frac{1}{y + \sqrt{1 + y^2}}\left(1 + \frac{y}{\sqrt{1 + y^2}}\right)$$

$$= \frac{1}{\sqrt{1 + y^2}}.$$

The function arsinh is defined on $\mathbf{R}^1$.

To define the function inverse to $\cosh x$, we need to choose a branch so that if $y = \cosh x$, $x \geqslant 0$, then $\operatorname{arcosh} y := x$ (*area cosine*). Accordingly, let us take

$$y = \cosh x = \frac{e^x + e^{-x}}{2}.$$

Let $t = e^x$. We have the equation

$$t^2 - 2yt + 1 = 0, \quad t = y \pm \sqrt{y^2 - 1}, \quad e^x = y \pm \sqrt{y^2 - 1}.$$

In order to have $e^x \geqslant 1$ for $x \geqslant 0$, we need to take the plus sign here. We obtain

$$\operatorname{arcosh} y = \ln(y + \sqrt{y^2 - 1}) \quad \text{for } y \geqslant 1,$$

$$(\operatorname{arcosh} y)' = \left(\ln(y + \sqrt{y^2 - 1})\right)' = \frac{1}{\sqrt{y^2 - 1} + y}\left(1 + \frac{y}{\sqrt{y^2 - 1}}\right)$$

$$= \frac{1}{\sqrt{y^2 - 1}}.$$

IV. CONVEX SETS AND FUNCTIONS

Perhaps because it is so suggestive, convexity is one of the most important concepts in mathematics. We shall come across it in many chapters of this book. The first to make a thorough study of this concept was Hermann Minkowski (one of the co-founders of the special theory of relativity, a great friend of Hilbert's). On discovering profound relations for convex sets, he expressed his rapture in the famous saying, "Everything that is convex is of interest to me".

In the present chapter we shall give the definitions and several simple properties of convex functions. These straightforward results in themselves indicate the weight of this concept.

1. CONVEX SETS AND FUNCTIONS. TESTS FOR CONVEXITY

DEFINITION. Let X denote a vector space over the field of real numbers. A set $W \subset X$ is said to be *convex* when

$$\bigwedge_{x_1, x_2 \in W} \bigwedge_{\substack{\alpha_1, \alpha_2 \geqslant 0 \\ \alpha_1 + \alpha_2 = 1}} (\alpha_1 x_1 + a_2 x_2) \in W.$$

Remark. By definition it follows immediately that for $0 \leqslant t \leqslant 1$ we have $t x_1 + (1-t) x_1 \in W$. From analytical geometry, however, we know that the equation $p = t x_1 + (1-t) x_2$ is the parametric equation of a straight line passing through the points x_1 and x_2, and for $t \in [0, 1]$ this is the segment of the straight line lying between the points x_1 and x_2. A convex set may thus be defined as one such that for any two points in it the segment joining those points also belongs to that set.

Example. In the space $X = R^3$ the set $W = K(0, l)$ $(l > 0)$ is a convex set whereas the set $W' = R^3 - K(0, l)$ is not.

LEMMA IV.1.1. $(W\text{—convex set}) \Leftrightarrow \Big(\bigwedge_{n \geqslant 2} \bigwedge_{x_1, \ldots, x_n \in W} \bigwedge_{\substack{\alpha_1, \ldots, \alpha_n \geqslant 0 \\ \alpha_1 + \ldots + \alpha_n = 1}} (\alpha_1 x_1 + \ldots +$

$+ \alpha_n x_n) \in W \Big)$.

PROOF. $\Leftarrow$: Setting in particular $n = 2$, by hypothesis we get the definition of convexity. The set W thus is convex.

$\Rightarrow$: We use induction with respect to the number n. For $n = 2$ the theorem holds (by virtue of the definition of convex set). We make the inductive assumption that the theorem is valid for $k = n-1$. We have

$$\sum_{j=1}^{n} \alpha_j x_j = \Big(\sum_{j=1}^{n-1} \alpha_j x_j \Big) + \alpha_n x_n = \alpha_0 \Big(\sum_{j=1}^{n-1} \frac{\alpha_j}{\alpha_0} x_j \Big) + \alpha_n x_n,$$

where

$$\alpha_0 := \sum_{j=1}^{n-1} \alpha_j, \quad \text{whence} \quad \sum_{j=1}^{n-1} \frac{\alpha_j}{\alpha_0} = \frac{1}{\alpha_0} \sum_{j=1}^{n-1} \alpha_j = 1.$$

Let

$$x_0 := \left(\frac{\alpha_1}{\alpha_0} x_1 + \ldots + \frac{\alpha_{n-1}}{\alpha_0} x_{n-1} \right).$$

By the inductive hypothesis, $x_0 \in W$; thus we have

$$\sum_{j=1}^{n} \alpha_j x_j = (\alpha_0 x_0 + \alpha_n x_n) \in W,$$

since

$$\alpha_0 + \alpha_n = \left(\sum_{j=1}^{n-1} \alpha_j \right) + \alpha_n = \sum_{j=1}^{n} \alpha_j = 1. \quad \square$$

DEFINITION. Let X be a real vector space, and $W \subset X$ a convex set. The function $f \colon W \ni x \to f(x) \in R^1$ is said to be *convex* if

$$\bigwedge_{x_1, x_2 \in W} \bigwedge_{\substack{\alpha_1, \alpha_2 \geqslant 0 \\ \alpha_1 + \alpha_2 = 1}} f(\alpha_1 x_1 + \alpha_2 x_2) \leqslant \alpha_1 f(x_1) + \alpha_2 f(x_2).$$

When the strict inequality occurs for $x_1 \neq x_2$, the function is *strictly convex*.

Remark. For $t \in [0, 1]$ we have $f(tx_1 + (1-t)x_2) \leqslant tf(x_1) + (1-t)f(x_2)$.

If the letter F is used to denote a function of the form $F(x) = a + l(x)$ where l is a linear function defined on the interval connecting the points x_1 and x_2, then

$$F(tx_1 + (1-t)x_2) = tF(x_1) + (1-t)F(x_2).$$

If now we require that at the end-points of this interval the function F takes on the same values as f (i.e. that $F(x_1) = f(x_1)$, $F(x_2) = f(x_2)$), we get

$$f(tx_1 + (1-t)x_2) \leqslant tF(x_1) + (1-t)F(x_2) = F(tx_1 + (1-t)x_2).$$

A convex function may thus be defined as one which, along the entire interval between any two points, assumes values not greater than the function $a + l(\cdot)$ equal to it at the end-points of that interval.

LEMMA IV.1.2. (A set $W \subset X$ is convex, f is a convex function on W)

$$\Leftrightarrow \left(\bigwedge_{n \geqslant 2} \bigwedge_{x_1, \ldots, x_n \in W} \bigwedge_{\substack{\alpha_1, \ldots, \alpha_n \geqslant 0 \\ \alpha_1 + \ldots + \alpha_n = 1}} f\left(\sum_{j=1}^{n} \alpha_j x_j \right) \leqslant \sum_{j=1}^{n} \alpha_j f(x_j) \right)$$

(the strict inequality holds for a strictly convex function whenever not all x_j coincide).

The proof is similar to the proof of the preceding lemma.

LEMMA IV.1.3. (Functions $f_1, f_2, \ldots, f_n$ are convex on a convex set W; $c_1, c_2, \ldots, c_n \geqslant 0$) $\Rightarrow$ ($c_1 f_1 + c_2 f_2 + \ldots + c_n f_n$ is a function convex on the set W).

PROOF. If $c \geqslant 0$ and f is a convex function, cf is also a convex function since

$$cf(tx_1 + (1-t)x_2) \leqslant c[tf(x_1) + (1-t)f(x_2)]$$
$$= tcf(x_1) + (1-t)cf(x_2).$$

The sum of convex functions is a convex function; addition of the conditions

$$f_i(tx_1 + (1-t)x_2) \leqslant tf_i(x_1) + (1-t)f_i(x) \quad \text{for } i = 1, 2, \ldots, n$$

yields the convexity condition for $f = \sum_{i=1}^{n} f_i$.

It is seen, therefore, that any linear combination of convex functions with non-negative coefficients is also a convex function. $\square$

Remark. Frequently, a function satisfying the condition

$$f(\alpha_1 x_1 + \alpha_2 x_2) \geqslant \alpha_1 f(x_1) + \alpha_2 f(x_2)$$

$$\text{for } \alpha_1, \alpha_2 \geqslant 0, \quad \alpha_1 + \alpha_2 = 1,$$

is called a *concave function*. Plainly, when $c \leqslant 0$ and f is concave, $\tilde{f} = cf$ is convex and, conversely, when f is convex, cf is a concave function.

DEFINITION. If $m_1, m_2, \ldots, m_n \geqslant 0$ are masses at the points $x_1, x_2, \ldots, x_n$, respectively, the *centre of gravity* of such a system is the point X_M:

$$X_M = \frac{m_1 x_1 + m_2 x_2 + \ldots + m_n x_n}{m_1 + m_2 + \ldots + m_n} = \sum_{j=1}^{n} \frac{m_j}{M} x_j$$

(the symbol M stands for $M = \sum_{j=1}^{n} m_j$, that is, the mass of the entire system).

COROLLARY IV.1.4. If the set W is convex, then for any system of masses distributed at points belonging to W, the centre of gravity of the system belongs to W.

The proof is implied by the definition of convex sets (note that for numbers $\alpha_j = m_j/M$ we have $\alpha_1, \ldots, \alpha_n \geqslant 0$ and $\alpha_1 + \ldots + a_n = 1$). $\square$

Let us now formulate a necessary and sufficient condition for the convexity of a function of one variable.

THEOREM IV.1.5. (The function f is (strictly) convex on $W \subset R^1$) $\Leftrightarrow$ (For every $a \in W$, the function (*difference quotient*)

$$f_a : W - \{a\} \ni x \to f_a(x) = \frac{f(x) - f(a)}{x - a}$$

is a (increasing) non-decreasing function).

PROOF. $\Rightarrow$: Suppose $x > a$ and $x' = ta + (1-t)x$, $0 < t < 1$. Then

(1) $\qquad x' - a = (1-t)(x-a)$.

On making use of the fact that the function f is convex, we have

(2) $\qquad f(x') = f\big(ta + (1-t)x\big) \leqslant tf(a) + (1-t)f(x)$.

This yields

(3) $\qquad f(x') - f(a) \leqslant (1-t)\big(f(x) - f(a)\big)$,

whence

(4) $\qquad f_a(x') = \dfrac{f(x') - f(a)}{x' - a} \leqslant \dfrac{(1-t)\big(f(x) - f(a)\big)}{(1-t)(x-a)} = \dfrac{f(x) - f(a)}{x - a}$

$\qquad = f_a(x)$.

Since for any $x' \in \,]a, x[$ it is possible to find a $t \in \,]0, 1[$ such that $x' = ta + (1-t)x$, in the same manner we can prove for every such x' that $f_a(x') \leqslant f_a(x)$, that is, that f_a is a non-decreasing function (when the strict inequality holds, f_a is an increasing function). The proof is identical for $x < a$.

$\Leftarrow$: Let $0 < t < 1$. We then write $x' = ta + (1-t)x$; consequently, equation (1) holds. Since f_a is a non-decreasing function, inequality (4) is valid, that is

$$\frac{f(x') - f(a)}{x' - a} \leqslant \frac{(1-t)\big(f(x) - f(a)\big)}{(1-t)(x-a)}.$$

On taking (1) into account, we thus arrive at (2), and hence f is convex. $\square$

When a function is differentiable, the following extremely useful criterion is valid:

THEOREM IV.1.6. f is a function differentiable on a convex set $W \subset R^1$, (f' is (increasing) non-decreasing) $\Leftrightarrow$ (f is (strictly) convex).

PROOF. $\Rightarrow$ (a.a.): (We give only the proof of the implication $\Rightarrow$, i.e. proof of the sufficiency of the condition (f' — non-decreasing).) Suppose that f is not convex. In that event, by the preceding theorem, there exist a point $a \in W$ and points $x_1 > x_2 > a$ or points $a > x_1 > x_2$, such that $f_a(x_1) < f_a(x_2)$.

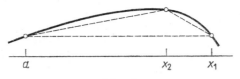

Fig. 2

Suppose, for instance, that the first case occurs (see Fig. 2):

$$\frac{f(x_1)-f(a)}{x_1-a} < \frac{f(x_2)-f(a)}{x_2-a} = l.$$

Consequently, $f(x_1) < f(a)+l(x_1-a)$ and $f(x_2) = f(a)+l(x_2-a)$. Therefore

$$\frac{f(x_1)-f(x_2)}{x_1-x_2} < \frac{f(a)+l(x_1-a)-f(x_2)}{x_1-x_2} = \frac{l(x_1-x_2)}{x_1-x_2} = l.$$

Hence we have obtained the inequality

(5) $$\frac{f(x_1)-f(x_2)}{x_1-x_2} < \frac{f(x_2)-f(a)}{x_2-a}.$$

From Lagrange's mean-value theorem of differential calculus (III.2.3), however, we know that there exist points ξ, η; $\xi \in]a, x_2[$; $\eta \in]x_2, x_1[$, such that

$$f'(\xi) = \frac{f(x_2)-f(a)}{x_2-a}, \quad f'(\eta) = \frac{f(x_1)-f(x_2)}{x_1-x_2}.$$

Consequently, it follows from (5) that $f'(\eta) < f'(\xi)$. But $\xi < \eta$, and thus we have a contradiction with the hypothesis that the derivative f' is non-decreasing. $\square$

THEOREM IV.1.7. The function f is twice-differentiable on a convex set $W \subset R^1$; (f is convex) $\Leftrightarrow$ ($f'' \geqslant 0$ on W).

PROOF. $\Rightarrow$: By Theorem IV.1.6 it follows that f' is a non-decreasing function, whence $(f')' = f'' \geqslant 0$.

$\Leftarrow$: $(f')' \geqslant 0$, so that consequently f' is a non-decreasing function. Thus, it follows from Theorem IV.1.6 that f is convex.

COROLLARY. IV.1.8. The exponential function is strictly convex over all of R^1.

PROOF. We have $(\exp x)'' = (\exp x)' = \exp x > 0$. $\square$

THEOREM IV.1.9 (On the Weighted Geometric Mean and the Weighted Arithmetic Mean).

Hypothesis: $0 < b_1, ..., b_n$; $0 < \alpha_1, ..., \alpha_n$; $\alpha_1 + ... + \alpha_n = 1$.

Thesis: $b_1^{\alpha_1} b_2^{\alpha_2} ... b_n^{\alpha_n} \leqslant \alpha_1 b_1 + \alpha_2 b_2 + ... + \alpha_n b_n$; the equality holding only for $b_1 = b_2 = ... = b_n$.

PROOF. Since $b_j > 0$, we can denote $a_j = \log b_j$ $(e^{a_j} = b_j)$. An exponential function, however, is convex, and hence $b_1^{\alpha_1} \cdot b_2^{\alpha_2} \cdot ... \cdot b_n^{\alpha_n}$ $= \exp(\alpha_1 a_1 + ... + \alpha_n a_n) \leqslant \alpha_1 e^{a_1} + ... + \alpha_n e^{a_n} = \alpha_1 b_1 + ... + \alpha_n b_n$. $\square$

COROLLARY IV.1.10 (On Arithmetic and Geometric Means). If $\alpha_i \underset{i}{\equiv} \dfrac{1}{n}$, then we get

$$\sqrt[n]{b_1 \cdot b_2 \cdot ... \cdot b_n} = b_1^{1/n} \cdot ... \cdot b_n^{1/n} \leqslant \frac{1}{n} b_1 + ... + \frac{1}{n} b_n$$

$$= \frac{b_1 + b_2 + ... + b_n}{n}.$$

Consequently, the arithmetic mean of numbers greater than zero is not smaller than their geometric mean.

2. CONVEXITY AND SEMICONTINUITY

To begin with, let us recall Corollary II.7.9: (the function $f: X \to R^1$ is continuous) $\Leftrightarrow$ $\left((i) \bigwedge_{c \in R^1} \{x \in X: f(x) > c\}$ is open and (ii) $\bigwedge_{c \in R^1} \{x \in X: f(x) < c\}$ is open$\right)$.

The proof followed immediately from the fact that the open half-lines $\{t \in R^1 : t > c\}$ and $\{t \in R : t < a\}$ form a *subbase of the topology* of R^1, i.e. that on taking (finite) intersections of these and then arbitrary set-theoretical sums, we can get an arbitrary open set in R^1. The continuity (in the neighbourhood sens) of a mapping $T: (X, \mathcal{T}) \to (Y, \mathcal{T}_1)$ may be formulated, after all, in the following form:

2. CONVEXITY AND SEMICONTINUITY

Let $\mathscr{B}_1$ be the subbase of the topology $\mathscr{T}_1$ (space Y). (The mapping $T: X \to Y$ is continuous) $\Leftrightarrow \left(\bigwedge_{\mathscr{U} \in \mathscr{B}_1} T^{-1}(\mathscr{U}) \text{ is open, i.e. } \subset \mathscr{T} \right)$.

1. Semicontinuity. If only part (i) (or (ii)) of the continuity condition in Corollary II.7.9 is preserved, we obtain a broader class of functions, that of lower (upper) semicontinuous functions.

DEFINITION. The function $f: X \to R^1$ is *lower (upper) semicontinuous* when $\bigwedge_{c \in R^1} \{x \in X: f(x) > c \ (< c)\}$ is open.

It is thus seen that the function $f: X \to R^1$ is continuous if and only if it is simultaneously lower and upper semicontinuous.

We shall refer to semicontinuous functions on many occasions in the sequel (in the theory of Radon integrals and in Chapter XII). At this point we shall merely note that their role in problems of extrema, e.g. in calculus of variations (which deals with the extrema of functions written as an integral; cf. Chapter VII), is justified by the following theorem:

THE GENERALIZED THEOREM OF WEIERSTRASS IV.2.1. A function which is lower (upper) semicontinuous on a compact space X attains its lower (upper) bound.

The proof can be found in Part II.

The weaker a topology on X, that is the fewer open sets there are, the more numerous the compact sets but the fewer continuous functions.

A natural question comes to mind in relation to problems of minima: to what extent should the initial topology be weakened so that many compact sets appear and the occurring functions be still lower semicontinuous ? This problem is particularly important in calculus of variations: the spaces X occurring there are (for the most part) spaces of functions which are continuous or continuously differentiable. However, a closed ball in $C([a, b])$, $d(f, g) := \|f - g\|$, where $\|f\| := \sup |f([a, b])|$, is not compact (a closed ball in a Banach space is compact only when that space is finite-dimensional). This ball is compact, on the other hand, in the so-called weak topology (cf. Part II). By the Generalized Theorem of Weierstrass, we see that in calculus of variations it is of paramount importance to investigate the weak-topology semicontinuity of functions of the form

$$C^1([a, b]) \ni x \rightarrow f(x) := \int_a^b L\left(t, x(t), \frac{dx}{dt}(t)\right) dt \in \mathbf{R}^1,$$

where L is a given function in three arguments. Functionals of this type keep cropping up in theoretical mechanics (there, L is called a *Lagrangian* (the difference between kinetic and potential energy)) and in differential geometry (cf. Chapter X and the final chapters of Part III) (the length of the arc of a curve, etc., cf. K. Maurin, *Methods of Hilbert Spaces*, Warszawa 1967).

A serious study of these topics will be possible only when a greater store of mathematical knowledge becomes available. The remarks presented above were intended to convince the reader that such a seemingly "farfetched" concept as semicontinuity does have a perfectly tangible "mechanical" meaning.

2. Convexity and Semicontinuity. Let $f: D \rightarrow \mathbf{R}^1$. Let

$$[f, D] := \{(x, y) \in D \times \mathbf{R}^1: y \geqslant f(x)\},$$

that is, let $[f, D]$ be the set of points of the product $\mathbf{R}^1 \times X$ which lie "over" the graph of f. It is easy to verify

PROPOSITION IV.2.2. If D is a convex subset of a real vector space, then (f is a convex function) $\Leftrightarrow$ (the set $[f, D]$ is convex).

PROPOSITION IV.2.3. Let D be a metric (or topological) space. Then (function f is lower semicontinuous) $\Leftrightarrow$ ($[f, D]$ is closed).

3. The Minkowski Functional. Minkowski found a close relationship between convex sets in a real vector space X and seminorms.

DEFINITION. A function $p: X \rightarrow \mathbf{R}_+$ is called a *seminorm* when it is
  (i) convex,
  (ii) homogeneous, i.e. $p(t \cdot x) = |t| p(x)$.
A set $M \subset X$ is said to be *absorbing* when $\bigwedge_x \bigvee_{\mathcal{N}} \bigwedge_{\mathcal{N} < a \in R} x \in aM$, where $aM = \{x \in X: x = am, m \in M\}$, and *balanced* when $(x \in M, |t| < 1) \Rightarrow (tx \in M)$.

Each balanced, convex, and absorbing set $M \subset X$ can be assigned a function $p_M(x) := \inf\{\lambda > 0: x \in \lambda M\}$.

LEMMA IV.2.4. p_M is a seminorm.

DEFINITION. The seminorm p_M is called the *Minkowski functional of the set M* (or the *indicatrix of the set M*).

Conversely as well, every seminorm p can be assigned a convex absorbing set $M_p := \{x \in X : p(x) < 1\}$.

It is left to the reader to prove the following relations between indicatrices and sets:

THEOREM IV.2.5.

(i) $(R^1 \ni a \neq 0) \Rightarrow (p_{aM} = |a|^{-1} p_M)$,

(ii) $(M_1 \subset M_2) \Leftrightarrow (p_{M_2} \leqslant p_{M_1})$,

(iii) $p_{M_1 \cap M_2} = \sup(p_{M_1}, p_{M_2})$.

4. Polar. Let X^* denote the set of linear forms on a vector space X, the so-called *space algebraically dual to X*.

Remark. When X is a topological vector space (i.e. algebraic operations are continuous), one may speak of continuous linear forms on X; the set of continuous linear forms on X is designated by X'. The concept of polars appeared in analytical geometry.

DEFINITION. Let $A \subset X$; the *polar* of the set A is the set $A^0 \subset X^*$ defined as

$$A^0 := \{x^* \in X^* : |\langle x, x^* \rangle| \leqslant 1 \text{ for } x \in A\},$$

where, as usual, we write $\langle x, x^* \rangle$ instead of $x^*(x)$.

Polars have many interesting properties and play a fundamental role in the theory of duality of locally convex spaces. The reader should verify the following

PROPOSITION. IV.2.6.

(i) A^0 is a balanced, convex set in X^*,

(ii) $(A \subset B) \Rightarrow (B^0 \subset A^0)$,

(iii) $(\lambda \neq 0) \Rightarrow ((\lambda A)^0 = \lambda^{-1} A^0)$,

(iv) $(\bigcup_i A_i)^0 = \bigcap_i A_i^0$,

(v) (M is a subspace or, more generally, a cone with vertex 0, that is $\bigwedge_{a \geqslant 0} aM \subset M) \Rightarrow (M^0 = M^\perp$, where $M^\perp = \{x^* \in X^* : \langle m, x^* \rangle = 0$ for every $m \in M\}$ is the set of all linear forms orthogonal to M).

5. Convex Cones. A subset S of a real vector space X is a *cone* when

$$\bigwedge_{\lambda \geqslant 0} (x \in S) \Rightarrow (\lambda x \in S).$$

Let S be a convex cone in X, then an order in X can be defined as follows: $x \geqslant y$ when $x - y \in S$. Thus S consists of all *positive* elements in X, i.e. $(x \geqslant 0) \Leftrightarrow (x \in S)$.

Convex cones play an important role in the so-called convex analysis where linear inequalities are considered very often. Here, we are going to give—as a sample—the famous

FARKAS LEMMA. Let $a_1, \ldots, a_m \in R^n$ be linear independent, $a_0 \in R^n$ and $S := \left\{ \sum_{i=1}^{m} \lambda_i a_i \in R^n : \lambda_i \geqslant 0 \right\}$; S is the convex cone spanned by vectors $a_1, \ldots, a_m$. For every $x \in (R^n)^*$

(*) $(\langle a_i, x \rangle \leqslant 0, i = 1, \ldots, m) \Rightarrow (\langle a_0, x \rangle \leqslant 0)$

if and only if $a_0 \in S$.

PROOF. Obviously, if $a_0 \in S$ then (*) holds for every $x \in (R^n)^*$.

Let $\{a_i\}_1^m \cup \{b_j\}_1^{n-m}$ be a basis in R^n and let $\{x_i\}_1^m \cup \{y_j\}_1^{n-m}$ be the dual basis in $(R^n)^*$, i.e.

$$\langle a_i, x_j \rangle = \delta_{ij}, \quad \langle b_i, x_j \rangle = 0,$$

$$\langle a_i, y_j \rangle = 0, \quad \langle b_i, y_j \rangle = \delta_{ij},$$

for all respective i's and j's.

Let $a_0 = \sum_{i=1}^{m} \lambda_i a_i + \sum_{j=1}^{n-m} \mu_j b_j$. Since $\langle a_i, -x_j \rangle = -\delta_{ij} \leqslant 0$, $i = 1$, $\ldots, m$, it follows from (*) that $-\lambda_j = \langle a_0, -x_j \rangle \leqslant 0$, i.e. $\lambda_j \geqslant 0$. Similarly, since $\langle a_i, \pm y_j \rangle = 0, i = 1, \ldots, m$, we have $\pm \mu_j = \langle a_0, \pm y_j \rangle \leqslant 0$, i.e. $\mu_j = 0$. So, we have proved that $a_0 \in S$. $\square$

V. THE TAYLOR FORMULA
THE CONVERGENCE OF SEQUENCES OF MAPPINGS
POWER SERIES

A differentiable function is one for which a linear function is a "good" approximation: the error is an infinitesimal of higher order with respect to increments of the argument. The Mean-Value Theorem allows an increment of a function to be expressed in terms of the (first) derivative of that function. The question is whether a function, possessing continuous derivatives up to the order $n+1$, can be "well" approximated by a polynomial of degree n. In this case "well" means: the difference is an infinitesimal of order higher than $|\Delta x|^n$. The famous Taylor Formula gives an answer in the affirmative. This formula plays an enormously important role in the examination of functions (extrema!).

The next question comes to mind at once: if a function f has derivatives of arbitrarily high order, does the sequence (f_n) of Taylor approximations "converge" to f? It is immediately seen that this question is premature: we know what it means that a sequence of points is convergent in metric space, or more generally in topological space, but we do not know how to construe the convergence of a sequence of functions on topological space. Thus, there arises the extremely important problem of introducing the concept of convergence (topology) in spaces of functions or, more generally, in spaces of maps.

In the present chapter we shall become acquainted with three kinds of convergence in spaces of maps: pointwise or simple convergence, uniform convergence, and, in the case of mappings $T_n: X \to Y$ where the space X is locally compact, compact convergence. Subsequent chapters will acquaint us with other types of convergence of function sequences. Accordingly, the present chapter is a first look at a vast field: the topology of function spaces (called functional analysis at one time).

1. GENERALIZATION OF THE MEAN-VALUE THEOREM OF INTEGRAL CALCULUS

LEMMA V.1.1 (Generalized Mean-Value Theorem of Integral Calculus).

$$(\varphi, \psi \in C([a, b]); \psi \geqslant 0) \Rightarrow \Big(\bigvee_{a \leqslant \xi \leqslant b} \int_a^b \varphi\psi = \varphi(\xi) \int_a^b \psi \Big).^1$$

Remark. On setting $\psi \equiv 1$, we get the ordinary Mean-Value Theorem of Integral Calculus.

---

[1] $C(A)$, where $A \subset R^1$, denotes the set of all continuous functions $f: A \to R^1$.

PROOF. We denote $m = \inf \varphi([a, b])$, $M = \sup \varphi([a, b])$. Then

$$[a, b] \ni x \to \varphi(x) \int_a^b \psi$$

is a continuous function. Since $m \leqslant \varphi(x) \leqslant M$, then

$$m \int_a^b \psi \leqslant \varphi(x) \int_a^b \psi \leqslant M \int_a^b \psi.$$

However,

$$m \int_a^b \psi = \int_a^b m\psi \leqslant \int_a^b \varphi\psi \leqslant \int_a^b M\psi = M \int_a^b \psi.$$

We know that a function which is continuous on an open connected set ranges over all values between the bounds. Hence, the function $\varphi(x) \int_a^b \psi$ assumes all values lying between $m \int_a^b \psi$ and $M \int_a^b \psi$; in particular, it assumes the value $\int_a^b \varphi\psi$, which means there exists a $\xi \in [a, b]$ such that

$$\varphi(\xi) \int_a^b \psi = \int_a^b \varphi\psi. \quad \square$$

2. THE TAYLOR FORMULA

LEMMA V.2.1. The functions $u, g \in C^n(]a, b[)$ satisfy the equation

$$u^{(n)}(t)g(t) + (-1)^{n-1}g^{(n)}(t)u(t)$$

$$= \frac{d}{dt}\left(\sum_{j=0}^{n-1}(-1)^j u^{(n-j-1)}(t)g^{(j)}(t)\right).$$

PROOF. We write out the sum on the right-hand side of the equation. We obtain

$$\frac{d}{dt}\left(u^{(n-1)}g - u^{(n-2)}g^{(1)} + u^{(n-3)}g^{(2)} + \ldots + (-1)^{(n-1)}ug^{(n-1)}\right).$$

When this expression is differentiated, all the terms vanish, with the exception of the first and last terms, which are equal to the left member of the equation. $\square$

This lemma directly implies a useful integration formula which is a generalization of the theorem on integration by parts (for $n = 1$). Namely, let us integrate both sides of the equation, in limits a and b. This gives us

$$\int_a^b u^{(n)}g + (-1)^{n-1} \int_a^b ug^{(n)} = \left[\sum_{j=0}^{n-1} (-1)^j u^{(n-j-1)} g^{(j)} \right]_a^b.$$

Taking the second term over to the right-hand side, we obtain

THEOREM V.2.2. The equation

$$\int_a^b u^{(n)}g = \left[\sum_{j=0}^{n-1} (-1)^j u^{(n-j-1)} g^{(j)} \right]_a^b + (-1)^{(n)} \int_a^b ug^{(n)}$$

holds.

THEOREM V.2.3 (The Taylor Formula).
Hypothesis: $f \in C^{n+1}(]a, b[)$, $a < x < b$.

Thesis: $f(x) = f(a) + \dfrac{x-a}{1!} f^{(1)}(a) + \dfrac{(x-a)^2}{2!} f^{(2)}(a) + \ldots +$

$$+ \frac{(x-a)^n}{n!} f^{(n)}(a) + \int_a^x \frac{(x-t)^n}{n!} f^{(n+1)}(t)\, dt .$$

PROOF. Let us make use of Theorem V.2.2, setting $g(t) = f'(t)$ and $u(t) = (t-x)^n/n!$. We calculate both sides of the equation in this theorem. Now

$$u^{(n)}(t) = \frac{d^n}{dt^n} \left(\frac{(t-x)^n}{n!} \right) \equiv 1 ,$$

and therefore

$$\int_a^x u^{(n)}g = \int_a^x 1 \cdot f' = f(x) - f(a).$$

Since $g^{(k)} = f^{(k+1)}$, then

$$(-1)^n \int_a^x u(t) g^{(n)}(t)\, dt = (-1)^n \int_a^x \frac{(t-x)^n}{n!} f^{(n+1)}(t)\, dt$$

$$= \int_a^x \frac{(x-t)^n}{n!} f^{(n+1)}(t)\, dt .$$

Moreover,

$$u^{(n-j-1)}(t) = \frac{(t-x)^{n-(n-j-1)}}{n!} n(n-1)(n-2) \dots$$

$$((n-(n-j-2)) = \frac{(t-x)^{j+1}}{(j+1)!},$$

and thus

$$(-1)^j u^{(n-j-1)}(t) g^{(j)}(t) = (-1)^j \frac{(t-x)^{j+1}}{(j+1)!} f^{(j+1)}(t)$$

$$= -\frac{(x-t)^{j+1}}{(j+1)!} f^{(j+1)}(t);$$

$$[(-1)^j u^{(n-j-1)}(t) g^{(j)}(t)]_a^x = -\frac{(x-x)^{j+1}}{(j+1)!} f^{(j+1)}(x) +$$

$$+\frac{(x-a)^{j+1}}{(j+1)!} f^{(j+1)}(a) = \frac{(x-a)^{j+1}}{(j+1)!} f^{(j+1)}(a).$$

Finally, on rearranging both sides of the equation from Theorem V.2.2, we obtain

$$f(x)-f(a) = \sum_{j=0}^{n-1} \frac{(x-a)^{j+1}}{(j+1)!} f^{(j+1)}(a) + \int_a^x \frac{(x-t)^n}{n!} f^{(n+1)}(t) dt.$$

We set $k = j+1$; since $j = 0, 1, 2, \dots, n-1$, therefore $k = 1, 2, \dots, n$. In this manner we finally arrive at

$$f(x) = f(a) + \sum_{k=1}^{n} \frac{(x-a)^k}{k!} f^{(k)}(a) + \int_a^x \frac{(x-t)^n}{n!} f^{(n+1)}(t) dt. \quad \square$$

Remark. If we denote $f^{(0)} := f$ (the zeroth derivative of a function is the function itself) and $0! := 1$, this formula may be rewritten (since

$f(a) = \dfrac{(x-a)^0}{0!} f^{(0)}(a)$) in a different form

$$f(x) = \sum_{k=0}^{n} \frac{(x-a)^k}{k!} f^{(k)}(a) + \int_a^x \frac{(x-t)^n}{n!} f^{(n+1)}(t) dt.$$

The last term is often called the *n-th remainder* and is denoted by $R_n(a, x)$. We shall prove that $R_n(x, a)$ can be written in still different forms (in the so-called *Lagrange* and *Cauchy forms*):

THEOREM V.2.4 (Lagrange).

$$\bigvee_{0 \leqslant \theta \leqslant 1} R_n(a, x) = \frac{(x-a)^{n+1}}{(n+1)!} f^{(n+1)}(a+\theta(x-a)).$$

This can, of course, be rewritten as

$$\bigvee_{\xi \in [a, x]} R_n(a, x) = \frac{(x-a)^{n+1}}{(n+1)!} f^{(n+1)}(\xi).$$

THEOREM V.2.5 (Cauchy).

$$\bigvee_{0 \leqslant \theta \leqslant 1} R_n(a, x) = \frac{(x-a)^{n+1}}{n!} (1-\theta)^n f^{(n+1)}(a+\theta(x-a)).$$

Remark. It is possible to demonstrate the existence of yet another form of the remainder (the *Schlömilch form*):

$$\bigwedge_{p > 0} \bigvee_{a < \xi < x} R_n(a, x) = \frac{(x-a)^p (x-\xi)^{n+1-p}}{pn!} f^{(n+1)}(\xi).$$

On setting $p = 1$, we get Cauchy's form for the remainder from this.

PROOF OF THEOREM V.2.4. By Theorem V.2.3 we have

$$R_n(a, x) = \int_a^x \frac{(x-t)^n}{n!} f^{(n+1)}(t) \, dt.$$

Let us employ Lemma V.1.1 (on the mean value), taking $\varphi = f^{(n+1)}$; $\psi = (x-t)^n/n!$. There exists a point $\xi \in [a, x]$ such that

$$R_n(a, x) = \int_a^x \psi\varphi = \varphi(\xi)\int_a^x \psi = f^{(n+1)}(\xi) \cdot \frac{1}{n!} \int_a^x (x-t)^n dt$$

$$= f^{(n+1)}(\xi) \frac{1}{n!} \left[(-1) \frac{1}{n+1} (x-t)^{n+1} \right]_a^x$$

$$= \frac{(x-a)^{n+1}}{(n+1)!} f^{(n+1)}(\xi). \quad \square$$

PROOF OF THEOREM V.2.5. Let us denote

$$R_n(a, x) = \int_a^x \frac{(x-t)^n}{n!} f^{(n+1)}(t)\,dt = \int_a^x \varphi(t)\,dt,$$

where $\varphi(t) := \dfrac{(x-t)^n}{n!} f^{(n+1)}(t).$

Next, we apply the elementary Mean-Value Theorem to the entire integrand. By virtue of this theorem, there exists a $\theta \in [0, 1]$ (or a $\xi \in [a, x]$; $\xi = a + \theta(x-a)$), such that

$$R_n(a, x) = \int_a^x \varphi(t)\,dt = (x-a) \cdot \varphi(\xi) = (x-a) \cdot \varphi\big(a + \theta(x-a)\big)$$

$$= (x-a) \frac{[x-a+\theta(a-x)]^n}{n!} f^{(n+1)}\big(a+\theta(x-a)\big)$$

$$= (x-a) \frac{[(1-\theta)(x-a)]^n}{n!} f^{(n+1)}\big(a+\theta(x-a)\big)$$

$$= \frac{(x-a)^{n+1}}{n!} (1-\theta)^n \cdot f^{(n+1)}\big(a+\theta(x-a)\big). \quad \square$$

Remark. Using Theorem V.2.4, when we replace a by x_0 and x by $x_0 + \Delta x$ (that is we replace $(x-a)$ by Δx), we obtain

$$f(x_0 + \Delta x) = \sum_{k=0}^{n} \frac{(\Delta x)^k}{k!} f^{(k)}(x_0) +$$

$$+ \frac{(\Delta x)^{n+1}}{(n+1)!} f^{(n+1)}\big(x_0 + \theta(\Delta x)\big), \qquad 0 \leqslant \theta \leqslant 1.$$

If the last term is denoted by $R_n(x_0 + \Delta x, x_0) =: r_n(x_0, \Delta x)$, it can be shown that the following lemma holds.

LEMMA V.2.6. $\dfrac{r_n(x_0, \Delta x)}{(\Delta x)^n} \xrightarrow[\Delta x \to 0]{} 0$, which means that the n-th remainder

tends to zero more rapidly than does the n-th power of the increment Δx.

PROOF.

$$\lim_{\Delta x \to 0} \frac{r_n(x_0, \Delta x)}{(\Delta x)^n} = \lim_{\Delta x \to 0} \left[\frac{\Delta x}{(n+1)!} f^{(n+1)}(x_0 + \theta(\Delta x)) \right]$$

$$= \left[\lim_{\Delta x \to 0} \frac{\Delta x}{(n+1)!} \right] \cdot \left[\lim_{\Delta x \to 0} f^{(n+1)}(x_0 + \theta(\Delta x)) \right]$$

$$= 0 \cdot f^{(n+1)}(x_0) = 0. \quad \square$$

Summarizing the results of this section, we may say that a function which is $(n+1)$ times differentiable in a connected neighbourhood of a point x_0, allows itself to be represented in that neighbourhood in the form

$$f(x_0 + \Delta x) = f(x_0) + \frac{\Delta x}{1!} f^{(1)}(x_0) + \frac{(\Delta x)^2}{2!} f^{(2)}(x_0) + \cdots +$$

$$+ \frac{(\Delta x)^n}{n!} f^{(n)}(x_0) + r_n(x_0, \Delta x).$$

An analogous formula can be obtained for functions of the class $C^n([a, b])$.

We have proved three forms for the n-th remainder R_n and shown that this remainder tends to zero more rapidly than does the n-th power of Δx.

Example. Let $f(x) = e^x$, and set $x_0 = 0$. The exponential function is differentiable an arbitrary number of times at zero, and thus we can write

$$e^x = 1 + \frac{x}{1!} + \frac{x^2}{2!} + \cdots + \frac{x^n}{n!} + \frac{x^{n+1}}{(n+1)!} e^{\theta x}, \quad 0 \leqslant \theta \leqslant 1.$$

Here we have made use of the fact that $f(0) = e^0 = 1$ and that $f^{(k)}(x) = e^x$. The remainder has been written in the Lagrange form. Note that

$$|r_n(0, x)| = \left| \frac{x^{n+1}}{(n+1)!} e^{\theta x} \right| \leqslant \frac{e^{|x|}}{(n+1)!} |x|^{n+1}.$$

We shall show that $a^n/n! \xrightarrow[n \to \infty]{} 0$ when $a > 0$. By n_0 we denote an arbitrary integer greater than a. Thus, for $n > n_0$ we have

$$\frac{a^n}{n!} = \frac{a^{n_0}}{n_0!} \cdot \frac{a}{n_0+1} \cdot \frac{a}{n_0+2} \cdot \ldots \cdot \frac{a}{n} < \frac{a^{n_0}}{n_0!} \cdot 1 \cdot 1 \cdot \ldots \cdot \frac{a}{n}$$

$$= \frac{a^{n_0+1}}{n_0!} \cdot \frac{1}{n}$$

and consequently

$$0 < \frac{a^n}{n!} < \frac{a^{n_0+1}}{n_0!} \cdot \frac{1}{n} \xrightarrow[n \to \infty]{} 0$$

(for $a^{n_0+1}/n_0!$ is a constant factor; if we denote it by A, then $A/n \xrightarrow[n \to \infty]{} 0$).
Hence the conclusion that $a^n/n! \xrightarrow[n \to \infty]{} 0$.

If now in our case we put $a = |x|$, we get

$$0 \leqslant |r_n(0, x)| \leqslant \frac{e^{|x|}}{(n+1)!} |x|^{n+1} \xrightarrow[n \to \infty]{} 0.$$

Hence

$$r_n(0, x) = \left(e^x - \sum_{k=0}^{n} \frac{x^k}{k!}\right) \xrightarrow[n \to \infty]{} 0.$$

Therefore

$$\lim_{n \to \infty} r_n(0, x) = \lim_{n \to \infty} e^x - \lim_{n \to \infty} \sum_{k=0}^{n} \frac{x^k}{k!} = e^x - \lim_{n \to \infty} \sum_{k=0}^{n} \frac{x^k}{k!} = 0.$$

Finally

$$e^x = \lim_{n \to \infty} \sum_{k=0}^{n} \frac{x^k}{k!} =: \sum_{k=0}^{\infty} \frac{x^k}{k!}. \quad \square$$

Incidentally we have defined the sum of an infinite number of terms a_k. Namely,

$$\sum_{k=0}^{\infty} a_k := \lim_{n \to \infty} \sum_{k=0}^{n} a_k.$$

If this limit does not exist, the given sequence (a_n) is said *not to have the sum*.

Remark. As is evident from this straightforward example, it is in general a tedious matter to estimate the remainder. It is therefore desirable to avoid this labour, and this fortunately is possible in the most important cases (cf. the chapter on holomorphic functions in Part II).

3. APPLICATIONS OF THE TAYLOR FORMULA

DEFINITION. Let (X, d) be a metric space; $f: X \ni x \to f(x) \in R^1$. A function f is said to have a *local maximum* (*minimum*) at a point x_0 when

$$\bigvee_{r>0} \bigwedge_{x \in K(x_0, r)} f(x) \leqslant f(x_0) \quad (f(x) \geqslant f(x_0)).$$

THEOREM V.3.1 (Necessary Condition for the Existence of an Extremum). $(f \in C^1(\Omega),\ \Omega$ is open $\subset R^1$, at a point $x_0 \in \Omega$ the function f has an extremum, i.e. has a local maximum or minimum) $\Rightarrow (f'(x_0) = 0)$.

PROOF. Let f have, say, a local minimum at the point x_0 (the proof for a maximum is similar). Then the difference quotient is

$$f_{x_0}(x) = \frac{f(x) - f(x_0)}{x - x_0} \begin{cases} \geqslant 0 \text{ for } x > x_0, \\ \leqslant 0 \text{ for } x < x_0. \end{cases}$$

Since the function is differentiable, if we take a positive sequence (ε_n), $\varepsilon_n \xrightarrow[n \to \infty]{} 0$, we obtain

$$f'(x_0) = \lim_{n \to \infty} \frac{f(x_0 + \varepsilon_n) - f(x_0)}{\varepsilon_n} \geqslant 0,$$

but we also get

$$f'(x_0) = \lim_{n \to \infty} \frac{f(x_0 - \varepsilon_n) - f(x_0)}{-\varepsilon_n} \leqslant 0.$$

Hence $0 \leqslant f'(x_0) \leqslant 0$, that is $f'(x_0) = 0$. $\square$

THEOREM V.3.2. (Sufficient Condition for the Existence of a Local Extremum). $(f \in C^{2k+1}(\Omega),\ \Omega - $ open $\subset R^1, k \in N,\ x_0 \in \Omega,\ f^{(1)}(x_0) = f^{(2)}(x_0) = \ldots = f^{(2k-1)}(x_0) = 0,\ 0 \neq f^{2k}(x_0) < 0\ (> 0)) \Rightarrow (f$ has a local maximum (minimum) at the point $x_0)$.

PROOF. Applying the Taylor Formula (we assume that $f^{2k}(x_0) > 0$), we have

$$f(x) - f(x_0) = (x - x_0)f^{(1)}(x_0) + \ldots + \frac{(x - x_0)^{(2k-1)}}{(2k-1)!} f^{2k-1}(x_0) +$$

$$+ \frac{(x - x_0)^{2k}}{(2k)!} f^{(2k)}(x_0) + r_{2k}(x_0, x - x_0)$$

$$= \frac{(x - x_0)^{2k}}{(2k)!} f^{(2k)}(x_0) + r_{2k}(x_0, x - x_0).$$

But, as we have proven, the remainder tends to zero more rapidly than $(x-x_0)^{2k}$ does. Hence, there exists a number $\varepsilon > 0$ such that

$$|r_{2k}(x_0, x-x_0)| \leqslant \left| \frac{(x-x_0)^{2k}}{(2k)!} f^{(2k)}(x_0) \right|$$

when $|x-x_0| \leqslant \varepsilon$. Thus, if $|x-x_0| \leqslant \varepsilon$, then $f(x)-f(x_0) \geqslant 0$ because

$$\frac{(x-x_0)^{2k}}{(2k)!} \geqslant 0 \quad \text{and} \quad f^{(2k)}(x_0) > 0.$$

Consequently f has a local minimum at x_0. If $f^{(2k)} < 0$, in the same way we get $f(x)-f(x_0) \leqslant 0$, and hence f has a local maximum at x_0. $\square$

DEFINITION. A function $f \in C^r(\Omega)$ is said to have at a point $x_0 \in \Omega$, *zero of order* k $(k \leqslant r)$ if

$$f(x_0) = f^{(1)}(x_0) = f^{(2)}(x_0) = \ldots = f^{(k-1)}(x_0) = 0,$$

whereas $f^{(k)}(x_0) \neq 0$.

We shall now prove two theorems which are frequently referred to as *de l'Hospital rules*.

THEOREM V.3.3.

Hypothesis: $f, g \in C^r(\Omega)$; $R^1 \supset \Omega$ open; $x_0 \in \Omega$; f has zero of order $k \leqslant r$ at x_0; g has zero of order $l \leqslant r$ at x_0; $\varphi(x) = \dfrac{f(x)}{g(x)}$ whenever defined.

Thesis:

(i) If $k > l$, then $\lim\limits_{x \to x_0} \varphi(x) = 0$,

(ii) if $k < l$, then $\lim\limits_{x \to x_0} |\varphi(x)| = \infty$,

(iii) if $k = l$, then $\lim\limits_{x \to x_0} \varphi(x) = \dfrac{f^{(k)}(x_0)}{g^{(x)}(x_0)}$.

PROOF. By the Taylor Formula, taking the remainder in the Lagrange form we have

$$f(x) = \frac{(x-x_0)^k}{k!} f^{(k)}(x_0 + \theta(x-x_0)),$$

$$g(x) = \frac{(x-x_0)^l}{l!} g^{(l)}((x_0 + \theta(x-x_0))$$

106

(the rest of the terms are zero), and hence

$$\lim_{x \to x_0} \varphi(x) = \frac{l!}{k!} \lim_{x \to x_0} \frac{(x-x_0)^k}{(x-x_0)^l} \cdot \lim_{x \to x_0} \frac{f^{(k)}(x_0+\theta(x-x_0))}{g^{(l)}(x_0+\theta(x-x_0))}$$

$$= \frac{l!}{k!} \cdot \left(\lim_{x \to x_0} \frac{(x-x_0)^k}{(x-x_0)^l} \right) \cdot \frac{f^{(k)}(x_0)}{g^{(l)}(x_0)};$$

and this precisely yields the thesis. $\square$

THEOREM V.3.4.

Hypothesis: $]a, b[\ni x \to \dfrac{f(x)}{g(x)}; f', g'$ continuous in $]a, b[$; $g(x) \nearrow \infty$

as $x \to b$ (the rising arrow indicates that g is non-decreasing, that is

$g' \geqslant 0$) and $\lim\limits_{x \to b} \dfrac{f'(x)}{g'(x)} = G$.

Thesis: The limit $\lim\limits_{x \to b} \dfrac{f(x)}{g(x)}$ exists and is equal to G.

PROOF. By definition we have

$$\bigwedge_{\varepsilon > 0} \bigvee_{c \in]a, b[} \bigwedge_{c < x < b} G - \varepsilon < \frac{f'(x)}{g'(x)} < G + \varepsilon.$$

Since $g'(x) \geqslant 0$, then $(G-\varepsilon)g'(x) < f'(x) < (G+\varepsilon)g'(x)$ for $c < x < b$.
An integral is a positive functional (which means that an integral of
a greater function is greater) and therefore

$$\int_c^y (G-\varepsilon)g'(x)\,dx < \int_c^y f'(x)\,dx < \int_c^y (G+\varepsilon)g'(x)\,dx$$

for $y \in]c, b[$.

Integration yields

$$(G-\varepsilon)\left(g(y)-g(c)\right) < f(y)-f(c) < (G+\varepsilon)\left(g(y)-g(c)\right),$$

$$(G-\varepsilon)\left(1-\frac{g(c)}{g(y)}\right) < \frac{f(y)}{g(y)} - \frac{f(c)}{g(y)} < (G+\varepsilon)\left(1-\frac{g(c)}{g(y)}\right),$$

$$(G-\varepsilon)\left(1-\frac{g(c)}{g(y)}\right) + \frac{f(c)}{g(y)} < \frac{f(y)}{g(y)}$$

$$< (G+\varepsilon)\left(1-\frac{g(c)}{g(y)}\right) + \frac{f(c)}{g(y)}.$$

As $y \to b$, $g(y) \to \infty$. Denoting the left member of the inequality by $L(y)$, and the right by $R(y)$, we have $G - \varepsilon \underset{b \leftarrow y}{\longleftarrow} L(y) \leqslant f(y)/g(y) \leqslant R(y) \underset{y \to b}{\longrightarrow} G + \varepsilon$. Consequently

$$\frac{f(y)}{g(y)} \underset{y \to b}{\longrightarrow} G$$

because ε was taken arbitrarily small. $\square$

Remark. The theorem remains valid if we set $b = \infty$, since we shall be integrating between the limits c and $y < \infty$ and then going to the limit $y \to \infty$. Accordingly, the proof does not change at all.

COROLLARY V.3.5. For $\alpha > 0$ we have

$$\frac{\log x}{x^\alpha} \underset{x \to \infty}{\longrightarrow} 0.$$

PROOF. By Theorem V.3.4

$$\frac{(\log x)'}{(x^\alpha)'} = \frac{1/x}{\alpha x^{\alpha-1}} = \frac{1}{\alpha x^\alpha} \underset{x \to 0}{\longrightarrow} 0. \ \square$$

THEOREM V.3.6. Let $a > 1$. Then the formula

$$\frac{a^x}{x^\alpha} \underset{x \to \infty}{\longrightarrow} \infty$$

is satisfied for any $\alpha \in R^1$.

PROOF. Let $\alpha = 1$. The hypotheses of Theorem V.3.4 are then satisfied and hence

$$\frac{(a^x)'}{(x)'} = (\log a) a^x \underset{x \to \infty}{\longrightarrow} \infty.$$

In the case $\alpha \neq 1$, but for $\alpha > 0$, we have

$$\varphi(x) = \left(\frac{a^x}{x^\alpha}\right)^{1/\alpha} = \frac{(a^{1/\alpha})^x}{x} \underset{x \to \infty}{\longrightarrow} \infty.$$

The power function is an increasing function for $\alpha > 0$, whereby

$$\frac{a^x}{x^\alpha} = (\varphi(x))^\alpha \underset{x \to \infty}{\longrightarrow} \infty.$$

For $\alpha = 0$ we have $a^x \to \infty$.

108

For $\alpha < 0$ we have $\dfrac{a^x}{x^\alpha} = x^{|\alpha|}a^x \xrightarrow[x\to\infty]{} \infty$, because this is the product of two factors, each of which tends to infinity. $\square$

Landau Notation. This section will be concluded with a presentation of the definition of notation introduced by Bachmann and Landau (so-called *Landau symbols*). Let f, g be functions defined in a neighbourhood of a point x_0 (x_0 may also be $\pm\infty$; in the latter case f, g may be sequences).

When $\lim\limits_{x\to x_0} \dfrac{f(x)}{g(x)} = 0$, we write

$$f(x) = o\big(g(x)\big).$$

When $x \to \dfrac{f(x)}{g(x)}$ is bounded in the neighbourhood of x_0, we write

$$f(x) = O\big(g(x)\big).$$

In the case of sequences the notation $f(n) = o(1)$ means that $f(n) \xrightarrow[n\to\infty]{} 0$ whereas $f(n) = O(1)$ denotes that the sequence $\big(f(n)\big)$ is bounded.

4. POINTWISE AND UNIFORM CONVERGENCE OF SEQUENCES OF MAPPINGS

Let (X, d), (Y, ϱ) denote metric spaces. Let T_0, T_n ($n = 1, 2, \ldots$) be mappings $X \to Y$.

DEFINITION. The sequence of mappings T_n is said to be *pointwise convergent* to the mapping T_0 (we write $T_n \xrightarrow[n\to\infty]{} T_0$) when

$$\bigwedge_{x\in X} Y \ni T_n(x) \xrightarrow[n\to\infty]{} T_0(x) \in Y.$$

Example. Let $X = [0, 1] \subset R^1$, $Y = R^1$. Take the sequence of mappings $X \ni x \to T_n(x) = x^n \in Y$.

Note that $T_n(1) \to 1$ for $x = 1$ (because $1^n \to 1$). For $0 < x < 1$ we have $x^n \xrightarrow[n\to\infty]{} 0$. Thus, if we define the mapping

$$T_0(x) = \begin{cases} 1 & \text{for } x = 1, \\ 0 & \text{for } 0 \leqslant x < 1, \end{cases}$$

we can write $T_n \to T_0$.

We have thus shown that a sequence of continuous functions may converge pointwise to a discontinuous function.

Let (X, d), (Y, ϱ) be metric spaces. We recall that a mapping $T \colon X \to Y$ is said to be *bounded* if

$$\bigvee_{y_0 \in Y} \bigvee_{R > 0} \bigwedge_{x \in X} \varrho\big(T(x), y_0\big) < R, \quad \text{that is} \quad T(x) \in K(y_0, R).$$

By $B(X, Y)$ let us denote all bounded mappings of the space (X, d) into the space (Y, ϱ). Now, we introduce a function of pairs of elements of the set $B(X, Y)$ and, as we shall prove, this function possesses the property of a metric.

DEFINITION. If $T_1, T_2 \in B(X, Y)$ then

$$\delta(T_1, T_2) := \sup_{x \in X} \varrho\big(T_1(x), T_2(x)\big).$$

The definition is a good one since $T_1(x) \in Y$, $T_2(x) \in Y$, and hence their distances ϱ, are calculable. The supremum of these distances does exist because both mappings are bounded.

LEMMA V.4.1. The space $\big(B(X, Y), \delta\big)$ is a metric space, that is $\delta(\cdot, \cdot)$ is a distance in the set $B(X, Y)$.

PROOF. Let us verify whether δ has all the properties of a distance. We have

(i) $\delta(T_1, T_2) = \sup_x \varrho\big(T_1(x), T_2(x)\big) = \sup_x \varrho\big(T_2(x), T_1(x)\big) = \delta(T_2, T_1)$;

we have here used the symmetry of the function ϱ.

(ii) $\delta(T_1, T_2) \geqslant 0$ because this is the supremum of non-negative numbers. If $\delta(T_1, T_2) = 0$, then

$$\sup_x \varrho\big(T_1(x), T_2(x)\big) = 0.$$

None of the numbers $\varrho\big(T_1(x), T_2(x)\big)$, however, is greater than the supremum (by the definition of supremum), and thus $\varrho\big(T_1(x), T_2(x)\big) \leqslant 0$. Since ϱ is a metric ($\varrho \geqslant 0$), we have $\varrho\big(T_1(x), T_2(x)\big) \underset{x}{\equiv} 0$. Consequently, $T_1(x) \equiv T_2(x)$, that is $T_1 = T_2$. Conversely, if $T_1 = T_2$, then plainly $\delta(T_1, T_2) = 0$.

(iii) $\delta(T_1, T_2) = \sup_x \varrho\big(T_1(x), T_2(x)\big) \leqslant \sup_x [\varrho\big(T_1(x), T_3(x)\big) + \varrho\big(T_3(x),$ $T_2(x)\big)] \leqslant \sup_x \varrho\big(T_1(x), T_3(x)\big) + \sup_x \varrho\big(T_3(x), T_2(x)\big) = \delta(T_1, T_3) + \delta(T_3, T_2)$.

Use has been made here of the fact that ϱ satisfies the triangle inequality and the fact that the supremum of a sum is not greater than the sum of suprema. $\square$

DEFINITION. Let T_0, T_n $(n = 1, 2, ...)$ be mappings from X to Y. We say that the sequence (T_n) *converges uniformly* to T_0 (we write $T_n \xrightarrow[n \to \infty]{\text{uniform}} T_0$), when

$$(1) \qquad \bigwedge_{\varepsilon > 0} \bigvee_{N(\varepsilon)} \bigwedge_{n > N(\varepsilon)} \sup_x \varrho(T_n(x), T_0(x)) \leqslant \varepsilon.$$

Since $\varrho(T_n(x), T_0(x)) \leqslant \sup_x \varrho(T_n(x), T_0(x))$, the definition of uniform convergence could be rewritten in a different form:

$$\bigwedge_{\varepsilon > 0} \bigvee_{N(\varepsilon)} \bigwedge_{n > N(\varepsilon)} \bigwedge_{x \in X} \varrho(T_n(x), T_0(x)) \leqslant \varepsilon.$$

If T_0 and T_n are elements of $B(X, Y)$, the uniform convergence is just the metric convergence in $B(X, Y)$.

LEMMA V.4.2. $\left(T_n \in B(X, Y); \ T_n \xrightarrow[n \to \infty]{\text{uniform}} T_0 \right) \Rightarrow \left(T_0 \in B(X, Y) \right)$.

PROOF. Take $\varepsilon > 0$ and N such that for any $n \geqslant N$ and $x \in X$ we have $\varrho(T_n(x), T_0(x)) < \varepsilon$. However, T_N is a bounded mapping, and thus there exist $y_0 \in Y$ and $R > 0$ such that for any x

$$\varrho(T_N(x), y_0) < R.$$

Accordingly, by the triangle inequality we have

$$\varrho(T_0(x), y_0) \leqslant \varrho(T_0(x), T_N(x)) + \varrho(T_N(x), y_0) < R_0 + \varepsilon$$

for all x. Thus, for any x, $T_0(x) \subset K(y_0, R + \varepsilon)$. $\square$

THEOREM V.4.3. (The space (Y, ϱ) is complete) $\Rightarrow \left((B(X, Y), \delta) \text{ is complete} \right)$.

PROOF. We want to prove that every Cauchy sequence (T_n) has a limit $T_0 \in B(X, Y)$. Let us write out what it means that (T_n) is a Cauchy sequence:

$$(2) \qquad \bigwedge_{\varepsilon > 0} \bigvee_{N(\varepsilon)} \bigwedge_{n, m > N(\varepsilon)} \bigwedge_{x \in X} \varrho(T_n(x), T_m(x)) \leqslant \varepsilon.$$

Note that for any $x \in X$, the sequence $(T_n(x))$ (where $T_n(x) \in Y$) is a Cauchy sequence. Since the space (Y, ϱ) is complete, this sequence has a limit. Let us denote it by $T_0(x)$, that is, $T_n(x) \to T_0(x)$ in the space (Y, ϱ). We shall prove that $T_n \to T_0$ uniformly. Given a number $\varepsilon > 0$. Take an $N(\varepsilon)$ such that the condition (2) speaks of its existence. Hence,

$\bigwedge\limits_{x \in X} \varrho\big(T_n(x), T_m(x)\big) \leqslant \varepsilon$ for $n, m > N(\varepsilon)$. The metric ϱ, however, is a continuous function of its arguments, and thus

$$\lim_{m \to \infty} \varrho\big(T_n(x), T_m(x)\big) = \varrho\big(T_n(x), \lim_{m \to \infty} T_m(x)\big)$$
$$= \varrho\big(T_n(x), T_0(x)\big).$$

Since the limit of numbers not greater than ε is not greater than ε, consequently for $n > N(\varepsilon)$ we have $\varrho\big(T_n(x), T_0(x)\big) \leqslant \varepsilon$ for all x. We have thus proved that

$$\bigwedge_{\varepsilon > 0} \bigvee_{N(\varepsilon)} \bigwedge_{n > N(\varepsilon)} \bigwedge_{x \in X} \varrho\big(T_n(x), T_0(x)\big) \leqslant \varepsilon, \quad \text{i.e.} \quad T_n \xrightarrow[n \to \infty]{\text{uniform}} T_0.$$

Lemma V.4.2 implies that the limit T_0 is bounded ($T_0 \in B(X, Y)$); thus for any convergent sequence (T_n) of elements $B(X, Y)$ there exists a $T_0 \in B(X, Y)$ such that $T_n \to T_0$ uniformly, and this precisely means that the space $B(X, Y)$ is complete. $\square$

EXAMPLE. $\mathbf{R}^1 \supset X := [0, \infty[\ni x \to f_n(x) := xe^{-nx}$. Plainly, $f_n \to 0$ pointwise, but $(xe^{-nx})' = e^{-nx} - nxe^{-nx}$ vanishes only for $x = 1/n$. Since $f_n(0) = 0$ and $f_n''(1/n) = -n/e < 0$, then f_n has a maximum at $x = 1/n$;

$$\sup f_n(x) = f_n\left(\frac{1}{n}\right) = \frac{1}{ne^n} \xrightarrow[n \to \infty]{} 0.$$

Thus, $f_n \to 0$ also uniformly.

By $C(X, Y)$ we denote the *set of continuous mappings of the metric space (X, d) into the metric space (Y, ϱ)*. Let (Y, ϱ) be complete. The following theorem then holds.

THEOREM V.4.4. The set $B(X, Y) \cap C(X, Y)$ with the metric δ, is a complete space.

PROOF. This is a metric space because the conditions imposed on a metric are satisfied by δ on the set $B(X, Y)$ and hence also on the smaller set $B(X, Y) \cap C(X, Y) \subset B(X, Y)$. Let us take an arbitrary Cauchy sequence (T_n), $T_n \in B(X, Y) \cap C(X, Y)$. Since $T_n \in B(X, Y)$, it follows from Theorem V.4.3 that a limit $T_0 \in B(X, Y)$ of this sequence exists. Accordingly, it is sufficient to prove that $T_0 \in B(X, Y) \cap C(X, Y)$, that is, that the uniform limit of continuous mappings is a continuous mapping. Given $\varepsilon > 0$ and let $x_0, x_1 \in X$. Then

$$(*) \qquad \varrho\big(T_0(x_0), T_0(x_1)\big)$$
$$\leqslant \varrho\big(T_0(x_0), T_n(x_0)\big) + \varrho\big(T_n(x_0), T_n(x_1)\big) + \varrho\big(T_n(x_1), T_0(x_1)\big).$$

Since $T_n \to T_0$ uniformly, there exists an N such that for $n \geqslant N$

$$\varrho\big(T_n(x), T_0(x)\big) \leqslant \frac{1}{3}\varepsilon \quad \text{for any } x.$$

Since the mapping T_N is continuous, there exists a number $\gamma > 0$ such that for $d(x_1, x_0) < \gamma$

$$\varrho\big(T_N(x_1), T_N(x_0)\big) \leqslant \frac{1}{3}\varepsilon.$$

If we set in $(*)$ $n = N$, $x_1 \in K(x_0, \gamma)$, we obtain

$$\varrho\big(T_0(x_0), T_0(x_1)\big) \leqslant \frac{1}{3}\varepsilon + \frac{1}{3}\varepsilon + \frac{1}{3}\varepsilon = \varepsilon.$$

Thus $T_0 \in C(X, Y)$. $\square$

DEFINITION. A metric space (X, d) is said to be *locally compact* when the condition

$$\bigwedge_{x \in X}\ \bigvee_{X \supset K-\text{compact}}\ \bigvee_{r > 0} K(x, r) \subset K$$

is satisfied. This proposition may be recast in a somewhat different formulation: every point $x \in X$ has a compact neighbourhood.

Examples. 1. The compact space (X, d) is locally compact (then $K = X$).

2. The real axis R^1 (we take, say, $K = [x-r, x+r]$).

An example of a metric space which is not locally compact is the set Q of rational numbers with the metric $d(v_1, v_2) = |v_1 - v_2|$.

DEFINITION. Let the space (X, d) be locally compact, and the space (Y, ϱ), metric. The sequence (T_n) of mappings $T_n: X \to Y$ is said to be *convergent in the sense of compact convergence* (or *almost uniformly convergent*) to a mapping $T_0: X \to Y$ when for every compact set $K \subset X$ the mappings T_n, treated as the mappings $T_n: K \to Y$, converge uniformly to $T_0|_K$.

Example. Let $R^1 \supset X :=]0, 1[$; $f_n(x) := x^n$. It is easily seen that $f_n \to 0$ almost uniformly. These functions are not almost uniformly convergent on the set $[0, 1]$.

THEOREM V.4.5. (The sequence (T_n), $T_n \in C(X, Y)$, is almost uniformly convergent to $T_0: X \to Y$) $\Rightarrow$ ($T_0 \in C(X, Y)$).

113

PROOF. Take any compact set K. If we treat T_n as the mappings $K \to Y$, these will be bounded, continuous, uniformly convergent mappings. It thus follows from Theorem V.4.4 that their uniform limit $T_0|_K$ is continuous.

Take any $x \in X$. Choosing K to be a (compact) neighbourhood of x we see that T_0 is continuous at x. $\square$

LEMMA V.4.6. (The space (X, d) is locally compact, and a set $O \subset X$ is open) $\Rightarrow$ (the space (O, d) is locally compact).

PROOF. Take $x_0 \in O$. Since the set O is open,

$$\bigvee_{r_1 > 0} K(x_0, r_1) \subset O.$$

Since (X, d) is locally compact, then $\bigvee_{r_2} \bigvee_{K-\text{compact}} K(x_0, r_2) \subset K.$

Let us set $r = \frac{1}{2} \min(r_1, r_2)$. We then have $K' := \overline{K(x_0, r)} \subset K(x_0, r_1) \subset O$ and K' is compact as a closed subset of compact K. $\square$

THEOREM V.4.7 (Dini). (The space (X, d) is locally compact, $C(X) \ni f_n \searrow f \in C(X)) \Rightarrow (f_n \to f$ almost uniformly).

The sign $\searrow$ should be construed as the symbol for pointwise and monotonic convergence (that is, $f_n \geqslant f_m$ when $n \leqslant m$).

PROOF. Let $g_n := f_n - f$. By hypothesis we have $g_n \searrow 0$. It is sufficient to show that $g_n \to 0$ almost uniformly, i.e. uniformly on every compact K. Suppose that this is not so, i.e. suppose that

$$\bigvee_{X \supset K-\text{compact}} \bigvee_{\varepsilon > 0} \bigwedge_{N} \bigvee_{n > N} \bigvee_{x_n \in K} g_n(x_n) > \varepsilon.$$

Setting consecutively $N = 1, 2, \ldots$, we get a sequence $x_{n_1}, x_{n_2}, \ldots$ But since the set K is compact, a convergent subsequence $x_{n_{k_i}} \to x_0 \in K$ can be selected from it. Let us now take an arbitrary index m and i_0 such that $n_{k_{i_0}} \geqslant m$. Then, since the convergence is monotonic

$$g_m(x_{n_{k_i}}) \geqslant g_{n_{k_i}}(x_{n_{k_i}}) \geqslant \varepsilon \quad \text{for } i \geqslant i_0.$$

Since g_m is continuous, then

$$\varepsilon \leqslant \lim_{i \to \infty} g_m(x_{n_{k_i}}) = g_m(\lim_{i \to \infty} x_{n_{k_i}}) = g_m(x_0).$$

We have thus proved that for every m, $g_m(x_0) \geqslant \varepsilon$, which contradicts the assumed, pointwise convergence of g_m to zero. Hence, $g_m \to 0$ almost uniformly. $\square$

THEOREM V.4.8 $\left(\varphi_n \in C([a, b]), \varphi_n \to \varphi_0 \text{ uniformly}\right) \Rightarrow \left(\int_a^b \varphi_n \to \int_a^b \varphi_0\right)$.

PROOF. We have for any $\varepsilon > 0$ and sufficiently great n's

$$\left|\int_a^b \varphi_n - \int_a^b \varphi_0\right| \leqslant \int_a^b |\varphi_n - \varphi_0| \leqslant \int_a^b \varepsilon = \varepsilon |a - b|. \quad \square$$

Theorem V.4.9. $(C(\Omega) \ni f_n \to f_0$ almost uniformly in Ω; Ω is connected in R^1, $x_0 \in \Omega$; $\varphi_n(x) := \int_{x_0}^x f_n$, $\varphi_0(x) := \int_{x_0}^x f_0$ for $x \in \Omega) \Rightarrow (\varphi_n \to \varphi_0$ almost uniformly on Ω).

PROOF. Given a compact set K in Ω; hence K is compact in R^1 too. K is bounded, which means that there exists an R such that $|x - x_0| \leqslant R$ for $x \in K$. We have for any $\varepsilon > 0$ and sufficiently great n's

$$\left|\int_{x_0}^x f_n - \int_{x_0}^x f_0\right| \leqslant \int_{x_0}^x |f_n - f_0| < \varepsilon R \to 0. \quad \square$$

Remark. This theorem on the "passage to the limit under the integral sign" is not valid for the pointwise or the uniform convergence. For instance, let $f_n(x) := x + 1/n$, $f_0(x) := x$, $\Omega := R^1$, and $f_n \to f_0$ uniformly on Ω. But $\varphi_n(x) = \dfrac{1}{2} x^2 + \dfrac{1}{n} x$; $\varphi_0 = \dfrac{1}{2} x^2$, and consequently

$$\sup_x |\varphi_n(x) - \varphi_0(x)| = \sup_x \frac{1}{n} |x| = \infty.$$

This means that φ_n do not converge uniformly to φ_0.

THEOREM V.4.10. $(R^1 \supset \Omega$ open and connected; $g_n \in C'(\Omega)$; (i) $g_n' \to g^1 \in C(\Omega)$ almost uniformly; (ii) $\bigvee_{x_0 \in \Omega} g_n(x_0) \xrightarrow[n \to \infty]{} G) \Rightarrow ((i)$ g_n converges almost uniformly to a function g_0; (ii) $g_0' = g^1$).

PROOF. We have

$$g_n(x) = g_n(x_0) + \int_{x_0}^x g_n'.$$

Theorem V.4.9 implies that $\int_{x_0}^x g_n' \to \int_{x_0}^x g^1$ almost uniformly. Consequently, (since $g_n(x_0) \to G$) we have the convergence

$$g_n(x) \to G + \int_{x_0}^x g^1 =: g_0(x),$$

this convergence being almost uniform. By the definition of g_0 it follows that $g_0' = g^1$. $\square$

5. POWER SERIES

Given a metric vector space (X, d) (i.e. the operations of addition and multiplication by numbers are defined in it and these operations are continuous). Given also a sequence $(y_n)_0^\infty$, $y_n \in X$; let $S_n := \sum_{k=0}^{n} y_k$.

DEFINITION. A pair $\big((y_n)_0^\infty, (S_n)_0^\infty\big)$ is called a *series with the general term y_n*. If the sequence $(S_n)_0^\infty$ converges to $s \in X$, the element s is called the *sum of the series* $\big((y_n)_0^\infty, (S_n)_0^\infty\big)$, and is denoted by $\sum_{k=0}^{\infty} y_k$. Then we say that the *series* $\big((y_n)_0^\infty, (S_n)_0^\infty\big)$ or, more often but less correct, $\sum_{k=0}^{\infty} y_k$ *converges*. In this case we write also $\sum_{k=0}^{\infty} y_k < \infty$.

Remark. This definition could be extended to any topological vector space.

We shall see that the notion of sum (convergence) of a series can be introduced not only in the case of metric space.

Let F be the set of all mappings $T: X \to Y$; X, Y—metric vector spaces. The pointwise convergence could be defined in it. We define the operation of addition and the operation of multiplication by a number:

$$(T_1 + T_2)(x) := T_1(x) + T_2(x), \qquad (\lambda T)(x) := \lambda(T(x)).$$

Then the series $\big((T_n)_0^\infty, (S_n)_0^\infty\big)$ has a sum S if the sequence $S_n = \sum_{k=0}^{n} T_k$ is a sequence of mappings which converges pointwise to the mapping S. The series is then frequently said to be *pointwise convergent*. Instead of F we may take the space $B(X, Y)$ with the uniform convergence or the space $C(X, Y)$ with the almost uniform convergence. In those events we speak of the *uniform convergence of the series* $\big((T_n)_0^\infty, (S_n)_0^\infty\big)$, $T_n \in B(X, Y)$, when the sequence of mappings $S_n \in B(X, Y)$ converges uniformly to a mapping $S \in B(X, Y)$, or we speak of the *almost uniform convergence of the series* $\big((T_n)_0^\infty, (S_n)_0^\infty\big)$, $T_n \in C(X, Y)$ when the sequence of mappings $S_n \in C(X, Y)$ almost uniformly converges to a mapping $S \in C(X, Y)$.

In the next we shall also use the metric space (C^1, d), where

$$d(z_1, z_2) := |z_1 - z_2|, \quad z_1, z_2 \in C^1,$$

is the Pythagorean metric on the complex plane $C^1 \cong R^2$; $x + iy \leftrightarrow (x, y)$. Plainly, (C^1, d) is a complete space. This definition, in the case of the complex numbers with zero imaginary part, coincides with the definition of the metric in R^1.

DEFINITION. A series $((c_n)_0^\infty, (S_n)_0^\infty)$, where $c_n \in C^1$, is said to be *absolutely convergent* when the series $((|c_n|)_0^\infty, (S_n')_0^\infty)$ is convergent, what is written as $\sum\limits_{k=0}^{\infty} |c_k| < \infty$.

LEMMA V.5.1. Let $y_n \in C^1$, then

$$\left(\sum_{n=0}^{\infty} y_n < \infty\right) \Leftrightarrow \left(\bigwedge_{\varepsilon > 0} \bigvee_{N(\varepsilon)} \bigwedge_{N(\varepsilon) < m < n} |S_n - S_m| = \left|\sum_{k=m+1}^{n} y_k\right| < \varepsilon\right).$$

PROOF. $\Rightarrow$: Since $(S_n)_0^\infty$ converges it is a Cauchy sequence.
$\Leftarrow$: A Cauchy sequence in C^1 converges. $\square$

Taking $n = m+1$, we get $|y_n| < \varepsilon$, and thus we have the following

COROLLARY V.5.2. Let $y_n \in C^1$, then $\left(\sum\limits_{n=0}^{\infty} y_n < \infty\right) \Rightarrow (|y_n| \xrightarrow[n \to \infty]{} 0)$.

THEOREM V.5.3 (The Comparison Test for the Convergence of Series). Let $b_n \in C^1$, $c_n \in C^1$ for $n = 1, 2, \ldots$ Then

$$\left(\sum_{n=1}^{\infty} c_n < \infty \; ; \; \bigvee_i \bigwedge_{n > i} |b_n| \leqslant |c_n|\right)$$

$$\Rightarrow \left(\sum_{n=1}^{\infty} |b_n| < \infty\right).$$

PROOF. By Lemma V.2.1 we have

$$\sum_{k=m+1}^{n} |b_k| \leqslant \sum_{k=m+1}^{n} |c_k| < \varepsilon \quad \text{for } m, n \geqslant \max(N(\varepsilon), i). \; \square$$

An example of a convergent series is the geometric series $a_n = q^n$ $|q| < 1, q \in C^1$,

$$\sum_{n=0}^{\infty} a_n = \frac{1}{1-q}, \quad \text{since} \quad S_n = \sum_{k=0}^{n} a_k = \frac{1-q^n}{1-q} \xrightarrow[n \to \infty]{} \frac{1}{1-q}.$$

LEMMA V.5.4 (The Weierstrass Test). Let $f_n\colon X \to C^1$. Then

$$\left(\sum_{n=0}^{\infty} |c_n| < \infty; \bigwedge_{x\in X} \bigwedge_n |f_n(x)| \leqslant |c_n|\right) \Rightarrow \left(\sum_{n=0}^{\infty} f_n = f \in B(X, C^1)\right).$$

PROOF. By virtue of Theorem V.5.3 the sequence $S_n = \sum_{k=0}^{n} f_k$ is uniformly convergent, and thus Theorem V.4.3 on the completeness of the space $B(X, C^1)$ implies the existence of a limit f such that $S_n \to f$ uniformly. $\square$

We shall also give the following simple but often useful test for the convergence of numerical series.

PROPOSITION V.5.5 (Leibniz). $(R^1 \ni a_n \geqslant 0, n = 1, 2, \ldots$, and $a_n \to 0$ monotonically, that is, $a_n \geqslant a_m$ for $m \geqslant n) \Rightarrow \left(\text{There exists } \sum_{n=0}^{\infty} (-1)^n a_n\right).$

Remark. This series may not at all be absolutely convergent, just as for $a_n = 1/n$, where $(((-1)^n \cdot (1/n))_0^\infty, (S_n)_0^\infty)$ is convergent but not absolutely convergent.

PROOF. We denote

$$S_a = \sum_{k=0}^{n} (-1)^k a_k.$$

For even numbers we have

$$S_{2n} = (a_0 - a_1) + (a_2 - a_3) + \ldots + (a_{2n-2} - a_{2n-1}) + a_{2n}.$$

Since $a_n \geqslant a_m$, when $n \leqslant m$, then $S_{2n} \geqslant 0$ for every n. But

$$S_{2n} = a_0 - (a_1 - a_2) - (a_3 - a_4) - \ldots - (a_{2n-1} - a_{2n}).$$

If n increases, new negative terms are added, and hence the sequence (S_{2n}) is non-increasing, that is $S_{2n} \leqslant S_{2m}$ for $m < n$. But the non-increasing sequence, which is bounded from below (by 0), is convergent and hence there exists a $p \in R^1$ such that $S_{2n} \to p$.

In the same way for odd numbers we have

$$S_{2n+1} = (a_0 - a_1) + \ldots + (a_{2n} - a_{2n-1}),$$

and thus (S_{2n+1}) is non-decreasing and

$$S_{2n+1} = a_0 - (a_1 - a_2) - \ldots - (a_{2n-1} - a_{2n}) - a_{2n+1} \leqslant a_0.$$

Thus the sequence (S_{2n+1}), being non-decreasing and bounded from above, is convergent.

Let $S_{2n+1} \to s$, for example. But $|S_{2n+1} - S_{2n}| = |a_{n+1}| \xrightarrow[n \to \infty]{} 0$. Hence, these two sequences are equivalent to each other, and hence their limits are equal: $s = p$ and $\lim_{n \to \infty} S_n = s$. $\square$

DEFINITION. A *power series* is a series with terms y_n of the form $y_n = a_n(z - z_0)^n \in C^1$.

Treating y_n as a function of the variable z, one may speak of the convergence of this series being uniform, almost uniform, etc., on certain subsets of the plane C^1.

Obviously, there are three possible alternatives:

  (i) the series is divergent for $z \neq z_0$,

  (ii) the series is convergent everywhere (i.e. for any $z \in C^1$),

  (iii) $\bigvee_{z \in C^1} z \neq z_0 : \sum_{n=0}^{\infty} y_n(z) < \infty$, but there also exist z for which the series is divergent.

The Circle of Convergence of a Power Series. It will presently be seen that a circle, known as the *circle of convergence,* can be associated with every power series; the series is absolutely (and almost uniformly) convergent inside this circle and divergent outside it (beyond the closure of the circle). This follows immediately from the straightforward remark below:[1]

Remark. (a) (The series $\sum a_n(z - z_0)^n$ is convergent in $z = z_1 \neq z_0$) $\Rightarrow (\sum a_n(z - z_0)^n$ is absolutely convergent in the circle $K(z_0, |z_1 - z_0|))$;

  (b) $(\sum a_n(z - z_0)^n$ is divergent in $z = z_1 \neq z_0) \Rightarrow (\sum a_n(z - z_0)^n$ is divergent for every z such that $|z - z_0| > |z_1 - z_0|$, i.e. is divergent in $C^1 - \overline{K(z_0, |z_1 - z_0|)})$.

PROOF. (a) By Corollary V.5.2 $\bigvee_{M > 0} \bigwedge_n |a_n(z_1 - z_0)^n| < M$, and hence

$$|a_n| < \frac{M}{|z_1 - z_0|^n}. \text{ Since}$$

$$t := \frac{|z - z_0|}{|z_1 - z_0|} < 1 \quad \text{for } z \in K(z_0, |z_1 - z_0|),$$

then

$$\sum |a_n| \cdot |z - z_0|^n \leqslant M \sum t^n < \infty.$$

---

[1] In the next we shall write very often $\sum$ instead of $\sum_{n=0}^{\infty}$.

(b) follows from (a). $\square$

The foregoing remark enables us to formulate the following

COROLLARY V.5.6 (On the Existence of a Circle of Convergence). Let Z be the set of points of convergence of the series $\sum a_n(z-z_0)^n$, that is,

$$Z = \left\{ z \in C^1 : \sum a_n(z-z_0)^n \text{ is convergent} \right\}.$$

Let

$$R := \sup_{z \in Z} |z-z_0|.$$

Then $K(z_0, R)$ is the circle of convergence of the series $\sum a_n(z-z_0)^n$, i.e. for z: $|z-z_0| > R$ the series $\sum a_n(z-z_0)^n$ is divergent, while on $K(z_0, R)$ the series $\sum a_n(z-z_0)^n$ is absolutely convergent.

The problem thus arises of determining the *radius of convergence* R. The solution yields the following important

THEOREM V.5.7 (Cauchy, Hadamard). Given a series with the general term $y_n = a_n(z-z_0)^n \in C^1$. Let

$$\limsup_{n \to \infty} \sqrt[n]{|a_n|} =: \varrho.$$

We denote $R := 1/\varrho$. Then the series $((y_n)_0^\infty, (S_n)_0^\infty)$ is absolutely convergent in the circle $|z-z_0| < R$ and divergent beyond its closure.

This formulation should be construed as follows:

(i) If $\varrho = 0$, then $R = \infty$ and the series is absolutely convergent for all z.

(ii) If $\varrho = \infty$, then $R = 0$ and the series is divergent everywhere except for the point $z = z_0$.

(iii) If $0 < \varrho < \infty$, the series is absolutely convergent in the circle $|z-z_0| < 1/\varrho = R$ and divergent for $|z-z_0| > R$.

PROOF. (i) Given an arbitrary point $z \in C^1$. We have

$$\lim_{n \to \infty} \left(\sup_{m \geqslant n} \sqrt[m]{|a_m|} \right) = 0,$$

which means that it is possible to find an n_0 such that

$$\sup_{m \geqslant n_0} \sqrt[m]{|a_m|} \leqslant \frac{1}{2|z-z_0|}.$$

Thus, for $m \geqslant n_0$

$$\sqrt[m]{|a_m|} \leqslant \frac{1}{2|z-z_0|}, \qquad |a_m| \leqslant \frac{1}{2^m|z-z_0|^m}$$

whence

$$|y_m| = |a_m| \cdot |z-z_0|^m \leqslant 1/2^m \qquad \text{for } m \geqslant n_0.$$

Therefrom, by Theorem V.5.3

$$\sum_{k=n_0}^{\infty} |y_k| = a < \infty.$$

Consequently,

$$\sum_{k=0}^{\infty} |y_k| = \left(\sum_{k=0}^{n_0-1} |y_k|\right) + a < \infty.$$

(ii) Given an arbitrary point $z \in C^1$, but $z \neq z_0$. Then

$$\lim_{n \to \infty} \left(\sup_{m \geqslant n} \sqrt[m]{|a_m|}\right) = \infty,$$

which means that it is possible to find a sequence (n_k) such that

$$\bigwedge_{n_k} \sup_{m \geqslant n_k} \sqrt[m]{|a_m|} \geqslant \frac{3}{|z-z_0|}.$$

Thus, for every n_k there exists an $m_k \geqslant n_k$ such that

$$\sqrt[m_k]{|a_{m_k}|} \geqslant \frac{2}{|z-z_0|}, \qquad |a_{m_k}| \geqslant \frac{2^{m_k}}{|z-z_0|^{m_k}}.$$

Hence, $|y_{m_k}| = |a_{m_k}| \cdot |z-z_0|^{m_k} \geqslant 2^{m_k}$, which means that the series is convergent, for if it were divergent, then we would have $y_{m_k} \to 0$; cf. Corollary V.5.2.

(iii) Take an arbitrary point $z \in C^1$ such that $0 < |z-z_0| =: r < R = 1/\varrho$. Consequently, $\varrho < 1/r$. Let us introduce the notation $\frac{1}{r} - \varrho$

$=: \varepsilon > 0.$

$$\lim_{n \to \infty} \left(\sup_{m \geqslant n} \sqrt[m]{|a_m|}\right) = \varrho,$$

and hence there exists an n_0 such that

$$\sup_{m \geqslant n_0} \sqrt[m]{|a_m|} \in [\varrho - \tfrac{1}{2}\varepsilon, \varrho + \tfrac{1}{2}\varepsilon],$$

121

and thus

$$\sup_{m \geqslant n_0} \sqrt[m]{|a_m|} \leqslant \varrho + \tfrac{1}{2}\varepsilon = 1/r - \tfrac{1}{2}\varepsilon.$$

Therefore

$$\sqrt[m]{|a_m|} \leqslant 1/r - \tfrac{1}{2}\varepsilon \quad \text{for } m \geqslant n_0$$

and

$$|a_m| \leqslant (1/r - \tfrac{1}{2}\varepsilon)^m, \quad |y_m| = |a_m||z - z_0|^m \leqslant (1 - r\tfrac{1}{2}\varepsilon)^m.$$

In view of this, however, the series with general term $|y_m|$ is dominated by the geometric series. Since $|1 - r\tfrac{1}{2}\varepsilon| < 1$, $\sum |y_n| < \infty$ and thus, by virtue of Theorem V.5.3, our series $\sum y_n$ is absolutely convergent. In a similar way, we prove that the series is divergent for $|z - z_0| > R$. $\square$

Remark. When $|z - z_0| = R$, the series is convergent sometimes, and divergent at other time (as we shall soon show with examples). In any event, if the series converges for a particular z_1 then it certainly converges absolutely for $|z - z_0| < |z_1 - z_0|$.

Examples. 1. $y_n = (z - z_0)^n$. This is a geometric series which converges, as we know, for $|z - z_0| < 1$. The same result is obtained by taking $\limsup\limits_{n \to \infty} \sqrt[n]{|a_n|} = 1$, since $a_n \equiv 1$ for all n.

This series is not convergent for $|z - z_0| = 1$, because $|y_n| \nrightarrow 0$.

2. $y_n = \dfrac{1}{n^2}(z - z_0)^n$. We have

$$\limsup_{n \to \infty} \sqrt[n]{\frac{1}{n^2}} = \lim_{n \to \infty} \frac{1}{\sqrt[n]{n^2}} = \lim_{x \to \infty} \frac{1}{\sqrt[x]{x^2}} = \lim_{x \to \infty} e^{\frac{-2\log x}{x}} = 1,$$

since $\dfrac{\log x}{x} \xrightarrow[x \to \infty]{} 0$ (we have incidentally proved that $\sqrt[n]{n^\alpha} \xrightarrow[n \to \infty]{} 1$ for $\alpha \in R^1$). Hence the series is absolutely convergent for $|z - z_0| < 1$. However, this series is also absolutely convergent for $|z - z_0| = 1$, because $\sum\limits_{n=0}^{\infty} (1/n^2) < \infty$ (we shall prove later that in general $\sum\limits_{n=0}^{\infty} (1/n^\alpha) < \infty$ for $\alpha > 1$).

3. $y_n = \dfrac{1}{n}(z - z_0)^n$. Here, too, the radius of convergence is $R = 1$.

The series diverges for $z = z_0 + 1$, because $\sum\limits_{n=1}^{\infty} (1/n) = \infty$. On the other hand, the series converges for $z = z_0 - 1$ because

$$\sum_{n=1}^{\infty} \frac{(-1)^n}{n} < \infty;$$

cf. Proposition V.5.5.

THEOREM V.5.8. A series whose general term is $a_n(z-z_0)^n$ in the circle of convergence $K(z_0, R)$ is absolutely, almost uniformly convergent, i.e. the continuous functions P_n and S_n,

$$S_n := \sum_{k=0}^{n} a_k(z-z_0)^k, \qquad P_k := \sum_{k=0}^{n} |a_k(z-z_0)^k|,$$

are almost uniformly convergent.

PROOF. Let a set K, $K \subset K(z_0, R)$, be compact; $K(z_0, R)$ is an open ball. Take the continuous function $l(z) := |z-z_0|$. This function attains its bounds on the compact set, and hence there exists a point $z_K \in K \subset K(z_0, R)$ such that

$$|z-z_0| \leqslant |z_K - z_0| < R, \qquad \text{for } z \in K.$$

But

$$|P_m - P_n| = \sum_{j=n+1}^{m} |a_j| \cdot |z-z_0|^j \leqslant \sum_{j=n+1}^{m} |a_j| \cdot |z_K - z_0|^j < \varepsilon$$

for $n, m > N(\varepsilon)$ and for $z \in K$, because the series converges absolutely at the point z_K. Since this is an estimate independent of $z \in K$, the functions P_n are uniformly convergent on every compact set $K \subset K(z_0, R)$ and thus are almost uniformly convergent on $K(z_0, R)$. $\square$

COROLLARY V.5.9. The function $f(z) := \sum\limits_{n=0}^{\infty} a_n(z-z_0)^n$, $z \in K(z_0, R)$, is continuous as the almost uniform limit of the continuous functions S_n.

6. ANALYTIC FUNCTIONS

DEFINITION. Let $\Omega \subset C$ (or R) be an open set and let $f: \Omega \to C$ (or R respectively); we say that the function f is *analytic* on Ω if for every $z_0 \in \Omega$ there exists an $r > 0$ such that for $z \in K(z_0, r) \subset \Omega$ the function

f is equal to the sum of a power series $\sum\limits_{n=0}^{\infty} a_n(z-z_0)^n$, $a_n \in C$ (or R). In other words, f admits local representiation as a power series.

In the sequel, we shall consider analytic functions of the complex variable.

The function $\varphi: C^1 \supset \Omega \ni z \to \varphi(z) \in C^1$ is said to be *differentiable* at a point $z_0 \in \Omega$, if it has a *derivative* at z_0, i.e. there exists

$$\varphi'(z_0) := \lim_{\Delta z \to 0} \frac{\varphi(z_0 + \Delta z) - \varphi(z_0)}{\Delta z}.$$

LEMMA V.6.1. (The radius of convergence for a series with the general term $a_n(z-z_0)^n$ is R) $\Rightarrow$ (The series $\sum\limits_{n=1}^{\infty} n \cdot a_n(z-z_0)^{n-1}$ has the same circle of convergence).

PROOF. We have

$$\limsup_{n \to \infty} {}^{n-1}\!\!\sqrt{|n \cdot a_n|} = \limsup_{n \to \infty} {}^{n-1}\!\!\sqrt{|a_n|} \cdot \lim_{n \to \infty} {}^{n-1}\!\!\sqrt{n} = R \cdot 1 = R. \quad \square$$

THEOREM V.6.2. An analytic function is *infinitely differentiable*, i.e. differentiable any arbitrary number of times.

Hypothesis: $K(z_0, R) \ni z \to f(z) = \sum\limits_{n=0}^{\infty} a_n(z-z_0)^n$ (i.e. this series is convergent in the circle $K(z_0, R)$).

Thesis: (i) The function $f(z)$ is differentiable in $K(z_0, R)$ and

$$f'(z) = \sum_{n=1}^{\infty} n \cdot a_n(z-z_0)^{n-1},$$

(ii) f is infinitely differentiable,
(iii) $f^{(k)}(z_0) = k! a_k$.

PROOF. (i) The reader will prove (by induction) the following inequality:

$$(z + \Delta z - z_0)^n - (z - z_0)^n - n(z - z_0)^{n-1}\Delta z|$$
$$\leqslant n^2(|z - z_0| + |\Delta z|)^{n-2}|\Delta z|^2, \quad \text{where } n = 2, 3, \dots$$

Hence, for $z \in K(z_0, R)$ and $\Delta z \neq 0$ such that $z + \Delta z \in K(z_0, R)$,

$$\left| \frac{1}{\Delta z} \left[\sum_{n=0}^{\infty} a_n(z + \Delta z - z_0)^n - \sum_{n=0}^{\infty} a_n(z - z_0)^n \right] - \right.$$

$$\left. - \sum_{n=0}^{\infty} n a_n (z-z_0)^{n-1} \right| \leqslant |\Delta z| \sum_{n=2}^{\infty} n^2 |a_n| (|z-z_0|+|\Delta z|)^{n-2}.$$

But $\sum_{n=2}^{\infty} n^2 |a_n| (\zeta-z_0)^{n-2}$ converges for $\zeta \in K(z_0, R)$; hence, taking $\zeta := z_0 + r, 0 \leqslant r < R$, we have

$$\sum_{n=2}^{\infty} n^2 |a_n| r^{n-2} < \infty \quad \text{for } 0 \leqslant r < R.$$

If for our fixed $z \in K(z_0, R)$ we take Δz very small, namely, $|\Delta z| \leqslant \varepsilon := \frac{1}{2}(R-|z-z_0|) > 0$, then $|z-z_0|+|\Delta z| \leqslant \frac{1}{2}(R+|z-z_0|) = R - \varepsilon < R$, and therefore

$$|\Delta z| \sum_{n=2}^{\infty} n^2 |a_n| (|z-z_0|+|\Delta z|)^{n-2}$$

$$\leqslant |\Delta z| \sum_{n=2}^{\infty} n^2 |a_n| (R-\varepsilon)^{n-2} \xrightarrow[z \to 0]{} 0,$$

what completes the proof of (i).

(ii) By virtue of (i), the function f' is also analytic and has a derivative equal to $\sum_{n=2}^{\infty} n(n-1) \cdot a_n (z-z_0)^{n-2}$. By induction we prove that every derivative of f is analytic and has a derivative and consequently f is infinitely differentiable.

(iii) We have

$$f^{(k)}(z) = \sum_{n=k}^{\infty} n(n-1) \quad (n-k+1) \cdot a_n (z-z_0)^{n-k},$$

$$f^{(k)}(z_0) = k(k-1) \dots (k-k+1) \cdot a_k (z_0-z_0)^0 = k! a_k. \quad \square$$

An example of an analytic function:

$$e^z := \sum_{n=0}^{\infty} \frac{z^n}{n!}.$$

In particular, when $\operatorname{Im} z = 0$ (which means $z \in R^1$), this definition coincides with the one given previously by the formula

$$e^x = \sum_{n=0}^{\infty} \frac{x^n}{n!}.$$

Let us examine the radius of convergence of this series. We have

$$\sum_{n=0}^{\infty} \left| \frac{z^n}{n!} \right| = \sum_{n=0}^{\infty} \frac{|z|^n}{n!},$$

and this series, as proven earlier, converges for any $|z|$, and thus for any z. Hence e^z is an analytic function in the entire complex plane. Accordingly, we can calculate its derivative:

$$(e^z)' = \sum_{n=1}^{\infty} \frac{n \cdot z^{n-1}}{n!} = \sum_{n=1}^{\infty} \frac{z^{n-1}}{(n-1)!}.$$

Let us take $n-1 = k$. Then

$$(e^z)' = \sum_{n-1=0}^{\infty} \frac{z^{n-1}}{(n-1)!} = \sum_{k=0}^{\infty} \frac{z^k}{k!} = e^z.$$

Thus we have a formula analogous to that for real numbers: $(e^z)' = e^z$. Plainly, $e^0 = 1+0+0+ \dots = 1$, just as for the function e^x.

7. THE TRIGONOMETRIC FUNCTIONS AND THEIR RELATION TO THE EXPONENTIAL FUNCTION

Just as in the case of the logarithm and the exponential function, we assume no knowledge of the trigonometric functions. We shall give definitions in terms of power series. These definitions have the following advantages:

(i) They are frequently "analytic", i.e. they do not resort to geometric pictures, as the "schoolbook" definition does.

(ii) These functions are defined immediately for complex arguments, which is of vast importance, for applications as well.

(iii) The definition in terms of power series "shows the analyticity" of the sine and cosine functions.

Let us emphasize at this juncture that the "schoolbook" definition is not precise: it operates with a concept of length of arc, without having previously defined what that "length" means. The reader could scarcely be expected to be a *tabula rasa*, and hence at the end of this section we shall compare the definitions given here with those learnt at school, after we have made the schoolbook definitions more precise, i.e. after introducing radian measure.

$$\cos z := \sum_{n=0}^{\infty} (-1)^n \frac{z^{2n}}{(2n)!}, \qquad R = \infty,$$

$$\sin z := \sum_{n=0}^{\infty} (-1)^n \frac{z^{2n+1}}{(2n+1)!}, \qquad R = \infty.$$

Both of these functions are analytic throughout the complex plane.
Now we evaluate their derivatives. We have

$$(\cos z)' = \sum_{n=1}^{\infty} (-1)^n \frac{2n}{(2n)!} z^{2n-1}$$

$$= - \sum_{n-1=0}^{\infty} (-1)^{n-1} \frac{z^{2(n-1)+1}}{(2(n-1)+1)!}.$$

We set $n-1 = k$; then

$$(\cos z)' = - \sum_{k=0}^{\infty} (-1)^k \frac{z^{2k+1}}{(2k+1)!} = -\sin z.$$

In identical fashion, we get

$$(\sin z)' = \sum_{n=0}^{\infty} (-1)^n \frac{(2n+1)}{(2n+1)!} z^{2n} = \sum_{n=0}^{\infty} (-1)^n \frac{z^{2n}}{(2n)!} = \cos z.$$

The sine and cosine functions are associated with the exponential function by a very important relation (*Euler's formula*):

$$e^{iz} = \sum_{n=0}^{\infty} \frac{i^n z^n}{n!} = \sum_{n=0}^{\infty} \frac{i^{2n} z^{2n}}{(2n)!} + \sum_{n=0}^{\infty} \frac{i^{2n+1} z^{2n+1}}{(2n+1)!}$$

$$= \sum_{n=0}^{\infty} (-1)^n \frac{z^{2n}}{(2n)!} + i \sum_{n=0}^{\infty} (-1)^n \frac{z^{2n+1}}{(2n+1)!} = \cos z + i \sin z.$$

Use has been made here of the fact that $i^{2n} = (i^2)^n = (-1)^n$.
The definition implies immediately that

$$\sin(-z) = -\sin z, \quad \cos(-z) = \cos z.$$

127

Accordingly

$$e^{-iz} = \cos z - i \sin z.$$

From this we can directly establish

$$\cos z = \frac{e^{iz} + e^{-iz}}{2},$$

$$\sin z = \frac{e^{iz} - e^{-iz}}{2i}.$$

Comparison of these formulae with the formulae defining the hyperbolic functions yields

$$\cos z = \cosh iz, \quad \cosh z = \cos iz,$$
$$\sin z = -i \sinh iz, \quad \sinh z = -i \sin iz.$$

We also have

$$e^{z_1} \cdot e^{z_2} = e^{z_1 + z_2};$$

both sides are analytic in z_1, coincide in $z_1 = 0$ and all their derivatives with respect to z_1 coincide in 0, so by Theorem V.6.2 (iii) they coincide for all z_1 (and z_2).

Proceeding from this formula, we get

$$\cos(z_1 + z_2) = \tfrac{1}{2}(e^{i(z_1+z_2)} + e^{-i(z_1+z_2)}) = \tfrac{1}{2}(e^{iz_1}e^{iz_2} + e^{-iz_1}e^{-iz_2})$$

$$= \tfrac{1}{2}[(\cos z_1 + i \sin z_1)(\cos z_2 + i \sin z_2) +$$

$$+ (\cos z_1 - i \sin z_1)(\cos z_2 - i \sin z_2)]$$

$$= \cos z_1 \cdot \cos z_2 - \sin z_1 \cdot \sin z_2.$$

In the same way we prove the formula

$$\sin(z_1 + z_2) = \sin z_1 \cos z_2 + \cos z_1 \sin z_2.$$

Setting $z_2 = -z_1$, we obtain

$$\cos(z_1 - z_1) = \cos 0 = 1 = (\cos z_1)^2 - \sin z_1 \sin(-z_1)$$

$$= (\cos z_1)^2 + (\sin z_1)^2.$$

Thus, we have the highly important formula (the *Pythagoras Theorem*)

$$(\sin z)^2 + (\cos z)^2 \equiv 1 \quad \text{for all } z.$$

LEMMA V.7.1. For $x \in R^1$, $\cos x$ is a decreasing function on the interval $]0, 2[$.

PROOF. We know that $(\cos x)' = -\sin x$. But

128

$$\sin x = x - \frac{x^3}{3!} - \ldots = x\left(1 - \frac{x^2}{3!}\right) + \frac{x^5}{5!}\left(1 - \frac{x^2}{6\cdot 7}\right) +$$

$$+ \frac{x^9}{9!}\left(1 - \frac{x^2}{10\cdot 11}\right) + \ldots > 0 \quad \text{for } x \in \,]0, 2[,$$

because this is a sum of positive terms. Hence, $(\cos x)' = -\sin x < 0$, for $x \in \,]0, 2[$, and thus $\cos x$ is a decreasing function in this interval. $\square$

LEMMA V.7.2. The cosine function has only one zero (i.e. there exists only one x such that $\cos x = 0$) in the interval $]0, 2[$.

PROOF. We have

$$\cos 0 = 1 > 0,$$

$$\cos 2 = 1 - \frac{2^2}{2!} + \frac{2^4}{4!} - \frac{2^6}{6!} + \ldots$$

$$= 1 - \frac{4}{2} + \frac{16}{24} + \text{negative number}$$

$$= -\frac{1}{3} + \text{negative number} < 0,$$

and thus the function is positive at the beginning of the interval, negative at the end, and decreasing. Since it is continuous, it has a zero, but only one zero. $\square$

To denote the point at which the cosine function becomes zero let us use the symbol $\frac{1}{2}\pi \in \,]0, 2[$:

$$\cos \tfrac{1}{2}\pi = 0.$$

Hence, by the Pythagoras Theorem, $(\sin \tfrac{1}{2}\pi)^2 = 1$. However, as we have shown, $\sin x > 0$ for $x \in \,]0, 2[$. Consequently, $\sin \tfrac{1}{2}\pi = 1$ (and not -1).

Now let us demonstrate that the sine and cosine functions are periodic; we have

$$\sin(2\pi) = \sin\pi\cos\pi + \cos\pi\sin\pi = 2\sin\pi\cos\pi$$

$$= 2\cdot 2\sin\tfrac{1}{2}\pi\cos\tfrac{1}{2}\pi\cos\pi = 0,$$

because $\cos\tfrac{1}{2}\pi = 0$.

In identical manner we prove that $\cos 2\pi = 1$. Hence

$$\sin(z+2\pi) = \sin z\cos 2\pi + \cos z\sin 2\pi = \sin z,$$

$$\cos(z+2\pi) = \cos z\cos 2\pi - \sin z\sin 2\pi = \cos z.$$

129

It may be shown by induction that

$$\begin{aligned} \sin(z+n\cdot 2\pi) &= \sin z, \\ \cos(z+n\cdot 2\pi) &= \cos z, \end{aligned} \qquad \text{for } n = 0, \pm 1, \pm 2, \dots$$

Accordingly

$$e^{z+n2\pi i} = e^z \cdot e^{n2\pi i} = e^z,$$

because $e^{n2\pi i} = 1$.

Using the trigonometric functions derived above, we may now try to give a parametric description of the circumference of a circle.

Let

(3) $\qquad x(t) = \cos t, \quad y(t) = \sin t, \quad t \in [0, 2\pi[.$

Clearly, $x^2 + y^2 = 1$, and hence for an arbitrary t the point $\big(x(t), y(t)\big)$ $\in R^2$ lies on the circumference of the unit circle. Since the trigonometric functions are periodic with a period 2π, it is sufficient to take the interval $t \in [0, 2\pi[$ (for other t, our points will be repeated at intervals of 2π). Knowing the behaviour of the trigonometric functions, we can easily show that for every point (x, y) lying on the unit circle there exists a $t \in [0, 2\pi[$ such that $x = \cos t$, $y = \sin t$, and that different points are obtained for different t.

A mapping $t \to \big(x(t), y(t)\big) \in R^2$ (defined on a connected subset of R^1) which is one-to-one is called the *parametric description of a curve* in R^2. Curves may, of course, be written in spaces of three or more dimensions. For instance, the equations $x = at$, $y = bt$, $z = ct$, as we recall, represent a straight line in R^3 for $t \in]-\infty, \infty[$.

Hence, equations (3) constitute the parametric description of the unit circle. Any circle of radius R and centre at a point $(a, b) \in R^2$ can be parametrized in the same way. Namely

(*) $\qquad x(t) = a + R\cos t, \quad y(t) = b + R\sin t, \quad t \in [0, 2\pi[.$

The equation of the circle $(x-a)^2 + (y-b)^2 = R^2$ is then satisfied. Now let us define the length of a curve in R^n.

DEFINITION. Let $R^1 \ni t \to \big(\varphi_1(t), \varphi_2(t), \dots, \varphi_n(t)\big)$ be the parametric description for a curve in R^n. The *length of the curve* from the point $\big(\varphi_1(t_0), \varphi_2(t_0), \dots, \varphi_n(t_0)\big)$ to the point $\big(\varphi_1(t_1), \varphi_2(t_1), \dots, \varphi_n(t_1)\big)$ is the number

$$S = \left| \int_{t_0}^{t_1} \sqrt{\sum_{i=1}^{n} (\varphi_i'(t))^2 dt} \right|.$$

(It is assumed that φ_i is differentiable, $i = 1, \ldots, n$.)

By way of example, let us calculate the circumference of the circle (*) of radius R:

$$S = \int_0^{2\pi} \sqrt{(-R\sin t)^2 + (R\cos t)^2} dt = R \int_0^{2\pi} \sqrt{1} dt = 2\pi R.$$

Thus, the length of the arc (circumference) between $(x(t), y(t))$ and $(x(t+h), y(t+h))$ is h. Accordingly, the length of arc, or *radian measure* is the parameter in the functions $\sin t$ and $\cos t$.

It turns out (this will be seen in Part II) that for complex functions the requirement that f be analytic is equivalent to the requirement

that it be differentiable in the sense of the space C^1 $\Big($i.e. that there exists

a $\lim\limits_{\Delta z \to 0} \dfrac{f(z+\Delta z)-f(z)}{\Delta z}\Big)$. Thus, the condition of differentiability in the

sense of the space C^1 is a very strong condition. After all, differentiability in the sense of the space R^1 does not at all ensure analyticity. Moreover, we shall demonstrate that even differentiability infinite in the sense of the space R^1 does not ensure analyticity.

Example. Consider the function

$$f(x) = \begin{cases} e^{-1/x^2} & \text{for } x > 0, \\ 0 & \text{for } x \leqslant 0. \end{cases}$$

Clearly, f is continuous. We have

$$f'(x) = \begin{cases} \dfrac{2}{x^3} e^{-1/x^2} \to 0 & \text{for } 0 < x \searrow 0, \\ 0 \to 0 & \text{for } 0 > x \nearrow 0, \end{cases}$$

and consequently f' is continuous ($f'(0) = 0$). We have

$$f''(x) = \begin{cases} -\dfrac{6}{x^4} e^{-1/x^2} + \dfrac{4}{x^6} e^{-1/x^2} \to 0 & \text{for } 0 < x \searrow 0, \\ 0 \to 0 & \text{for } 0 > x \nearrow 0. \end{cases}$$

Thus f'' is continuous ($f''(0) = 0$). In identical manner it is proved that all derivatives of this function at zero are continuous and equal to zero (this is due to the factor e^{-1/x^2} which tends to zero very rapidly).

Consequently, f is infinitely differentiable. However, f is not analytic for its Taylor series at zero is equal to 0 ($\sum 0 = 0$) and thus does not coincide with the function f in any neighbourhood of zero. In summary, we may say that:

(i) analytic functions are infinitely differentiable,

(ii) functions infinitely differentiable in the sense of the space R^1 are in general not analytic,

(iii) complex functions which are differentiable even once in the sense of the space C^1 are not only infinitely differentiable but even analytic.

We shall show that the logarithmic function is analytic and we shall find its power series at the point 1. Obviously, this could be done by the Taylor Formula, by showing that $r_n(x_0, \Delta x) \xrightarrow[n \to \infty]{} 0$. This can, however, be done in a different way:

$$\log(1+x) = \int_1^{1+x} \frac{dt}{t} = \int_0^x \frac{d\tau}{1+\tau},$$

but

$$\frac{1}{1+\tau} = 1 - \tau + \tau^2 - \tau^3 + \ldots = \sum_{n=0}^{\infty} (-\tau)^n.$$

This (geometric) series converges almost uniformly on $K(0, 1)$ which means that

$$\sum_{k=0}^{n} (-\tau)^k = S_n(\tau) \to \frac{1}{1+\tau},$$

almost uniformly on $K(0, 1)$.

On invoking Theorem V.4.9, we have

$$\int_0^x S_n(\tau)d\tau \xrightarrow[n \to \infty]{} \log(1+x) \quad \text{almost uniformly on } K(0, 1).$$

But

$$\int_0^x S_n(\tau)d\tau = \sum_{k=0}^{n} (-1)^k \frac{x^{k+1}}{k+1} \xrightarrow[n \to \infty]{} \sum_{k=0}^{\infty} (-1)^k \frac{x^{k+1}}{k+1}$$

$$= \sum_{k=1}^{\infty} (-1)^{k+1} \frac{x^k}{k}.$$

Finally, we obtain

$$\log(1+x) = \sum_{k=1}^{\infty} (-1)^{k+1} \frac{x^k}{k} = x - \frac{x^2}{2} + \frac{x^3}{3} - \frac{x^4}{4} + \frac{x^5}{5} \ldots$$

for $-1 < x < 1$.

In accordance with what we have said above about analytic functions of a real variable, it will be natural to define the analytic function $\log z$ for $z \in K(0, 1)$ by means of a power series which is convergent in this circle:

$$\log(1+z) = \sum_{n=1}^{\infty} (-1)^{n+1} \frac{z^n}{n} = z - \frac{z^2}{2} + \frac{z^3}{3} - \frac{z^4}{4} + \ldots$$

for $z \in K(0, 1)$.

Of course, the equality

$$\log x = \log z \Big|_{\substack{\mathrm{Im}\, z = 0 \\ \mathrm{Re}\, z = x}}$$

holds.

Remark. Analytic (holomorphic) functions are treated at length in Part II.

VI. INTEGRALS ON SOME NON-COMPACT SETS

1. INTEGRALS ON SOME NON-COMPACT SETS

The notion of Riemann integrability, as defined in Chapter III, concerns only bounded functions defined on bounded closed intervals. A general method of extension of integral to a wider class of domains and functions is the subject of Chapter XII (Part II). We shall give here a modest extension of that notion, which is, however, very useful and does not need any complicated construction.

Let Ω be a subset of R^1 and f be a real function on Ω. We consider the family $\mathscr{R}$ of all subsets of Ω which are finite sums of closed proper (bounded and not one-point) intervals. Note that $\mathscr{R}$ with the relation $K_{\alpha_1} \prec K_{\alpha_2}$, when $K_{\alpha_1} \subset K_{\alpha_2}$ is a directed set. Suppose that Ω is exhausted by the sets K_α, i.e. $\bigwedge_{x \in \Omega} \bigvee_{K_\alpha \in \mathscr{R}} x \in K_\alpha$; this is written as $K_\alpha \nearrow \Omega$. Suppose, moreover, that f is integrable on each K_α.

DEFINITION. Let Ω and f satisfy the above assumptions. If the net $K_\alpha \to I(K_\alpha) = \int_{K_\alpha} f$ is convergent, f is called *integrable* on Ω and the limit of the net is called an *improper integral of the function f on the set Ω*.

PROPOSITION VI.1.1. If $g \geqslant f \geqslant 0$, g is integrable on Ω, f is integrable on each K_α, then f is integrable on Ω and $\int_\Omega f \leqslant \int_\Omega g$.

PROOF. By assumption

$$\bigwedge_{\varepsilon > 0} \bigvee_{K_0} \bigwedge_{K_1, K_2 \succ K_0} \left| \int_{K_1} g - \int_{K_2} g \right| < \varepsilon.$$

When K_1 is used to denote the set for which the integral of f is larger, we have

$$\left| \int_{K_1} f - \int_{K_2} f \right| = \int_{K_1} f - \int_{K_2} f \leqslant \int_{K_1 \cup K_2} f - \int_{K_2} f$$

$$= \int_{K_1 \cup K_2 - K_2} f \leqslant \int_{K_1 \cup K_2 - K_2} g = \int_{K_1 \cup K_2} g - \int_{K_2} g$$

$$= \left| \int_{K_1 \cup K_2} g - \int_{K_2} g \right| < \varepsilon,$$

since $K_1 \cup K_2 \succ K_1 \succ K_0$. Hence the net $\int_{K_\alpha} f$ has a limit, which means that f is integrable on Ω. $\square$

DEFINITION. A function f is *absolutely integrable on Ω* if f is integrable on Ω and if $|f|$ is integrable on Ω.

Note that we have the following formulae, analogous to those in the theory of integrals on intervals:

THEOREM VI.1.2.

(i) $\int_{\Omega} (\alpha f + \beta g) = \alpha \int_{\Omega} f + \beta \int_{\Omega} g$ for $\alpha, \beta \in C^1$, f, g integrable on Ω;

(ii) an integrable function f is always absolutely integrable and

$$\left| \int_{\Omega} f \right| \leq \int_{\Omega} |f|;$$

(iii) $\int_{\Omega_1} f + \int_{\Omega_2} f = \int_{\Omega_1 \cup \Omega_2} f$, when $\Omega_1 \cap \Omega_2 = \varnothing$ or when $\Omega_1 \cap \Omega_2 = \{x_0\}$.

PROOF. (i) Trivial because the equality holds on every set K_α, and hence the theorems on the limits of unions and intersections imply that it is also valid on Ω.

(ii) Take $\varepsilon > 0$. Since f is integrable on Ω, we have

(1) $$\bigwedge_{\varepsilon > 0} \bigvee_{K_0 \in \mathscr{R}} \bigwedge_{K} \left| \int_{K \cup K_0} f - \int_{K_0} f \right| < \frac{\varepsilon}{4}.$$

Suppose that there exists K_α such that $K_\alpha \cap K_0 = \varnothing$ and

$$\left| \int_{K} |f| - \int_{K_0} |f| \right| \geq \varepsilon.$$

Then

$$\int_{K_\alpha} |f| \geq \varepsilon.$$

From Theorem III.4.5 (iv) we know that the functions $f_+ = \dfrac{|f| + f}{2}$ and $f_- = \dfrac{|f| - f}{2}$ are integrable on K_α. Since $|f| = f_+ + f_-$ we conclude that at least one of the integrals $\int_{K_\alpha} f_+$, $\int_{K_\alpha} f_-$ is not less than $\varepsilon/2$. Without loss of generality we assume that $\int_{K_\alpha} f_+ \geq \varepsilon/2$. This means that there exists a partition $\pi = \{P_i\}_{i=1,\dots,n}$ of K_α such that

(2) $$S(f_+, \pi) = \sum_{i=1}^{n} \inf f(P_i) \cdot |P_i| \geq \frac{\varepsilon}{4}.$$

By K_β we denote the sum of closures of those intervals P_i for which the summands in (2) do not vanish. The function f is, therefore, positive on K_β, i.e. $f = f_+$ on K_β. This way we have

$$\int_{K_0 \cup K_\beta} f - \int_{K_0} f = \int_{K_\beta} f = \int_{K_\beta} f_+ \geqslant S(f_+, \pi) \geqslant \frac{\varepsilon}{4},$$

which contradicts (1).

(iii) It must be proved that the assumed integrability on $\Omega_1 \cup \Omega_2$ implies integrability on Ω_1 and on Ω_2. Suppose that f is not integrable on Ω_1 (for example); then

$$(1) \qquad \bigvee_\varepsilon \bigwedge_{K_0 \subset \Omega_1} \bigvee_{K_1, K_2 \succ K_0} \left| \int_{K_1} f - \int_{K_2} f \right| \geqslant \varepsilon.$$

Let us take an arbitrary set $\tilde{K}_0 \subset \Omega_1 \cup \Omega_2$, $\tilde{K}_0 \in \mathcal{R}$; $\tilde{K}_0 = \tilde{K}_0 \cap \cap (\Omega_1 \cup \Omega_2) = (\tilde{K}_0 \cap \Omega_1) \cup (\tilde{K}_0 \cap \Omega_2)$. We write

$$\tilde{K}_0 \cap \Omega_1 = K_0, \qquad \tilde{K}_0 \cap \Omega_2 = K;$$

without loss of generality assume that K_0, K are finite sums of intervals.

We now take two sets $K_1, K_2 \subset \Omega_1$, whose existence is guaranteed by formula (1) (i.e. $K_1, K_2 \succ K_0$). Then $\tilde{K}_1 = K_1 \cup K$; $\tilde{K}_2 = K_2 \cup K$ satisfy the condition $\tilde{K}_1 \succ \tilde{K}_0$; $\tilde{K}_2 \succ \tilde{K}_0$. But

$$\left| \int_{\tilde{K}_1} f - \int_{\tilde{K}_2} f \right| = \left| \int_{K_1} f + \int_K f - \int_{K_2} f - \int_K f \right| = \left| \int_{K_1} f - \int_{K_2} f \right| > \varepsilon.$$

We have made use here of the fact that for intervals

$$\int_{K_1} f + \int_K f = \int_{K_1 \cup K} f,$$

when $K_1 \cap K$ is an empty or one-element set. It has thus been shown that

$$\bigvee_\varepsilon \bigwedge_{\tilde{K}_0 \subset \Omega_1 \cup \Omega_2} \bigvee_{\substack{\tilde{K}_1, \tilde{K}_2 \\ \tilde{K}_1, \tilde{K}_2 \succ \tilde{K}_0}}^{\Omega_1 \cup \Omega_2} \left| \int_{\tilde{K}_1} f - \int_{\tilde{K}_2} f \right| \geqslant \varepsilon;$$

and, hence, f is not integrable on $\Omega_1 \cup \Omega_2$.

Since this contradicts our hypothesis, f must be integrable on Ω_1 and on Ω_2. The theorem on the limit of the sum implies immediately

$$\int_{\Omega_1} f + \int_{\Omega_2} f = \int_{\Omega_1 \cup \Omega_2} f. \quad \square$$

139

These facts and Proposition VI.1.1 may be used to construct the following comparison test for integrability:

THEOREM VI.1.3.

Hypothesis: (i) $0 \leqslant f$, g—functions defined on Ω and integrable on every subset $K_\alpha \subset \Omega$;

(ii) there exist constants $c, d > 0$ such that $c \leqslant \dfrac{f(x)}{g(x)} \leqslant d$ for $x \in \Omega$.

Thesis: (f is integrable on Ω) $\Leftrightarrow$ (g is integrable on Ω).

PROOF. $\Rightarrow$: We have

$$c \cdot g(x) \leqslant f(x),$$

and hence by Proposition VI.1.1 it follows that the function $c \cdot g$ is integrable on Ω. By Theorem VI.1.2 the function $\dfrac{1}{c} \cdot (c \cdot g) = g$ is also integrable on Ω.

$\Leftarrow$: The proof is identical, because $f(x) \leqslant d \cdot g(x)$. $\square$

Let us now consider special cases of improper integrals which frequently occur in applications.

I. Let the set Ω be unbounded and let $\Omega = [a, \infty[, a > 0$.

1. Take $\mu \neq 1$. We have

$$\int_a^b \frac{1}{x^\mu}\,dx = \frac{1}{1-\mu} x^{1-\mu}\Big|_a^b = \frac{1}{1-\mu}(b^{1-\mu}-a^{1-\mu}).$$

If we put $K_b = [a, b]$, then $K_b \nearrow \Omega$ for $b \to \infty$. Hence

$$\lim_{b \to \infty} \int_a^b \frac{1}{x^\mu}\,dx = \frac{1}{\mu-1}a^{1-\mu} + \frac{1}{1-\mu}\lim_{b \to \infty} b^{1-\mu}.$$

It is seen that in order for the sequence of integrals to be convergent, it is necessary and sufficient that $\mu > 1$.

When $\mu = 1$, then

$$\int_a^b \frac{dx}{x} = (\log b - \log a) \xrightarrow[b \to \infty]{} \infty.$$

Thus we have the following tests for integrability:

(a) $$\left(0 \leqslant f \leqslant C\frac{1}{x^\mu};\ \mu > 1\right) \Rightarrow \left(\int_{a>0}^\infty f < \infty\right);$$

(b) $\qquad \left(0 < C\dfrac{1}{x^\mu} \leqslant f; \mu \leqslant 1\right) \Rightarrow \left(\displaystyle\int\limits_{a}^{\infty} f = +\infty\right).$

Note that if $\mu > 1$, then $\mu - 1 = \alpha > 0$ and hence

$$\frac{1}{x^\mu} = \frac{1}{x} \cdot \frac{1}{x^\alpha}.$$

We see, therefore, that the function $1/x$ itself is not integrable on Ω, but once it has been "damped" by the factor $1/x^\alpha$ it becomes integrable for arbitrarily small α.

As we recall, for $\alpha > 0$

$$\lim_{x \to \infty} \frac{\log x}{x^\alpha} = 0,$$

and thus the factor $\log x$ tends to infinity even more slowly than x^α does. The question is what happens if we use precisely $1/\log x$, a much weaker damping factor than $1/x^\alpha$, to "damp" the function $1/x$.

2. Let $a > 1$. We substitute $\log x = t$, $\dfrac{dt}{dx} = \dfrac{1}{x}$, and get

$$\int\limits_{a}^{b} \frac{dx}{x(\log x)^\mu} = \int\limits_{\log a}^{\log b} \frac{dt}{t^\mu}$$

$$= \begin{cases} \dfrac{1}{1-\mu}\left[(\log b)^{1-\mu} - (\log a)^{1-\mu}\right] & \text{for } \mu \neq 1, \\ (\log\log b - \log\log a) & \text{for } \mu = 1. \end{cases}$$

It is seen that in order that these integrals converge for $b \to \infty$ it is necessary and sufficient that $\mu > 1$. Accordingly we have the following integrability tests, finer than the previous ones:

(a) $\qquad \left(0 \leqslant f \leqslant C\dfrac{1}{x(\log x)^\mu}; \mu > 1\right) \Rightarrow \left(\displaystyle\int\limits_{1 < a}^{\infty} f < \infty\right);$

(b) $\qquad \left(0 < C\dfrac{1}{x(\log x)^\mu} \leqslant f; \mu \leqslant 1\right) \Rightarrow \left(\displaystyle\int\limits_{a}^{\infty} f = +\infty\right).$

Remark. The functions considered in Examples 1 and 2 have a constant sign. The nets $\int\limits_{K_\alpha} f$ are then monotonic, whence

$$\lim_{K \,\in\,\mathfrak{R}} \int\limits_{K_\alpha} f = \lim_{b \to \infty} \int\limits_{a}^{b} f.$$

141

The following example shows that not every integral can be treated that way.

3. Let us consider $\int_\Omega \dfrac{\sin x}{x}\, dx$ where $\Omega = [a, \infty[,\ a > 0$. From Theorem VI.1.2 (ii) we conclude that it suffices to study the convergence of the integral $\int_a^\infty \left|\dfrac{\sin x}{x}\right| dx$ which gives a monotonic net and can, therefore, be evaluated by taking the limit

$$\lim_{b \to \infty} \int_a^b \left|\frac{\sin x}{x}\right| dx .$$

Since

$$\int_{n\pi}^{(n+1)\pi} \left|\frac{\sin x}{x}\right| dx \geqslant \frac{1}{(n+1)\pi} \int_{n\pi}^{(n+1)\pi} |\sin x|\, dx = \frac{2}{\pi(n+1)}$$

and the series $\displaystyle\sum_{n=n_0}^{\infty} \frac{2}{(n+1)\pi}$ diverges, the integral $\int_a^\infty \left|\dfrac{\sin x}{x}\right| dx$ diverges as well as $\int_a^\infty \dfrac{\sin x}{x}\, dx$.

Note that the limit $\lim\limits_{b \to \infty} \int_a^b \dfrac{\sin x}{x}\, dx$ exists (it can be proved by a similar calculation as above and using arguments similar to those which were applied in the proof of Proposition V.5.5).

II. The set Ω is bounded and not compact. Let $\Omega = \,]a, b]$, for example. Then

$$\int_{a+\varepsilon}^b \frac{dx}{(x-a)^\alpha} = \frac{1}{1-\alpha}\left[(b-a)^{1-\alpha} - \varepsilon^{1-\alpha}\right] \quad \text{for } \alpha \neq 1.$$

If the sequence of integrals is to converge for $0 < \varepsilon \to 0$, it is necessary and sufficient that $\alpha < 1$. For $\alpha = 1$

$$\int_{a+\varepsilon}^b \frac{dx}{x-a} = \left(\log(b-a) - \log\varepsilon\right)\underset{\varepsilon \to 0}{\longrightarrow} \infty.$$

Thus we have the following tests:

(a) $\quad \left(0 \leqslant f(x) \leqslant C\,\dfrac{1}{(x-a)^\alpha}\,; \alpha < 1\right) \Rightarrow \left(\int\limits_a^b f < \infty\right);$

(b) $\quad \left(0 < C\,\dfrac{1}{(x-a)^\alpha} \leqslant f; \alpha \geqslant 1\right) \Rightarrow \left(\int\limits_a^b f = +\infty\right).$

Examples. 1. *Euler's Beta Function.* Let $]0,1[\,\ni t \to f(t) = t^p(1-t)^q$; then

(2) $\qquad B(p+1, q+1) := \int\limits_0^1 t^p(1-t)^q\,dt.$

For $-1 < p, q$ the integral is convergent because, for $t \leqslant \dfrac{1}{2}$, for example,

$$t^p(1-t)^q \leqslant t^p\left(\frac{1}{2}\right)^{-q} = (2)^q\frac{1}{t^{-p}}, \qquad -p < 1,$$

whereas for $t \geqslant \dfrac{1}{2}$,

$$t^p(1-t)^q \leqslant \left(\frac{1}{2}\right)^{-p}(1-t)^q = (2)^p\frac{1}{(1-t)^{-q}}, \qquad -q < 1.$$

Thus the function $B(v, w)$ is well-defined by formula (2) for $v, w > 0$.

2. *Euler's Gamma Function.* Let $]0, \infty[\,\ni t \to f(t) = t^{z-1}e^{-t}$; then

(3) $\qquad \Gamma(z) := \int\limits_0^\infty t^{z-1}e^{-t}\,dt.$

When $t \to \infty$, $f(t)$ tends to zero more rapidly than any power of the variable t. Hence, the integral $\int\limits_0^\infty f(t)$ is convergent. When $t \to 0$, on the other hand, $e^{-t} \to 1$ and $f(t)$ behaves as t^{z-1}. If formula (3) is to define the function the requirement is that $z > 0$ (test 1(a)).

We shall prove interesting properties of the function $\Gamma(z)$. Let us integrate formula (3) by parts, treating the first factor as the derivative and the second as the function:

$$\int\limits_0^\infty t^{z-1}e^{-t}dt = \frac{t^z}{z}e^{-t}\Big|_0^\infty + \frac{1}{z}\int\limits_0^\infty t^z e^{-t}dt.$$

But $\dfrac{t^z}{z} e^{-t} \Big|_0^\infty = 0$ and hence $\Gamma(z) = \dfrac{1}{z} \Gamma(z+1)$, whereby

$$\Gamma(z+1) = z\Gamma(z).$$

LEMMA VI.1.4. The formula

$$\Gamma(n+1) = n!$$

holds.

PROOF (by induction). (a) $\Gamma(2) = 1!$; $\Gamma(1) = 1 = 0!$.

(b) Suppose that the lemma is valid for $n = k$, that is, $\Gamma(k+1) = k!$. Then

$$\Gamma(k+2) = (k+1)\Gamma(k+1) = (k+1)!,$$

or, in other words, the lemma holds for $n = k+1$. Hence we conclude that the lemma is true for any n. $\square$

Thus we have obtained a natural extension of the function $n!$ to arbitrary real numbers contained in the set $]-1, +\infty[$, that is,

$$x! = \Gamma(x+1).$$

It turns out that there is a very interesting relationship between the convergence of series and the convergence of integrals on a non-compact set. This is dealt with by the following theorem.

THEOREM VI.1.5 (The Cauchy–Maclaurin Integral Test).
Hypothesis: $[a, \infty[\ni x \to f(x) \geqslant 0$; $f(x) \searrow 0$ for $x \to \infty$.

Thesis: $\left(\sum\limits_{n=k}^{\infty} f(n) < \infty \right) \Leftrightarrow \left(\int\limits_{k}^{\infty} f(x)\,dx < \infty \right)$.

PROOF. Let us define the "step" functions $v(x)$ and $w(x)$:

$$v(x) := f(n) \quad \text{for } x \in [n, n+1[,$$
$$w(x) := f(n) \quad \text{for } x \in [n-1, n[.$$

Clearly, we have the inequalities

$$(4) \qquad 0 \leqslant w(x) \leqslant f(x) \leqslant v(x)$$

and

$$\int\limits_{k}^{p} v(x)\,dx = \sum\limits_{n=k}^{p-1} f(n); \qquad \int\limits_{k}^{p} w(x)\,dx = \sum\limits_{n=k+1}^{p} f(n).$$

Therefore the convergence of the series $\sum\limits_{n=k}^{\infty} f(n)$ is equivalent to the integrability of the functions $v(x)$ and $w(x)$. But by Proposition VI.1.1 the latter is equivalent to the integrability of the function $f(x)$. $\square$

Examples. 1. The harmonic series $\displaystyle\sum_{n=1}^{\infty} \frac{1}{n} = \infty$.

PROOF. The Cauchy–Maclaurin test implies that

$$\int_1^p \frac{1}{x}\,dx = \log p \xrightarrow[p\to\infty]{} \infty.$$

2. $\displaystyle\sum_{n=1}^{\infty} \frac{1}{n^{1+\sigma}} < \infty$ for $\sigma > 0$, because $\displaystyle\int_1^\infty \frac{1}{x^{1+\sigma}}\,dx < \infty$.

Remark. The sequence (d_p), where

$$d_p := \sum_{n=k}^p f(n) - \int_k^p f(x)\,dx = \int_k^{p+1} v(x)\,dx - \int_k^p f(x)\,dx,$$

is convergent for integrable functions, since

$$\lim_{p\to\infty}\left(\int_k^{p+1} v(x)\,dx - \int_k^p f(x)\,dx\right) = \sum_{n=k}^\infty f(n) - \int_k^\infty f(x)\,dx = d.$$

Thus $d_p \xrightarrow[p\to\infty]{} d$.

Let us see what the case is for functions which are not integrable but do satisfy the hypotheses of Theorem VI.1.5. We have

$$d_{p+1} - d_p = \sum_{n=k}^{p+1} f(n) - \int_k^{p+1} f - \sum_{n=k}^p f(n) - \int_k^p f = f(p+1) - \int_p^{p+1} f$$

$$= \int_p^{p+1} [f(p+1) - f(x)]\,dx < 0,$$

because $f(p+1) - f(x) < 0$ for $x \in [p, p+1]$, and thus $d_{p+1} < d_p$ (the sequence is decreasing). However,

$$d_p = \sum_{n=k}^p f(n) - \int_k^p f(x)\,dx = \int_k^{p+1} v(x)\,dx - \int_k^p f(x)\,dx$$

$$= \int_k^p (v-f)\,dx + \int_p^{p+1} v\,dx \geqslant 0,$$

since $v - f \geqslant 0$; $v \geqslant 0$. Accordingly, (d_p) is a decreasing sequence, bounded from below, and hence convergent:

$$d_p \to d = \lim_{p \to \infty} \left(\sum_{n=k}^{p} f(n) - \int_{k}^{p} f(x)\,dx \right).$$

This is an extremely interesting result, since the two sequences

$$S_p = \sum_{n=k}^{p} f(n) \quad \text{and} \quad I_p = \int_{k}^{p} f(x)\,dx$$

do not converge $(S_p \to \infty, I_p \to \infty)$, whereas their difference tends to a constant, $(S_p - I_p) \to d$.

Example. We have

$$\lim_{p \to \infty} \left(\sum_{n=1}^{p} \frac{1}{n} - \int_{1}^{p} \frac{1}{x}\,dx \right) := C = 0.5772 \ldots$$

This is *Euler's constant*.

VII. BANACH SPACES
DIFFERENTIATION OF MAPPINGS
EXTREMA OF FUNCTIONS AND FUNCTIONALS

In Chapter III we considered the differential calculus of functions of one variable. It will now be seen that the definition given there for a derivative carries over "unaltered" to mappings of Banach spaces: all that needs to be done is to replace the absolute value by a norm. The reasons underlying the treatment of this general situation are:

(i) The case of a function of two variables (the mappings $R^2 \to R^1$) entails the same difficulties as the general case does.

(ii) The general treatment makes it possible to bring out the "geometric" meaning of a number of constructions and imparts clarity to them.

(iii) As we know, many function spaces are Banach spaces. Thus variational calculus problems concern the examination of functions defined on Banach spaces; for example,

$$\varphi \to f(\varphi) = \int_a^b L\left(t, \varphi(t), \frac{d\varphi}{dt}(t)\right) dt.$$

Just as in the case of one variable, the important thing in the investigation of such functionals is to calculate their derivative and to consider the zero of the derivative, etc. It was precisely the variational calculus which was one of the main stimuli in the birth of the theory of Banach spaces and the differentiation of mappings of these spaces. We shall discuss some variational calculus problems further on when we shall present them as examples of more general theories.

In the present chapter we shall concentrate entirely on mappings ot vector spaces. The reader is assumed to be familiar with the concep: of vector space.

1. NORMED SPACES AND BANACH SPACES

DEFINITION. Let X be a vector space over the field $R^1(C^1)$. The function $||\cdot||$ defined on X, that is, $X \ni x \to ||x|| \in R^1 (C^1)$, is called a *norm* if

(i) $||x|| \geqslant 0$ for every $x \in X$,

(ii) $||\alpha x|| = |\alpha| \cdot ||x||$ for every $\alpha \in R^1 (C^1)$ and every $x \in X$,

(iii) $||x_1 + x_2|| \leqslant ||x_1|| + ||x_2||$ for every pair $x_1, x_2 \in X$ (this is the *triangle inequality*),

(iv) $||x|| = 0 \Leftrightarrow x = 0$, that is, $||\cdot||$ vanishes only on the null vector.

149

A function $p(\cdot)$ satisfying only (i), (ii), and (iii) is called a *seminorm*. The concept of seminorm will be needed in the theory of integration. Note furthermore that (ii) and (iii) imply that both seminorms and norms are convex functions.

DEFINITION. A pair $(X, ||\cdot||)$ is called a *normed space*. A pair $(X, p(\cdot))$ is called a *seminormed space*.

The structure of a metric space can be introduced in a normed space. A *distance* is defined in terms of the norm as follows:

$$d(x_1, x_2) := ||x_1 - x_2||, \quad x_1, x_2 \in X.$$

We verify whether this definition is correct:

(i) $\left(d(x_1, x_2) = 0\right) \Rightarrow (||x_1 - x_2|| = 0) \Rightarrow (x_1 - x_2 = 0) \Rightarrow (x_1 = x_2)$-

(ii) The symmetry is obvious.

(iii) The triangle inequality: $d(x_1, x_3) = ||x_1 - x_3|| \leqslant ||x_1 - x_2|| + ||x_2 - x_3|| = d(x_1, x_2) + d(x_2, x_3)$.

The distance so defined determines, in the familiar manner, a convergence (topology) in a normed space. Henceforth in speaking of normed space we shall assume that its topology (convergence) is determined by the norm in the manner described above. Thus we arrive at the concept of a Banach space.

DEFINITION. A normed and complete space is called a *Banach space*.

A Hausdorff theorem implies that every metric space can be completed. Strictly speaking, for every metric space X there exists a complete metric space $\tilde{X}$ such that $X \subset \tilde{X}$ and a natural mapping $j: X \to \tilde{X}$ is an isometry. Let us recall that the space $\tilde{X}$ is the set of classes of equivalent Cauchy sequences in the space X. If X is a normed space, then in the set of classes we have a vector structure and a norm:

$$\alpha[(x_n)] + \beta[(y_n)] := [(\alpha x_n + \beta y_n)], \quad [(x_n)], [(y_n)] \in \tilde{X}, \quad \alpha, \beta \in R^1,$$

$$||[(x_n)]|| := \lim_{n \to \infty} ||x_n||.$$

Clearly, the distance determined by this norm coincides with the distance determined by the Hausdorff theorem. Thus we infer that every normed space $(X, ||\cdot||)$ is embeddable in the Banach space $(\tilde{X}, ||\cdot||)$.

Once we have one norm in a vector space, we can multiply it by a positive number to obtain a new norm in this space. It may also be that other norms can be introduced in this space. We shall compare particular norms, comparing the topologies (convergences) which they determine. The concept of "convergence in the sense of norm" will henceforth be taken to mean convergence in the metric determined by that norm.

DEFINITION. The norm $||\cdot||_2$ is *stronger* than the norm $||\cdot||_1$ if for every $x \in X$ and for every sequence convergent $x_n \to x$ in the sense of the norm $||\cdot||_2$ we have $\lim\limits_{n\to\infty} x_n = x$ in the sense of the norm $||\cdot||_1$.

DEFINITION. Two norms are *equivalent* if the first is stronger than the second, and the second is stronger than the first.

It is easily verified that this is an equivalence relation.

LEMMA VII.1.1. (The norm $||\cdot||_2$ is stronger than the norm $||\cdot||_1$) $\Leftrightarrow$ (there exists an $a > 0$ such that $||x||_1 \leqslant a \cdot ||x||_2, x \in X$).

PROOF. $\Leftarrow$: Let $x_n \to x$ in the norm $||\cdot||_2$. Then $(x_n - x) \to 0$ in the norm $||\cdot||_2$ and because of the inequality $||x_n - x||_1 \leqslant a \cdot ||x_n - x||_2$ also $||x_n - x||_1 \to 0$, or $x_n \to x$ in the norm $||\cdot||_1$.

$\Rightarrow$: (a.a.) Suppose that for every $n > 0$ there exists an x_n such that $||x_n||_1 > n \cdot ||x_n||_2$, whence

$$\frac{||x_n||_2}{||x_n||_1} < \frac{1}{n}, \quad \text{and hence} \quad \frac{x_n}{||x_n||_1} \to 0$$

in the norm $||\cdot||_2$. On the other hand

$$\left|\left|\frac{x_n}{||x_n||_1}\right|\right|_1 = \frac{||x_n||_1}{||x_n||_1} = 1 \nrightarrow 0.$$

In this way we have obtained a sequence which converges to zero in the norm $||\cdot||_2$ but not in the norm $||\cdot||_1$. This contradicts the hypothesis. $\square$

COROLLARY. VII.1.2 Two norms $||\cdot||_1$ and $||\cdot||_2$ are equivalent if and only if there exist numbers $a > 0$ and $b > 0$ such that

$$||x||_1 \leqslant a \cdot ||x||_2, \quad ||x||_2 \leqslant b \cdot ||x||_1$$

for every $x \subset X$.

Examples of Norms. 1. R^1—the set of real numbers; $||x|| = |x|$.

2. R^n is the n-dimensional arithmetic space. The norms can be defined as follows:

(a) $\quad ||x||_1 = \left(\sum_{i=1}^{n} (x^i)^2 \right)^{1/2},$

(b) $\quad ||x||_2 = \sum_{i=1}^{n} |x^i|, \quad$ where $x = (x^1, x^2, ..., x^n),$

(c) $\quad ||x||_3 = \max_{i=1,...,n} |x^i|.$

The following inequalities hold:

$$||x||_3 \leqslant ||x||_2 \leqslant n||x||_3,$$
$$||x||_3 \leqslant ||x||_1 \leqslant n||x||_3,$$
$$\frac{1}{n}||x||_2 \leqslant ||x||_1 \leqslant ||x||_2.$$

All of these norms thus are equivalent.

The completeness of the space R^n follows from the completeness of the space R^1 and the fact that the Cartesian product of complete spaces is a complete space. This topic will be discussed in greater detail in the next example.

3. Let $(X, || \cdot ||)$ and $(Y, || \cdot ||)$ be normed spaces. On the Cartesian product $X \times Y$ we can introduce the norms:

(a) $\quad ||(x, y)||_1 = ((||x||)^2 + (||y||)^2)^{1/2};$

(b) $\quad ||(x, y)||_2 = ||x|| + ||y||;$

(c) $\quad ||(x, y)||_3 = \max(||x||, ||y||), \quad x \in X, \ y \in Y.$

Thus the Cartesian product of two normed spaces is a normed space. To show that it is a complete space, let us assume that (x_n, y_n) is a Cauchy sequence in one of the norms defined above. Then the sequences (x_n) and (y_n) are Cauchy sequences in the spaces X and Y, respectively. If X and Y are Banach spaces, then there exist vectors $x \in X$ and $y \in Y$ such that $x_n \to x$ and $y_n \to y$. Hence, the sequence $(x_n, y_n) \to (x, y)$. It has thus been shown that the Cartesian product of two Banach spaces is a Banach space. Plainly, this is true for the product of n spaces. The reasoning put forward in Example 2 demonstrates that norms defined in (a), (b), and (c) are equivalent.

4. Let X be a metric space, let Y be a normed space, and let $B(X, Y)$ be a set of bounded mappings of X into Y. This set in a natural manner becomes a vector space if the operations are defined as follows:

$$(f_1 + f_2)(x) := f_1(x) + f_2(x),$$
$$(\alpha \cdot f) : (x) = \alpha \cdot (f(x)), \qquad f_1, f_2 \in B(X, Y), \ \alpha \in R^1(C^1).$$

We can also define the norm

$$\|f\| := \sup \|f(X)\|.$$

Thus $B(X, Y)$ becomes a normed space which, in general, is not a Banach space. On the other hand, if we confine ourselves to the case when Y is a Banach space, this set will be a Banach space (cf. Theorem V.4.3).

5. Let $C([0, 1])$ be the space of functions continuous on the interval $[0, 1]$. We define the norm

$$\|f\|_1 := \int_0^1 |f(x)| \, dx.$$

The space $C([0, 1])$ with the norm so introduced is not a complete space. To see that this is the case we take the sequence of functions

$$f_n(x) := \begin{cases} 1 & \text{for } 0 \leqslant x \leqslant \dfrac{1}{2} - \dfrac{1}{n}, \\[2mm] -nx + \dfrac{n}{2} & \text{for } \dfrac{1}{2} - \dfrac{1}{n} < x < \dfrac{1}{2}, \ n = 2, 3, \ldots, \\[2mm] 0 & \text{for } \dfrac{1}{2} \leqslant x \leqslant 1; \end{cases}$$

it is easily verified that $f_n \to f$, where f is a discontinuous function:

$$f(x) := \begin{cases} 1 & \text{for } 0 \leqslant x < \dfrac{1}{2}, \\[2mm] 0 & \text{for } \dfrac{1}{2} \leqslant x \leqslant 1. \end{cases}$$

The completion of the space $C([0, 1])$ in the norm $\| \cdot \|_1$ is called the *space of functions integrable on* $[0, 1]$ and is denoted by $L^1([0, 1])$. This space will be discussed more thoroughly in Chapter XIII.

2. CONTINUOUS LINEAR MAPPINGS OF BANACH SPACES

Let $(X, \| \cdot \|_1)$ and $(Y, \| \cdot \|_2)$ be normed spaces both real or both complex. A mapping $L: X \to Y$ is said to be *linear* if

$$L(\alpha x_1 + \beta x_2) = \alpha L(x_1) + \beta L(x_2), \qquad x_1, x_2 \in X, \ \alpha, \beta \in R^1(C^1).$$

We shall frequently write Lx or $L \cdot x$ instead of $L(x)$ for linear mappings. It should be recalled that a mapping $L: X \to Y$ is said to be *continuous at the point* $x \in X$ if for every sequence $x_n \to x$ we have $\lim_{n \to \infty} Lx_n = Lx$.

In the sequel we shall deal with continuous linear mappings, a set of which we denote by $L(X, Y)$. The following theorem characterizes continuous linear mappings.

THEOREM VII.2.1. Let $(X, \| \cdot \|)$ and $(Y, \| \cdot \|)$ be normed spaces and L a linear mapping of X into Y. The four conditions below are then equivalent:

(i) L maps sequences convergent to zero upon bounded sequences, i.e. for every sequence $x_n \to 0$ there exists a $C > 0$ such that

$$\|Lx_n\| \leqslant C, \quad n = 1, 2, \ldots$$

(ii) L is continuous at a particular point, e.g. at zero.

(iii) There exists an $A > 0$ such that for every $x \in X$ we have $\|Lx\| \leqslant A\|x\|$.

(iv) L is continuous at every point.

PROOF. The scheme of the proof is:

$$\text{(i)} \Rightarrow \text{(ii)} \Rightarrow \text{(iii)} \Rightarrow \text{(iv)} \Rightarrow \text{(i)}.$$

(i) $\Rightarrow$ (ii) (a.a.): Suppose that L is not continuous at zero. There thus exists a sequence $x_n \to 0$ such that $Lx_n \nrightarrow L(0) = 0$. Hence, it is possible to extract from the sequence (x_n) a subsequence (x_{n_k}) such that $\|Lx_{n_k}\| > C$ for a particular $C > 0$. Let $b_k := (\|x_{n_k}\|)^{-1/2}$. It is seen that $b_k \nearrow \infty$. On the other hand, the sequence $d_k := b_k x_{n_k} \to 0$. For we have

$$\|d_k\| = b_k\|x_{n_k}\| = \frac{\|x_{n_k}\|}{\|x_{n_k}\|^{1/2}} = \|x_{n_k}\|^{1/2} \to 0.$$

Therefore

$$\|Ld_k\| = \|L(b_k x_{n_k})\| = b_k\|Lx_{n_k}\| \geqslant b_k C \nearrow \infty.$$

This, however, contradicts condition (i).

(ii) $\Rightarrow$ (iii). Let L be continuous at the point $x_0 \in X$. For every $\varepsilon > 0$ there exists a $\delta > 0$ such that $\|L(x-x_0)\| < \varepsilon$, provided that $\|x-x_0\| < \delta$. If we denote $x-x_0 = u$, we have $\|Lu\| < \varepsilon$, provided that $\|u\| < \delta$. Now, let us take an arbitrary $x \neq 0$. Then

$$\|Lx\| = \left\| L\left(\frac{x}{\|x\|} \cdot \frac{\delta}{2} \right) \right\| \cdot \frac{2\|x\|}{\delta}.$$

2. CONTINUOUS LINEAR MAPPINGS

We apply the inequality given above to the right-hand side of the foregoing equation:

$$||Lx|| = \left\|L\left(\frac{x}{||x||} \cdot \frac{\delta}{2}\right)\right\| \cdot \frac{2||x||}{\delta} \leqslant \frac{2\varepsilon}{\delta}||x|| = A||x||.$$

Thus, we get $||Lx|| \leqslant A||x||$ for every $x \in X$.

(iii) $\Rightarrow$ (iv): Suppose that the inequality $||L(x)|| \leqslant A||x||$ holds for all $x \in X$. If x_0 is an arbitrary vector of the space X, and the sequence $x_n \to x_0$, then

$$||Lx_n - Lx_0|| = ||L(x_n - x_0)|| \leqslant A||x_n - x_0|| \to 0.$$

Consequently, $\lim_{n \to \infty} Lx_n = Lx_0$. This proves the continuity of the mapping L at every point.

(iv) $\Rightarrow$ (i): If $x_n \to x_0$, then by virtue of the continuity we have $Lx_n \to Lx_0$. We know, however, that a convergent sequence is bounded. $\square$

DEFINITION. A linear mapping possessing the property

$$||Lx|| \leqslant A \cdot ||x|| \qquad \text{for every } x \in X \text{ and a particular } A \geqslant 0$$

is said to be *bounded*.

Theorem VII.2.1 shows that a continuous linear mapping is bounded and, conversely, a bounded linear mapping is continuous. For this reason, in the case of linear mappings we shall use the designations "continuous" and "bounded" without invoking Theorem VII.2.1. In general, discontinuous linear mappings exist in Banach spaces. On the other hand, every linear mapping of finite-dimensional Banach spaces into any Banach space is continuous. This is related to the fact that all norms are equivalent in a finite-dimensional Banach space.

Let $(X, ||\cdot||)$ be an n-dimensional (say real) Banach space. Next, we choose some base $e_1, e_2, \ldots, e_n$ in X. We define a new norm

$$||x||_1 = \max_{i=1,\ldots,n} |x^i|,$$

where the numbers x^i $(i = 1, 2, \ldots, n)$ are coordinates of the vector x in the base $e_1, e_2, \ldots, e_n$; $x = \sum_{i=1}^{n} x^i e_i$.

By the inequality $||x|| \leqslant \max |x^i| \max ||e_i|| \cdot n$ the norm $||\cdot||_1$ is stronger than the norm $||\cdot||$. The converse theorem must now be shown

155

to hold. To this end, we consider the mapping $F: X \to R^n$, $X \ni x \to F(x) = (x^1, x^2, \ldots, x^n) \in R^n$; this mapping is clearly linear, and hence we want to demonstrate that it is continuous in the norm $||\cdot||$. Suppose that F is not continuous. Hence there exists a sequence $x_n \to 0$ such that $F(x_n) \not\to 0$. It is thus possible to extract a subsequence x_{n_k} such that $||F(x_{n_k})|| > a$ for some $a > 0$. Let $y_k = x_{n_k}(||F(x_{n_k})||)^{-1}$. We then have $y_k \to 0$ and $||F(y_k)|| = 1$. But the unit sphere in R^n is compact, and hence from the sequence $(F(y_k))$ we can select a subsequence convergent to the element $z \in R^n$, $||z|| = 1$. Bear in mind, however, that earlier we demonstrated the continuity of the mapping F^{-1}. Thus

$$F^{-1}(z) = 0, \text{ but on the other hand } F^{-1}(z) = \sum_{i=1}^{n} z^i e_i. \text{ By the uniqueness}$$

of the null-vector decomposition we have $z^i = 0$ ($i = 1, 2, \ldots, n$), which is inconsistent with the equation $||z|| = 1$. This contradiction proves the mapping F to be continuous. Hence there exists an $A > 0$ such that

$$||F(x)|| = \max_{i=1,\ldots,n} |x^i| \leqslant A||x||.$$

This inequality shows that the norm $||\cdot||$ is stronger than the norm $||\cdot||_1$. The equivalence of the two norms has thus been demonstrated.

The set of continuous linear mappings of a Banach space X into a Banach space Y is itself a Banach space. The algebraic operations are defined as follows:

$$(\alpha L)(x) := \alpha(Lx),$$

$$(L_1 + L_2)(x) := L_1 x + L_2 x,$$

$$x \in X, \quad \alpha \in R^1(C^1), \quad L_1, L_2 \in L(X, Y).$$

Item (iii) of Theorem VII.2.1 implies that for $L \in L(X, Y)$ there exists an $A > 0$ such that $||Lx|| \leqslant A||x||$. The smallest $A \geqslant 0$ satisfying this condition for all $x \in X$ is called the *norm of the mapping* L:

$$||L|| := \inf\{A: ||Lx|| \leqslant A||x||, \; x \in X\}.$$

We shall now show that $||L|| = \sup_{||x|| \leqslant 1} ||Lx||$. To this end let us assume that $x \neq 0$ and divide the inequality $||Lx|| \leqslant ||L|| \cdot ||x||$ by $||x||$; we obtain

$$\left|\left| L \frac{x}{||x||} \right|\right| \leqslant ||L||.$$

On taking the upper bound of both sides, we have

$$\sup_{x \neq 0} \left\| L \frac{x}{\|x\|} \right\| \leqslant \|L\|;$$

next we shall show that the strict inequality cannot hold. Suppose that

$$A = \sup_{x \neq 0} \left\| L \frac{x}{\|x\|} \right\| < \|L\|,$$

then

$$\left\| L \frac{x}{\|x\|} \right\| \leqslant \sup_{x \neq 0} \left\| L \frac{x}{\|x\|} \right\| = A < \|L\|,$$

whence

(1) $\|Lx\| \leqslant A\|x\|.$

This, however, is inconsistent with the definition of $\|L\|$ as the smallest constant to satisfy (1). Thus we have the equality:

$$\|L\| = \sup_{x \neq 0} \left\| L \frac{x}{\|x\|} \right\| = \sup_{\|x\| \leqslant 1} \|Lx\|.$$

By means of this formula we shall show that $\| \cdot \|$ is a norm:

(i) Obvious, because $\|L\| \geqslant 0$, $L \in L(X, Y)$,

(ii) $\|\alpha L\| = \sup_{\|x\| \leqslant 1} \|\alpha Lx\| = |\alpha| \cdot \sup_{\|x\| \leqslant 1} \|Lx\| = |\alpha| \cdot \|L\|,$

(iii) $\|L_1 + L_2\| = \sup_{\|x\| \leqslant 1} \|L_1 x + L_2 x\| \leqslant \sup_{\|x\| \leqslant 1} \|L_1 x\| + \sup_{\|x\| \leqslant 1} \|L_2 x\|$

$= \|L_1\| + \|L_2\|,$

(iv) If $\|L\| = 0$, then $Lx = 0$ for every $x \in X$, and hence $L = 0$.

THEOREM VII.2.2. The space $L(X, Y)$ is a Banach space.

PROOF (it is sufficient to demonstrate completeness). Let $L_n \in L(X, Y)$, $n = 1, 2, \ldots$, and let L_n be a Cauchy sequence in $L(X, Y)$, that is, for every $\varepsilon > 0$ there exists an $N > 0$ such that $\|L_n - L_m\| < \varepsilon$ for $n, m > N$. In view of the inequality $\|L_n x - L_m x\| \leqslant \|L_n - L_m\| \cdot \|x\|$ $< \varepsilon \cdot \|x\|$, the sequence $L_n x$ is a Cauchy sequence in Y. The completeness of the space Y implies that there exists a point $y \in Y$ such that $y = \lim_{n \to \infty} L_n x$. We define the mapping

$$Lx := \lim_{n \to \infty} L_n x.$$

Clearly the mapping L is linear; we need to show that it is continuous. The modified triangle inequality

$$| \, ||L_n|| - ||L_m|| \, | \leqslant ||L_n - L_m|| < \varepsilon \quad \text{for } n, m > N$$

shows that the sequence $(||L_n||)$ is a Cauchy sequence of real numbers. Let $A := \lim_{n \to \infty} ||L_n||$. We have

$$||Lx|| = || \lim_{n \to \infty} L_n x|| = \lim_{n \to \infty} ||L_n x|| \leqslant \lim_{n \to \infty} ||L_n|| \cdot ||x||$$
$$= A \cdot ||x||.$$

We have used the fact that a norm is continuous. Thus we have demonstrated that the mapping L is continuous.

It will now be shown that $L_n \to L$. We have

$$||(L - L_n)x|| = ||Lx - L_n x|| \leqslant ||Lx - L_m x|| + ||L_m x - L_n x||,$$

but

(2) $\qquad ||L_m x - L_n x|| < \varepsilon \quad \text{for } ||x|| \leqslant 1, \ n, m > N.$

Simultaneously for every $\varepsilon > 0$ and every $x \in X$ there exists an $M > 0$ such that

(3) $\qquad ||Lx - L_m x|| < \varepsilon \quad \text{for } m > M.$

Taking $m > \max(M, N)$ and making use of (2) and (3), for $x \in K(0, 1)$ we have

$$||(L - L_n)x|| < 2\varepsilon \quad \text{for } n > N.$$

Note that N does not depend on x; consequently we have

$$||(L - L_n)x|| < 2\varepsilon \quad \text{for } n > N \text{ and } x \in K(0, 1).$$

Hence

$$||L - L_n|| = \sup_{x \in K(0,1)} ||(L - L_n)x|| \leqslant 2\varepsilon \quad \text{for } n > N,$$

that is $L_n \to L$. $\square$

A continuous linear mapping of a real Banach space X into R^1 is called a *linear functional continuous on X*. The set of all linear functionals continuous on X is plainly a Banach space. This space is said to be the *conjugate* (*dual*) to the space X and is denoted by X^*:

$$X^* = L(X, R^1).$$

2. CONTINUOUS LINEAR MAPPINGS

The value of a functional $x^* \in X^*$ on a vector $x \in X$ is written as

$$x^*(x) := \langle x, x^* \rangle.$$

The elements of the space X are called *vectors*, and those of the space X^*, *covectors*.

Remark. Some authors (e.g. N. Bourbaki) denote the space X^* by the symbol X', reserving the symbol X^* for the set of all linear functionals (not necessarily continuous) on X. For certain reasons we shall employ the traditional notation, X^*.

Examples of Continuous (Linear) Mappings. 1. The dual to the n-dimensional arithmetic space $(R^n)^* = L(R^n, R^1)$. Let $e_1, e_2, \ldots, e_n$ be the canonical basis in R^n. An arbitrary vector $x \in R^n$ has a unique representation in this basis $x = \sum_{i=1}^{n} x^i e_i$. If $L \in L(R^n, R^1)$, then

$$L(x) = L\left(\sum_{i=1}^{n} x^i e_i \right) = \sum_{i=1}^{n} x^i L e_i = \sum_{i=1}^{n} x^i a_i,$$

where $a_i = Le_i$, $i = 1, 2, \ldots, n$. A sequence of n real numbers $a_1, a_2, \ldots, a_n$ corresponds to every mapping L. Thus, as is easily verified, we have an algebraic isomorphism of the space R^n and the space $L(R^n, R^1)$. We shall show this isomorphism to be an isometry:

$$\|L\| = \sup_{\|x\| \leqslant 1} \|Lx\| = \sup_{\|x\| \leqslant 1} \sum_{i=1}^{n} x^i a_i$$

$$\leqslant \sup_{\|x\| \leqslant 1} \left(\sum_{i=1}^{n} (x^i)^2 \right)^{1/2} \left(\sum_{i=1}^{n} (a_i)^2 \right)^{1/2}$$

$$= \left(\sum_{i=1}^{n} a_i^2 \right)^{1/2} = \|a\|.$$

Here, use has been made of the *Schwarz inequality*, which is familiar from algebra:

$$\sum_{i=1}^{n} x^i a_i \leqslant \left(\sum_{i=1}^{n} (x^i)^2 \right)^{1/2} \left(\sum_{i=1}^{n} (a_i)^2 \right)^{1/2}.$$

Thus, we have $\|L\| \leqslant \|a\|$. If we put $x = a/\|a\|$, the Schwarz inequality becomes an equality and we have

$$\|L\| = \|a\|.$$

We write

$$L(R^n, R^1) \cong R^n \quad \text{or} \quad R^n \cong (R^n)^*.$$

2. The space $L(R^n, R^m)$. As is known from algebra, a matrix corresponds to every mapping $L \in L(R^n, R^m)$. If $e_1, e_2, \ldots, e_n$ is the canonical basis in R^n and $f_1, f_2, \ldots, f_m$ a canonical basis in R^m, then the matrix $[L_i^j]$ corresponding to the mapping L operates as follows:

$$Le_i = \sum_{j=1}^{m} L_i^j f_j.$$

It is easily shown that

$$\|L\| \leqslant \left(\sum_{i,j} (L_i^j)^2 \right)^{1/2}.$$

3. Let X be some real Banach space. We consider the space $L(R^1, X)$. Let $L \in L(R^1, X)$. The linearity of L implies that $L(a) = L(a \cdot 1) = a \cdot L(1) = a \cdot x$ where $x = L(1) \in X$, $a \in R^1$. Thus an element of the space X corresponds to every linear mapping R^1 in X. This relationship is plainly an algebraic isomorphism. We shall demonstrate that it is an isometry. We have

$$\|L\| = \sup_{|a| \leqslant 1} \|La\| = \sup_{|a| \leqslant 1} (\|aL(1)\|) = \sup_{|a| \leqslant 1} (|a| \|L(1)\|)$$

$$= \|L(1)\| = \|x\|,$$

and hence the relation $L \to x$ is a topological isomorphism. This enables us to identify the spaces X and $L(R^1, X)$; we may write $X \cong L(R^1, X)$. Analogous reasoning shows that the spaces $L(R^n, X)$ and $X \times X \times \ldots \times X$ are isomorphic.

3. DIFFERENTIATION OF MAPPINGS OF BANACH SPACES

In the rest of the present chapter we shall consider only real vector (normed, Banach) spaces, unless otherwise stated.

The definition given in Chapter III for the derivative of a function carries over immediately to mappings of open sets of Banach spaces; the absolute value $|\cdot|$ should be replaced by the norm $\|\cdot\|$, and consequently we have:

DEFINITION. Let U be an open set in a Banach space X. The mapping T of the set U into a Banach space Y is said to be *differentiable at a point*

$x_0 \in U$, if there exists a continuous linear mapping L_{x_0} of the space X into the space Y such that

$$T(x_0+h)-T(x_0) = L_{x_0}(h)+r(x_0; h),$$

where

$$\lim_{h \to 0} \frac{\|r(x_0; h)\|}{\|h\|} = 0.$$

It is seen that the fact the mapping T is differentiable at the point x_0 means that the difference $T(x_0+h)-T(x_0)$ decomposes into a part linear in h and a remainder which is an infinitesimal of order higher than h. The mapping $L_{x_0} \in L(X, Y)$ is called the *derivative of the mapping T at the point x_0* and is denoted by $T'(x_0)$. The value of the mapping L_{x_0} at a given $h \in X$ is called the *differential of the mapping T* for a given increment h and is denoted by $dT(x_0, h)$. When T is differentiable at every point of the set U, we say that T is *differentiable on U* and we call the mapping $U \ni x \to T'(x) \in L(X, Y)$ the *derivative of the mapping T* and denote it by T'. If the derivative is a continuous mapping, then T' is called *continuously differentiable on V*.

The definition of the differentiability of a mapping implies immediately that a mapping which is differentiable at a point x_0 is continuous at that point, and a mapping differentiable on U is continuous on U.

LEMMA VII.3.1. If a mapping T has a derivative at a point $x_0 \in U$, then it is unique.

PROOF. (a.a.) Suppose that there exist two decompositions

$$T(x_0+h)-T(x_0) = L_1(h)+r_1(h),$$
$$T(x_0+h)-T(x_0) = L_2(h)+r_2(h).$$

On subtracting both sides, we have

$$(L_1-L_2) \cdot h = r_1(h)-r_2(h).$$

Division of both sides of this equation by $\|h\|$ ($h \neq 0$) yields

$$\left\| (L_1-L_2)\frac{h}{\|h\|} \right\| = \frac{\|r_1(h)-r_2(h)\|}{\|h\|};$$

when $h \to 0$, we have

$$\left\| (L_1-L_2)\frac{h}{\|h\|} \right\| < \varepsilon \quad \text{for } \|h\| < \delta.$$

For any $x \in X$, such that $||x|| = 1$,

$$||(L_1 - L_2)x|| = \left\| (L_1 - L_2) \frac{x \cdot \frac{1}{2}\delta}{\frac{1}{2}\delta} \right\| < \varepsilon$$

and the arbitrariness of ε implies that

$$||(L_1 - L_2)x|| = 0,$$

whence $L_1 = L_2$. $\square$

The derivative (differential) defined above is frequently called the *strong derivative* (*strong differential*) or the *Fréchet derivative* (*differential*).

DEFINITION. Let U be an open subset of a Banach space X, and let T be a mapping of the set U into a Banach space Y. The mapping T is said to have a *derivative in the direction of the vector* $e \in X$ at a point $x_0 \in U$ if there exists a limit

$$\lim_{t \to 0} \frac{T(x_0 + te) - T(x_0)}{t}, \qquad t \in R^1.$$

The limit of this quotient (if it does exist) is denoted by the symbol $\nabla_e T(x_0)$ and is called the *directional derivative*. Of course, $\nabla_e T(x_0) \in Y$.

THEOREM VII.3.2. Let T be a mapping of an open subset of a Banach space X into a Banach space Y. If T is differentiable (strongly) at a point $x_0 \in U$, then the derivative exists in every direction and $\nabla_e T(x_0) = T'(x_0)e$.

PROOF. The differentiability of T implies that

$$T(x_0 + te) - T(x_0) = T'(x_0)(te) + r(te).$$

Dividing both sides by t and making use of the linearity of the derivative, we have

$$\frac{T(x_0 + te) - T(x_0)}{t} = T'(x_0)e + \frac{r(te)}{t}.$$

Passage to the limit and use of the fact that

$$\left\| \frac{r(te)}{t} \right\| = \frac{||r(te)|| \cdot ||e||}{||te||} \xrightarrow[t \to 0]{} 0,$$

yields

$$\nabla_e T(x_0) = T'(x_0)e. \quad \square$$

The converse theorem is not in general true. This is due to the fact that the existence of a directional derivative (even in all directions) is a much weaker property than differentiability; it does not even entail continuity. Examples will be given in subsequent sections.

Note that if the strong derivative exists at a point x_0, then the mapping $X \ni e \to \nabla_e T(x_0) \in Y$ is linear and continuous. In general, the directional derivative is neither linear nor continuous with respect to e. If the directional derivative is linear and continuous with respect to e, the mapping $X \ni e \to \nabla_e T(x_0) \in Y$ is said to be the *weak* or *Gateaux derivative* of the mapping T at the point x_0. A subsequent section in this chapter will be devoted to weak derivatives.

Let us now consider a differentiable mapping F of an open subset U of the space R^n into R^m. As we know, the derivative of this mapping is represented by the matrix $[F_j'^i]$.

Let $e_1, e_2, ..., e_n$ be a canonical basis in R^n, and let $k_1, k_2, ..., k_m$ be a canonical basis in R^m. In these bases, the mapping F corresponds to m functions $U \ni x \to f^i(x) \in R^1$, $i = 1, 2, ..., m$, such that

$$F(x) = \sum_{i=1}^{m} f^i(x) \cdot k_i.$$

This equation is simply the decomposition of the vector $F(x) \in R^m$ into basis vectors. We may thus write

$$(1) \qquad F(x_0+h) - F(x_0) = \sum_{i=1}^{m} f^i(x_0+h) \cdot k_i - \sum_{i=1}^{m} f^i(x_0) \cdot k_i.$$

By the differentiability of F we have

$$(2) \qquad F(x_0+h) - F(x_0) = F'(x_0) \cdot h + r(h).$$

Making use of the linearity of the derivative, we have

$$(3) \qquad F'(x) \cdot h = \sum_{i,j} F_j'^i(x_0) h^j \cdot k_i, \qquad \text{where } h = \sum_{j=1}^{n} h^j \cdot e_j.$$

The decomposition of the remainder into basis vectors gives

$$(4) \qquad r(h) = \sum_{i=1}^{m} r^i(h) \cdot k_i.$$

163

From (2)–(4) we have

$$\sum_{i=1}^{m} \left(f^i(x_0+h)-f^i(x_0)\right) \cdot k_i = \sum_{i=1}^{m} \left(\sum_{j=1}^{n} \left(h^j F_j^{\prime i}(x_0)+r^i(h)\right)\right) k_i.$$

Making use in turn of the linear independence of the basis vectors, we obtain

$$f^i(x_0+h)-f^i(x_0) = \sum_{j=1}^{n} \left(h^j F_j^{\prime i}(x_0)\right)+r^i(h).$$

We have thus shown that the differentiability of the mapping F implies the differentiability of its coordinates f^i $(i = 1, 2, ..., m)$ and the relation

$$f^{i\prime}(x_0) \cdot h = \sum_{j=1}^{n} \left(h^j F_j^{\prime i}(x_0)\right).$$

But

$$f^{i\prime}(x_0) \cdot h = \sum_{j=1}^{n} h^j f^{i\prime}(x_0) \cdot e_j = \sum_{j=1}^{n} h^j \nabla_{e_j} f^i(x_0),$$

whence

$$F_j^{\prime i}(x_0) = \nabla_{e_j} f^i(x_0).$$

The derivative in the direction of the j-th basis vector is denoted by

$$\nabla_{e_j} f^i(x_0) \overset{\text{df}}{=} \frac{\partial f^i(x_0)}{\partial x^j}$$

or symbolically

$$\nabla_{e_j} = \frac{\partial}{\partial x^j}.$$

The deeper meaning of this symbol will be explained further, when partial derivatives in general will be discussed. In the meantime we have obtained the following theorem:

THEOREM VII.3.3. Let F be a differentiable mapping of an open subset U of the space R^n into the space R^m. Then the matrix $[F_j^{\prime i}]$ of the derivative of the mapping F consists of partial derivatives (derivatives in the direction of basis vectors of R^n) of the system of functions f^i $(i = 1, 2, ..., m)$

$$F_j^{\prime i}(x) = \nabla_{e_j} f^i(x) = \frac{\partial f^i(x)}{\partial x^j}$$

$$(i = 1, 2, ..., m; \quad j = 1, 2, ..., n).$$

164

4. FORMAL LAWS OF DIFFERENTIATION

Let us write this matrix out in explicit terms:

$$[F_j^{\prime i}(x)] = \begin{bmatrix} \dfrac{\partial f^1(x)}{\partial x^1} & \dfrac{\partial f^1(x)}{\partial x^2} & \cdots & \dfrac{\partial f^1(x)}{\partial x^n} \\[2mm] \dfrac{\partial f^2(x)}{\partial x^1} & \dfrac{\partial f^2(x)}{\partial x^2} & \cdots & \dfrac{\partial f^2(x)}{\partial x^n} \\[2mm] \cdots\cdots\cdots\cdots\cdots\cdots \\[2mm] \dfrac{\partial f^m(x)}{\partial x^1} & \dfrac{\partial f^m(x)}{\partial x^2} & \cdots & \dfrac{\partial f^m(x)}{\partial x^n} \end{bmatrix}.$$

This matrix is called a *Jacobi matrix* and its determinant, a *Jacobian* when $m = n$. In the general case, when f is a real-valued function i.e. a mapping of the open subset U of the space R^n into R^1, the deriva tive is represented by a row matrix

$$f'(x) = \begin{bmatrix} \dfrac{\partial f(x)}{\partial x^1}, & \dfrac{\partial f(x)}{\partial x^2}, & \ldots, & \dfrac{\partial f(x)}{\partial x^n} \end{bmatrix}.$$

We know, however, that there exists an isomorphism of the space $(R^n)^*$ and R^n, $i\colon (R^n)^* \to R^n$. The derivative of the function f can then have a vector field $R^n \supset U \ni x \to (i \circ f')(x) \in R^n$ associated with it. This vector field is referred to as the gradient of the function f.

The gradient is denoted by the symbol grad, and the gradient of a function f by grad$f$.

4. THE FORMAL LAWS OF DIFFERENTIATION

The definition of the derivative of a mapping immediately implies the following theorem:

THEOREM VII.4.1. Let F and G be mappings of an open subset of a Banach space X into a Banach space Y. If F and G are differentiable at a point $x_0 \in U$, then the mapping $a \cdot F + b \cdot G$ is also differentiable at x_0, and

$$(a \cdot F + b \cdot G)'(x_0) = a \cdot F'(x_0) + b \cdot G'(x_0) \qquad (a, b \in R^1).$$

PROOF. We have

$$(a \cdot F + b \cdot G)(x_0 + h) - (a \cdot F + b \cdot G)(x_0)$$
$$= a \cdot (F(x_0 + h) - F(x_0)) + b \cdot (G(x_0 + h) - G(x_0))$$
$$= a \cdot F'(x_0) \cdot h + b \cdot G'(x_0) \cdot h + a \cdot r_1(h) + b \cdot r_2(h).$$

165

The increment has thus been decomposed into a linear part and a remainder. It is seen that the remainder is an infinitesimal of order higher than h. The theorem has been proved. $\square$

The theorem above tells us that differentiation is a linear operation.

Now we shall consider the differentiation of a composition of mappings. The theorems we get here are similar to those concerning the derivative of a composition of real-valued functions.

THEOREM VII.4.2. Let G be a mapping of an open subset U of a Banach space X into a Banach space Y, and let F be a mapping of an open set $V \subset Y$ into a Banach space Z. If $x_0 \in U$ and $G(x_0) \in V$ and G is differentiable at the point x_0, while F is differentiable at the point $y_0 := G(x_0)$, then the mapping $F \circ G$ is differentiable at $x_0 \in U \subset X$ and

$$(F \circ G)'(x_0) = F'(G(x_0)) \circ G'(x_0).$$

PROOF (cf. proof of Theorem III.2.7, p. 63). We calculate the increment:

$$(F \circ G)(x_0 + h) - (F \circ G)(x_0) = F(G(x_0 + h)) - F(G(x_0))$$
$$= F(G(x_0) + G'(x_0) \cdot h + r_1(x_0; h)) - F(G(x_0))$$
$$= F'(G(x_0))(G'(x_0) \cdot h) + F'(G(x_0))(r_1(x_0; h)) +$$
$$+ r_2(G(x_0), G'(x_0) \cdot h + r_1(x_0; h))$$
$$= (F'(G(x_0))G'(x_0)) \cdot h + R(x_0; h).$$

Next we examine the remainder:

$$\frac{\|R(x_0; h)\|}{\|h\|}$$
$$= \frac{\|F'(G(x_0))(r_1(x_0; h)) + r_2(G(x_0); G'(x_0) \cdot h + r_1(x_0; h))\|}{\|h\|}$$
$$\leqslant \frac{\|F'(G(x_0))(r_1(x_0; h))\|}{\|h\|} +$$
$$+ \frac{\|r_2(G(x_0), G'(x_0) \cdot h + r_1(x_0; h))\| \cdot \|G'(x_0) \cdot h + r_1(x_0; h))\|}{\|G'(x_0) \cdot h + r_1(x_0; h)\| \cdot \|h\|}$$
$$\leqslant \frac{\|F'(G(x_0))\| \cdot \|r_1(x_0; h)\|}{\|h\|} +$$

166

$$+ \frac{\|r_2(G(x_0); G'(x_0) \cdot h + r_1(x_0; h))\|}{\|G'(x_0) \cdot h + r_1(x_0, h)\|} \left(\|G'(x_0)\| + \right.$$

$$\left. + \frac{\|r_1(x_0; h)\|}{\|h\|} \right).$$

It is evident that this expression tends to zero as $h \to 0$. $\square$

In the particular case when $X = R^n$, $Y = R^m$, and $Z = R^r$, a set of m functions corresponds to the mapping G, and a set of r functions to the mapping F:

$$G(x) \leftrightarrow (g^1(x), g^2(x), \ldots, g^m(x)),$$

$$F(y) \leftrightarrow (f^1(y), f^2(y), \ldots, f^r(y)).$$

Associated with G and F are the matrices

$$G'(x) \leftrightarrow [G_j'^i(x)] = \left[\frac{\partial g^i(x)}{\partial x^j} \right],$$

$$F'(y) \leftrightarrow [F_i'^l(y)] = \left[\frac{\partial f^l(y)}{\partial y^i} \right],$$

$i = 1, 2, \ldots, m, j = 1, 2, \ldots, n, l = 1, 2, \ldots, r$. The derivative of the composition of the mappings is equal to the composition of the derivatives and, as is known, the matrix product corresponds to the composition of linear mappings.

$$(F \circ G)'(x_0) = F'(G(x_0)) \circ G'(x_0),$$

$$(F \circ G)_j'^l(x) = \sum_{i=1}^m \frac{\partial f^l(y)}{\partial y^i} \cdot \frac{\partial g^i(x)}{\partial x^j},$$

where the substitution $x = x_0, y = G(x_0), j = 1, 2, \ldots, n, l = 1, 2, \ldots$, r has been made.

Let us now prove a specially important property of the gradient.

DEFINITION. The *level surface* of a function $f: U \to R^1$ is the name given the set

$$f^{-1}(c) := \{x \in U: f(x) = c\}.$$

The level surface $f^{-1}(c)$ is denoted by P_c.

Let $R^1 \supset]a, b[\ni t \to x(t)$ be a differentiable curve on the level surface P_c, i.e. for every $t \in]a, b[$ we have $x(t) \in P_c$. The vector corresponding to the derivative of the mapping $t \to x(t)$ at a point t_0 is called the *tangent vector* at the point $x(t_0)$ to the given curve. In the case when $U \subset R^n$, we have

$$\frac{dx(t)}{dt} = \begin{bmatrix} \dfrac{dx^1(t)}{dt} \\ \vdots \\ \dfrac{dx^n(t)}{dt} \end{bmatrix}.$$

For every $t \in \;]a, b[$ we have $f(x(t)) = c$, whence

$$\frac{df(x(t))}{dt} = 0, \quad \text{or} \quad (f' \circ x')(t) = 0,$$

and thus

$$\left[\frac{\partial f(x)}{\partial x^1}, \;\ldots, \;\frac{\partial f(x)}{\partial x^n} \right]_{x=x(t)} \begin{bmatrix} \dfrac{dx^1(t)}{dt} \\ \vdots \\ \dfrac{dx^n(t)}{dt} \end{bmatrix} = 0.$$

The left-hand side is equal to the scalar product of the gradient at the point $x(t)$ and the tangent vector to the curve $x(\cdot)$ at this point. The gradient of the function f is thus perpendicular to every curve lying on the level surface of that function. We say that the gradient is perpendicular to the level surface.

This situation is frequently encountered in physics, for example in electrostatics, the electric field E is proportional to the gradient of the potencial V, that is, $E = -\text{grad} V$. Our theorem expresses a familiar physical fact, viz. that the field lines are prependicular to the equipotential surface.

Examples of Derivatives. 1. Let X and Y be Banach spaces;

$$T(x) := Ax, \quad \text{where } A \in L(X, Y).$$

We have

$$T(x+h) - T(x) = A(x+h) - Ax$$
$$= Ax + Ah - Ax = Ah.$$

This equation demonstrates the differentiability of the mapping T; moreover $T'(x) = A$.

2. If T is a mapping of an open set $U \subset R^1$ into a Banach space Y, differentiable at a point $x_0 \in U \subset R^1$, then

(1) $T(x_0+h)-T(x_0) = T'(x_0) \cdot h + r(h),$

where $T'(x_0) \in L(R^1, Y)$. We know, however, that $L(R^1, Y) \cong Y$, and hence there exists an $y \in Y$ such that for every $h \in R^1$

(2) $T'(x_0) \cdot h = h \cdot y.$

From (1) and (2) we obtain

$$T(x_0+h)-T(x_0) = h \cdot y + r(h).$$

Dividing both sides of this equation by h and going to zero as h does, we get

$$y = \lim_{h \to 0} \frac{T(x_0+h)-T(x_0)}{h}.$$

We write

$$y = \left. \frac{dT(x)}{dx} \right|_{x=x_0} = T'(x_0).$$

Thus, in the case of the mapping T of the set of real numbers into a Banach space (also called a *vector-valued function*) the derivative can be found through the limit of the difference quotient just as for numerical functions.

3. A *unitary space* U is the vector space over the field of complex (real) numbers, in which is defined the form

$$U \times U \ni (x, y) \to (x|y) \in C^1 \ (R^1)$$

which satisfies the following conditions:

I. $(x|y) \geqslant 0,$
II. $(x|y) = \overline{(y|x)},$
III. $(x_1+x_2|y) = (x_1|y)+(x_2|y),$
IV. $(ax|y) = a \cdot (x|y), \ a \in C^1 \ (R^1),$
V. $((x|x) = 0) \Leftrightarrow (x = 0).$

A space which satisfies conditions I–IV is called a *pre-Hilbert space.*

Properties II, III, and IV imply that $(x|ay) = \bar{a}(x|y)$. This form is said to be *linear in the first argument*, and *antilinear in the second*. If we consider only the space over R^1, the form is clearly bilinear and sym-

169

metric. A form which satisfies conditions I–V is called a *scalar product* in the space U. The norm

$$||x|| := ((x|x))^{1/2}$$

can be introduced in a unitary space. Of the four conditions which the norm must satisfy, only the triangle inequality is difficult to demonstrate. This can be done by employing the *Schwarz inequality*

$$|(x|y)| \leqslant ||x|| \cdot ||y||.$$

First of all, let us demonstrate this inequality. Observe that for $y = 0$ the equality is obvious. For any vectors x and y and for every $a \in C^1$ (R^1), we have

$$(x-ay|x-ay) \geqslant 0,$$
$$(x|x)-a(y|x)-\bar{a}(x|y)+a\bar{a}(y|y) \geqslant 0.$$

Now we put $a = \dfrac{(x|y)}{(y|y)}$ for $y \neq 0$ and we obtain

$$\frac{|(x|y)|^2}{(y|y)} \leqslant (x|x), \quad \text{whence } |(x|y)| \leqslant ||x|| \cdot ||y||.$$

Having the Schwarz inequality, we prove the triangle inequality:

$$||x+y||^2 = (x+y|x+y) = (x|x)+(x|y)+(y|x)+(y|y)$$
$$= ||x||^2+2\,\mathrm{Re}(x|y)+||y||^2;$$

but

$$2\,\mathrm{Re}(x|y) \leqslant 2|(x|y)| \leqslant 2||x|| \cdot ||y||,$$

whence

$$||x+y||^2 \leqslant ||x||^2+2||x|| \cdot ||y||+||y||^2$$

and consequently

$$||x+y|| \leqslant ||x||+||y||.$$

The triangle inequality has thus been demonstrated.

A complete unitary space is called a *Hilbert space*. Obviously, a Hilbert space is a Banach space.

In our subsequent considerations we shall concern ourselves with the real Hilbert space H.

Let us examine the differentiability of a function $H \ni x \to ||x||^2 \in R^1$. We evaluate the increment

$$||x+h||^2-||x||^2 = (x+h|x+h)-(x|x)$$
$$= (x|x)+(x|h)+(h|x)+(h|h)-(x|x) = 2(x|h)+(h|h).$$

Thus we see that $(||x||^2)' \cdot h = 2(x|h)$; by the usual identification of the vector h and the functional $(\cdot |h)$ we can write $(||x||^2)' = 2x$.

To find the derivative of the function $H \ni x \to ||x|| \in R^1$, we make use of the identity: $||x|| = (||x||^2)^{1/2}$. Accordingly

$$(||x||)' = ((||x||^2)^{1/2})' = \frac{(||x||^2)'}{2||x||} = \frac{x}{||x||}.$$

Finally,

$$(||x||)' = \frac{x}{||x||}.$$

The norm is thus a function differentiable everywhere apart from zero.

Now let us evaluate the derivative of the function

$$H \ni x \to f(x) := \frac{c}{||x||};$$

we have

$$f'(x) = -\frac{c}{||x||^2}(||x||)' = -\frac{c}{||x||^2} \cdot \frac{x}{||x||} = -\frac{c \cdot x}{||x||^3}.$$

Therefore

$$f'(x) = -\frac{c \cdot x}{||x||^3}, \quad f'(x) \cdot h = -\frac{c \cdot (x|h)}{||x||^3}.$$

In the particular case when $H = R^3$, the Coulomb law follows from these formulae.

4. Let X and Y be Banach spaces, and let F be a mapping of X into Y. We say that F is a positive homogeneous mapping of degree $\alpha \neq 0$, if

$$F(t \cdot x) = t^\alpha \cdot F(x)$$

for every $x \in X$ and every $t > 0$. Suppose that F is differentiable in some neighbourhood of a fixed point $x_0 \in X$. Then we have

$$\frac{dF(t \cdot x_0)}{dt} = F'(t \cdot x_0) \cdot x_0,$$

whereas on the other hand

$$\frac{dF(t \cdot x_0)}{dt} = \alpha t^{\alpha-1} \cdot F(x_0).$$

When we put $t = 1$ these equations yield

$$F'(x_0) \cdot x_0 = \alpha \cdot F(x_0).$$

This equation is known as the *Euler theorem for homogeneous mappings*.

5. Let be given a set of n^2 differentiable functions

$$R^1 \in t \to x_j^i(t) \in R^1, \quad i, j = 1, 2, \ldots, n.$$

We now compute the derivative of the function $t \to \det[x_j^i(t)]$, where det denotes the determinant

$$\det[x_j^i(t)] = \begin{vmatrix} x_1^1(t) \ldots x_1^n(t) \\ x_2^1(t) \ldots x_2^n(t) \\ \cdots\cdots\cdots \\ x_n^1(t) \ldots x_n^n(t) \end{vmatrix}.$$

Using the chain rule we get

$$\frac{d}{dt}\left(\det[x_j^i(t)]\right) = \sum_{i,j=1}^{n} \frac{\partial\left(\det[x_j^i(t)]\right)}{\partial x_j^i} \cdot \frac{dx_j^i(t)}{dt}.$$

However, it is not difficult to see, upon expanding the determinant in the row (or column) containing the element $x_j^i(t)$, that

$$\frac{\partial\left(\det[x_j^i(t)]\right)}{\partial x_j^i} = D_j^i(t),$$

where $D_j^i(t)$ is the cofactor of the element $x_j^i(t)$. Thus, we have finally

$$\frac{d\det[x_j^i(t)]}{dt} = \sum_{i,j=1}^{n} D_j^i(t)\frac{dx_j^i(t)}{dt}.$$

6. An example of a function possessing the directional derivative bu not the Fréchet derivative. Let

$$f: R^2 \to R^1;$$

$$f(x_1, x_2) = \begin{cases} 0, & \text{when } x_1 = x_2 = 0, \\ \dfrac{x_1 x_2^2}{x_1^2 + x_2^2}, & \text{when } (x_1, x_2) \neq 0. \end{cases}$$

We calculate the derivative in the direction of the vector $h = (h_1, h_2)$ at the point $(x_1, x_2) = 0$:

$$\frac{f(x_1 + th_1, x_2 + th_2) - f(x_1, x_2)}{t} = \frac{f(th_1, th_2)}{t} = \frac{t^3 h_1 h_2^2}{t^3(h_1^2 + h_2^2)}.$$

Passing with t to zero, we obtain

$$\nabla_h f(0, 0) = \frac{h_1 h_2^2}{h_1^2 + h_2^2}.$$

172

The directional derivative does exist but is not linear in h. This shows that there is no strong derivative. For if there were one, the relation

$$f'(0,0) \cdot h = \nabla_h f(0,0)$$

would hold.

5. THE MEAN-VALUE THEOREM

THEOREM VII.5.1 (The Mean-Value Theorem for Functions). Let U be an open set in a Banach space X and let f be a real-valued function defined on U. If f is differentiable at every point of the interval $[x, x+ +h] \subset U$, where

$$[x, x+h] :=$$

$$\{y \in X: y = (1-a) \cdot x + a \cdot (x+h), 0 \leqslant a \leqslant 1\},$$

then there exists a $0 < \theta < 1$ such that

$$f(x+h) - f(x) = f'(x+\theta h) \cdot h.$$

PROOF. We introduce the auxiliary function g:

$$R^1 \supset [0, 1] \ni t \rightarrow g(t) := f(x+th).$$

Clearly, g is differentiable on $]0, 1[$, continuous on $[0, 1]$, and

$$g'(t) = f'(x+th) \cdot h.$$

The function g thus satisfies the hypotheses of the classical mean-value theorem (Theorem III.2.3) and hence there exists a $0 < \theta < 1$ such that

$$g(1) - g(0) = g'(\theta),$$

whence

$$f(x+h) - f(x) = f'(x+\theta h) \cdot h. \quad \square$$

With a view to generalizing Theorem VII.5.1 to mappings of Banach spaces, we first make several remarks. We have proved in Example 3 of the previous section that the scalar product $(\cdot \mid \cdot)$ in a unitary space satisfies the Schwarz inequality

$$|(x|y)| \leqslant ||x|| \cdot ||y||.$$

This inequality implies that

$$\sup_{||y|| \leqslant 1} |(x|y)| \leqslant ||x||.$$

However, on putting $y = x/||x||$, we obtain

$$||x|| = \sup_{||y|| \leqslant 1} |(x|y)|.$$

For any Banach space, this equation assumes the form

$$||x|| = \sup_{||x^*|| \leqslant 1} |\langle x, x^* \rangle|, \quad x^* \in X^*.$$

We give this fact without proof, merely pointing out that it is a corollary to the Hahn–Banach theorem (cf. Chapter XIII, Section 15) which is so important in functional analysis.

THEOREM VII.5.2 (The Mean-Value Theorem for Mappings). Let F be a mapping of an open subset U of a Banach space into a Banach space Y. If F has a continuous derivative (i.e. the mapping $X \ni x \to F'(x) \in L(X, Y)$ is continuous) at every point of the interval $[x, x+h]$, then

$$||F(x+h) - F(x)|| \leqslant ||h|| \sup_{0 \leqslant \theta \leqslant 1} ||F'(x+\theta h)||.$$

PROOF. We introduce a family of auxiliary functions

$$X \supset U \ni x \to f_{y*}(x) := \langle F(x), y^* \rangle \in R^1, \quad y^* \in Y^*.$$

It is easily verified that each of these functions satisfies Theorem VII.5.1, and their derivatives affect the increment h in the following manner:

$$f'_{y*}(x) \cdot h = \langle F'(x) \cdot h, y^* \rangle.$$

Thus, there exists a $0 < \theta_{y*} < 1$ such that

$$f_{y*}(x+h) - f_{y*}(x) = f'_{y*}(x+\theta_{y*}h) \cdot h = \langle F'(x+\theta_{y*}h) \cdot h, y^* \rangle,$$

whereas on the other hand

$$f_{y*}(x+h) - f_{y*}(x) = \langle F(x+h) - F(x), y^* \rangle;$$

we also have

$$||F(x+h) - F(x)|| = \sup_{||y^*|| \leqslant 1} |\langle F(x+h) - F(x), y^* \rangle|.$$

These three equations yield

$$||F(x+h) - F(x)|| = \sup_{||y^*|| \leqslant 1} |\langle F'(x+\theta_{y*} \cdot h) \cdot h, y^* \rangle|,$$

and further rearrangement gives

$$||F(x+h) - F(x)|| \leqslant \sup_{||y^*|| \leqslant 1} ||F'(x+\theta_{y*} \cdot h)|| \cdot ||h|| \cdot ||y^*||,$$

whence

$$\|F(x+h)-F(x)\| \leqslant \|h\| \cdot \sup_{0 \leqslant \theta \leqslant 1} \|F'(x+\theta h)\|.$$

The theorem has thus been proved. $\square$

In contradistinction to Theorem VII.5.1, Theorem VII.5.2 is in the form of an inequality. It is thus seen that only in the special case, when the mapping is a function, the Mean-Value Theorem is expressed by an equation.

Now we present a theorem which follows directly from the Mean-Value Theorem.

THEOREM VII.5.3. Let U be an open, connected set in a Banach space X and let T be a mapping of U into a Banach space Y which is differentiable on U. If $T'(x) = 0$ for every $x \in U$, then T is a constant mapping.

PROOF. Let x_0 be an arbitrary point of the set U. The set A is defined to be

$$A := \{x \in U: T(x) = T(x_0)\}.$$

The continuity of the mapping T implies that A is a closed set in U. The set U is open, and thus for every $x \in U$ there exists a $K(x, r) \subset U$. The ball is a convex set, i.e. together with the points x and $x+h$ it contains the interval $[x, x+h]$. By the mean-value theorem, therefore, we have

$$\|T(x+h)-T(x)\| \leqslant \|h\| \sup_{0 \leqslant \theta \leqslant 1} \|T'(x+\theta h)\| = 0,$$

whence

$$T(x+h) = T(x).$$

It has thus been shown that $K(x, r) \subset A$. The set A thus is open in U. We know, however, that the only sets which are both open and closed in a connected set U are the empty set and U itself. The set A is not empty since it contains the point x_0. Hence $A = U$. $\square$

LEMMA VII.5.4. Let X and Y be Banach spaces, and let T be a differentiable mapping of an open neighbourhood U of the interval $[x, x+ +h] \subset X$ into Y. Then for every $z \in U \subset X$ we have

$$\|T(x+h)-T(x)-T'(z) \cdot h\| \leqslant \|h\| \cdot \sup_{v \in [x, x+h]} \|T'(v)-T'(z)\|.$$

175

Proof. We introduce the mapping F:

$$X \supset U \ni x \to F(x) := T(x) - T'(z) \cdot x.$$

The mapping F is differentiable and

$$F'(x) = T'(x) - T'(z).$$

The Mean-Value Theorem is applied to the mapping F:

$$\|F(x+h) - F(x)\| = \|T(x+h) - T'(z)(x+h) - T(x) + T'(z) \cdot x\|$$

$$\leqslant \|h\| \sup_{v \in [x,\, x+h]} \|F'(v)\|$$

$$= \|h\| \sup_{v \in [x,\, x+h]} \|T'(v) - T'(z)\|.$$

Finally, we have

$$\|T(x+h) - T(x) - T'(z) \cdot h\| \leqslant \|h\| \sup_{v \in [x,\, x+h]} \|T'(v) - T'(z)\|. \quad \Box$$

Remark. The supremum is finite whenever the derivative T' is continuous on the interval $[x, x+h]$.

In the particular case, when we set $x = z$, we obtain an evaluation of the remainder in the definition of the derivative. We then have

$$\|r(x\,;\,h)\| = \|T(x+h) - T(x) - T'(x) \cdot h\|$$

$$\leqslant \|h\| \sup_{v \in [x,\, x+h]} \|T'(v) - T'(x)\|.$$

6. PARTIAL DERIVATIVES

It is known that the Cartesian product of Banach spaces X_1 and X_2 is also a Banach space. The algebraic operations are defined in the natural manner:

$$(x_1, x_2) + (y_1, y_2) := (x_1 + y_1, x_2 + y_2),$$
$$a \cdot (x_1, x_2) := (a \cdot x_1, a \cdot x_2),$$
$$(x_1, x_2), (y_1, y_2) \in X_1 \times X_2, \quad a \in R^1.$$

A norm, as we know (cf. Section VII.1), can be introduced in any of several ways. Henceforth we shall employ a norm defined as follows:

$$X_1 \times X_2 \ni (x_1, x_2) \to \|(x_1, x_2)\| := \|x_1\| + \|x_2\|.$$

The space $X_1 \times X_2$ is a Banach space in this norm. These properties carry over onto the case of the Cartesian product of n Banach spaces. The norm then is of the form

$$X_1 \times X_2 \times \ldots \times X_n \ni (x_1, x_2, \ldots, x_n)$$

$$\to \|(x_1, x_2, \ldots, x_n)\| := \sum_{i=1}^{n} \|x_i\|.$$

We remind that a differentiable mapping T of an open subset U of a Banach space X into a Banach space Y is said to be *continuously differentiable on U* if the mapping

$$X \supset U \ni x \to T'(x) \in L(X, Y)$$

is continuous on U.

DEFINITION. Let V be an open subset of the Cartesian product $X_1 \times X_2$ of two Banach spaces X_1 and X_2 and let T be a mapping of the set V into a Banach space Y. The mapping T is said to have a *partial derivative with respect to X_1 at a point* $(x_1, x_2) \in V \subset X_1 \times X_2$ if there exists a decomposition

$$T(x_1 + h_1, x_2) - T(x_1, x_2) = \underset{1}{L_{(x_1, x_2)}}(h_1) + r_1\big((x_1, x_2); h_1\big)$$

where

$$\underset{1}{L_{(x_1, x_2)}} \in L(X_1, Y) \quad \text{and} \quad \frac{\|r_1((x_1, x_2); h_1)\|}{\|h_1\|} \xrightarrow{h_1 \to 0} 0.$$

The linear continuous mapping $\underset{1}{L_{(x_1, x_2)}}$ is called the *partial derivative of the mapping T with respect to X_1 at the point* (x_1, x_2) and is denoted by the symbol $T'_{X_1}(x_1, x_2)$.

Similarly, by the decomposition

$$T(x_1, x_2 + h_2) - T(x_1, x_2) = \underset{2}{L_{(x_1, x_2)}}(h_2) + r_2\big((x_1, x_2); h_2\big),$$

where

$$\underset{2}{L_{(x_1, x_2)}} \in L(X_2, Y) \quad \text{and} \quad \frac{\|r_2((x_1, x_2); h_2)\|}{\|h_2\|} \xrightarrow{h_2 \to 0} 0,$$

we define the partial derivative of T with respect to X_2 at the point (x_1, x_2), which we denote by the symbol $T'_{X_2}(x_1, x_2)$.

If the partial derivative with respect to X_1 (with respect to X_2) exists at every point of the set V, the mapping T is said to be *differentiable with respect to X_1 (X_2) on V* and the mappings

$$X_1 \times X_2 \supset V \ni (x_1, x_2) \to T'_{X_1}(x_1, x_2) \in L(X_1, Y),$$

$$X_1 \times X_2 \supset V \ni (x_1, x_2) \to T'_{X_2}(x_1, x_2) \in L(X_2, Y)$$

are called *partial derivatives of the mapping T with respect to X_1 and X_2*, respectively.

The definition of a partial derivative carries over immediately onto the case of the direct (Cartesian) product of n Banach spaces.

DEFINITION. Let X_i ($i = 1, 2, ..., n$) be Banach spaces and let V be an open subset of the Cartesian product $X_1 \times X_2 \times ... \times X_n$. A mapping T of V into a Banach space Y has a *partial derivative with respect to X_i at a point* $x = (x_1, x_2, ..., x_n)$ if

$$T(x_1, ..., x_{i-1}, x_i+h_i, x_{i+1}, ..., x_n) -$$
$$- T(x_1, ..., x_{i-1}, x_i, x_{i+1}, ..., x_n)$$
$$= T'_{X_i}(x_1, ..., x_n) \cdot h_i + r_i((x_1, ..., x_n); h_i),$$

where $T'_{X_i}(x_1, ..., x_n) \in L(X_i, Y)$ and the remainder is an infinitesimal of order higher than h_i, that is,

$$\lim_{h_i \to 0} \frac{||r_i((x_1, ..., x_n); h_i)||}{||h_i||} = 0.$$

If T has a partial derivative with respect to X_i at every point of V, it is said to be *differentiable with respect to X_i on V* and the mapping

$$X_1 \times ... \times X_n \supset V \ni (x_1, ..., x_n) \to T'_{X_i}(x_1, ..., x_n) \in L(X_i, Y)$$

is called the *partial derivative of T with respect to X_i*.

In the particular case when $X_i = R^1$, we have $X_1 \times ... \times X_n = R^n$. If F is a mapping of an open set $U \subset R^n$ into a Banach space Y, then

(1) $$F(x_1, ..., x_{i-1}, x_i+h_i, x_{i+1}, ..., x_n) -$$
$$- F(x_1, ..., x_{i-1}, x_i, x_{i+1}, ..., x_n)$$
$$= F'_i(x_1, ..., x_n) \cdot h_i + r_i(h_i);$$
$$F'_i(x_1, ..., x_n) \in L(R^1, Y) \cong Y.$$

Hence, there exists an $y \in Y$ such that

(2) $$F'_i(x_1, ..., x_n) \cdot h_i = y \cdot h_i.$$

178

When the limiting process $h \to 0$ is carried out, equations (1) and (2) yield the formula

$$F_i'(x_1, \ldots, x_n) = y$$

$$= \lim_{h_i \to 0} \frac{1}{h_i} [F(x_1, \ldots, x_{i-1}, x_i + h_i, x_{i+1}, \ldots, x_n) -$$

$$- F(x_1, \ldots, x_{i-1}, x_i, x_{i+1}, \ldots, x_n)].$$

This expression is seen to coincide with the formula for the i-th directional derivative (the derivative in the direction of the i-th canonical basis vector R^n) of the mapping F. Thus we have

$$F_i'(x) = \nabla_{e_i} F(x).$$

As we recall, the notation we used earlier was

$$\nabla_{e_i} F(x) = \frac{\partial F(x)}{\partial x^i}.$$

Finally, we have

$$F_i'(x) = \frac{\partial F(x)}{\partial x^i} = \nabla_{e_i} F(x).$$

The terminology adopted in Section 3 thus is consistent with that used in the present section.

As for differentiation of functions defined on open subsets of R^n (also referred to as *functions of n variables*), it is best to use a formula which defines the partial derivative in terms of the limit of a difference quotient:

$$\frac{\partial f(x)}{\partial x^i} = \lim_{h_i \to 0} \frac{f(x_1, \ldots, x_i + h_i, \ldots, x_n) - f(x_1, \ldots, x_i, \ldots, x_n)}{h_i}.$$

Accordingly, we differentiate with respect to one particular variable, while treating the others as constant parameters.

Examples. 1. $f(x_1, x_2) := x_1^2 + x_2^2$;

$$\frac{\partial f(x_1, x_2)}{\partial x_1} = 2x_1, \qquad \frac{\partial f(x_1, x_2)}{\partial x_2} = 2x_2.$$

2. $f(x_1, x_2) := \sin x_1 x_2$;

$$\frac{\partial f(x_1, x_2)}{\partial x_1} = x_2 \cos x_1 x_2, \qquad \frac{\partial f(x_1, x_2)}{\partial x_2} = x_1 \cos x_1 x_2.$$

179

3. $f(x_1, x_2, x_3) := x_1 \sin x_2 x_3;$

$$\frac{\partial f(x_1, x_2, x_3)}{\partial x_1} = \sin x_2 x_3,$$

$$\frac{\partial f(x_1, x_2, x_3)}{\partial x_2} = x_1 x_3 \cos x_2 x_3,$$

$$\frac{\partial f(x_1, x_2, x_3)}{\partial x_3} = x_1 x_2 \cos x_2 x_3.$$

Suppose that T is a mapping of an open subset V of the product $X_1 \times X_2$ of two Banach spaces X_1 and X_2 into a Banach space Y. Then T may be differentiable at a point $(x_1, x_2) \in V$. On the other hand, T may have a partial derivative at that point. The question is how these things are related. Let us provide the answer.

LEMMA VII.6.1. Let V be an open subset of the Cartesian product $X_1 \times X_2$ of two Banach spaces and let T be a mapping of V into a Banach space Y. If T has a derivative at a point $(x_1, x_2) \in V$, then both partial derivatives exist at that point, and

$$T'(x_1, x_2) \cdot (h_1, h_2) = T'_{X_1}(x_1, x_2) \cdot h_1 + T'_{X_2}(x_1, x_2) \cdot h_2.$$

PROOF. The differentiability of T at the point $(x_1, x_2) \in V$ implies that

$$T(x_1 + h_1, x_2) - T(x_1, x_2) = T'(x_1, x_2) \cdot (h_1, 0) + r(h_1, 0)$$

and

$$T(x_1, x_2 + h_2) - T(x_1, x_2) = T'(x_1, x_2) \cdot (0, h_2) + r(0, h_2).$$

Both partial derivatives are thus seen to exist and

$$T'_{X_1}(x_1, x_2) = T'(x_1, x_2) \circ i_1,$$
$$T'_{X_2}(x_1, x_2) = T'(x_1, x_2) \circ i_2,$$

where i_1, i_2 are natural embeddings of the spaces X_1 and X_2 into $X_1 \times X_2$:

$$X_1 \ni x_1 \to i_1(x_1) := (x_1, 0) \in X_1 \times X_2,$$
$$X_2 \ni x_2 \to i_2(x_2) := (0, x_2) \in X_1 \times X_2.$$

Thus we have

$$T'(x_1, x_2)(h_1, h_2) = T'_{X_1}(x_1, x_2)h_1 + T'_{X_2}(x_1, x_2)h_2. \quad \square$$

Lemma VII.6.1 shows that the existence of partial derivatives is implied by the differentiability of a mapping. A result which in a certain degree is the converse of that above is given by the following theorem.

THEOREM VII.6.2. Let X_1, X_2, and Y be Banach spaces and let T be a mapping of an open subset V of the product $X_1 \times X_2$ into Y. In order for T to be continuously differentiable on V (i.e. for the mapping $V \ni (x_1, x_2) \to T'(x_1, x_2) \in L(X_1 \times X_2, Y)$ to be continuous on V), it is necessary and sufficient that T be continuously differentiable with respect to X_1 and X_2 (i.e. that the mappings $V \ni (x_1, x_2) \to T'_{X_i}(x_1, x_2)$ $\in L(X_i, Y)$, $i = 1, 2$, be continuous on V).

PROOF. If a derivative exists at a point $(x_1, x_2) \in V$, then partial derivatives also exist (cf. Lemma VII.6.1) and

$$T'_{X_1}(x_1, x_2) = T'(x_1, x_2) \circ i_1, \qquad T'_{X_2}(x_1, x_2) = T'(x_1, x_2) \circ i_2,$$

where i_1, i_2 are natural injections of spaces X_1 and X_2 into $X_1 \times X_2$. The continuity of the mapping

$$X_1 \times X_2 \supset V \ni (x_1, x_2) \to T'(x_1, x_2) \in L(X_1 \times X_2, Y)$$

implies the continuity of the mapping

$$X_1 \times X_2 \supset V \ni (x_1, x_2) \to T'_{X_1}(x_1, x_2) \in L(X_1, Y),$$

for we have

$$\|T'_{X_1}(x'_1, x'_2) - T'_{X_1}(x_1, x_2)\|_{L(X_1, Y)}$$
$$= \|T'(x'_1, x'_2) \circ i_1 - T'(x_1, x_2) \circ i_1\|_{L(X_1, Y)}$$
$$= \sup_{\|h_1\| \leqslant 1} \|(T'(x'_1, x'_2) - T'(x_1, x_2)) \circ i_1 \cdot h_1\|_Y$$
$$\leqslant \sup_{\|h_1\|, \|h_2\| \leqslant 1} \|(T'(x'_1, x'_2) - T'(x_1, x_2))(h_1, h_2)\|_Y$$
$$= \|T'(x'_1, x'_2) - T'(x_1, x_2)\|_{L(X_1 \times X_2, Y)} < \varepsilon$$
$$\text{for } \|((x'_1, x'_2) - (x_1, x_2))\| < \delta.$$

The continuity of the other partial derivative is proved in an analogous manner.

Conversely, if partial derivatives exist, the mapping is differentiable. From the continuity of the partial derivative with respect to X_1 it follows that

$$\|T'_{X_1}(x_1, x_2 + h_2) - T'_{X_1}(x_1, x_2)\| < \varepsilon \qquad \text{for } \|h_2\| < r,$$

whence

$$(3) \qquad \|T'_{X_1}(x_1, x_2 + h_2) \cdot h_1 - T'_{X_1}(x_1, x_2) \cdot h_1\| \leqslant \|h_1\| \cdot \varepsilon$$
$$\text{for } \|h_2\| < r.$$

By Lemma VII.5.4 we have

$$(4) \qquad \|T(x_1 + h_1, x_2 + h_2) - T(x_1, x_2 + h_2) - T'_{X_1}(x_1, x_2 + h_2) \cdot h_1\|$$

$$\leqslant ||h_1|| \sup_{||z|| \leqslant ||h_1||} ||T'_{X_1}(x_1+z, x_2+h_2) - T'_{X_1}(x_1, x_2+h_2)||$$

$$\leqslant ||h_1|| \sup_{||z|| \leqslant ||h_1||} ||T'_{X_1}(x_1+z, x_2+h_2) - T'_{X_1}(x_1, x_2)|| +$$

$$+ ||h_1|| \cdot ||T'_{X_1}(x_1, x_2) - T'_{X_1}(x_1, x_2+h_2)|| \leqslant 2\varepsilon ||h_1||$$

for $\max(||h_1||, ||h_2||) < r$.

The existence of the partial derivative with respect to X_2 implies

(5) $\qquad ||T(x_1, x_2+h_2) - T(x_1, x_2) - T'_{X_2}(x_1, x_2) \cdot h_2|| \leqslant ||h_2|| \cdot \varepsilon$

for $||h_2|| < r'$;

on adding both sides of (3), (4), and (5), we obtain

$$||T(x_1+h_1, x_2+h_2) - T(x_1, x_2) - T'_{X_1}(x_1, x_2) \cdot h_1 -$$

$$- T'_{X_2}(x_1, x_2) \cdot h_2||$$

$$\leqslant \varepsilon(3||h_1|| + ||h_2||) \leqslant 3\varepsilon(||h_1|| + ||h_2||)$$

for $||h_1||, ||h_2|| < \min(r, r')$.

We have thus proved the existence of the derivative. Its continuity follows from the fact that it may be written in the form

$$T'(x_1, x_2) = T'_{X_1}(x_1, x_2) \circ \mathrm{pr}_1 + T'_{X_2}(x_1, x_2) \circ \mathrm{pr}_2,$$

where pr_1 and pr_2 are projections onto X_1 and X_2:

$$X_1 \times X_2 \ni (x_1, x_2) \to \mathrm{pr}_1(x_1, x_2) = x_1 \in X_1,$$

$$X_1 \times X_2 \ni (x_1, x_2) \to \mathrm{pr}_2(x_1, x_2) = x_2 \in X_2.$$

The continuity of the derivative is thus implied by the continuity of the partial derivatives and the continuity of the projections. $\square$

Note that in order to demonstrate the differentiability of the mapping T we needed the continuity of only one partial derivative. The continuity of the other partial derivative was required only in order to show the continuity of the derivative. With the weaker assumptions we thus have the following corollary:

COROLLARY VII.6.3. Let V be an open subset of the direct product $X_1 \times X_2$ of two Banach spaces and let T be a mapping of the set V into a Banach space Y. If both partial derivatives exist at a point (x_1, x_2) $\in V$ and one of them is continuous in some open neighbourhood $U \subset V$

of the point (x_1, x_2), then the mapping T is differentiable at (x_1, x_2) and

$$T'(x_1, x_2) = T'_{X_1}(x_1, x_2) \circ \mathrm{pr}_1 + T'_{X_2}(x_1, x_2) \circ \mathrm{pr}_2,$$

where pr_1 and pr_2 are the projections onto X_1 and X_2, respectively.

The results which we have obtained for the Cartesian product of two Banach spaces may be immediately generalized to the general case of n Banach spaces. Accordingly, we have

THEOREM VII.6.4. Let $X_1, X_2, \ldots, X_n$ and Y be Banach spaces and let T be the mapping of an open subset V of the product $X_1 \times X_2 \times \ldots \times X_n$ into Y. If the derivative of T exists at a point $x = (x_1, x_2, \ldots, x_n) \in V$, then partial derivatives with respect to $X_k, k = 1, 2, \ldots, n$, also exist at the point x and

$$T'_{X_k}(x_1, \ldots, x_n) = T'(x_1, \ldots, x_n) \circ i_k,$$

where i_k are the natural injections in $X_1 \times X_2 \times \ldots \times X_k$:

$$X_k \ni x_k \rightarrow i_k(x_k)$$
$$= (0, \ldots, 0, x_k, 0, \ldots, 0) \in X_1 \times X_2 \times \ldots \times X_n.$$

Conversely, if all partial derivatives with respect to $X_k, k = 1, 2, \ldots, n$, do exist at a point $(x_1, \ldots, x_n) \in V$ and $n-1$ of them are continuous in some open neighbourhood $U \subset V$ of the point $(x_1, \ldots, x_n)$, then the derivative of the mapping T exists at that point and

$$T'(x_1, \ldots, x_n) = \sum_{k=1}^{n} T'_{X_k}(x_1, \ldots, x_n) \circ \mathrm{pr}_k,$$

where pr_k are projections onto X_k:

$$X_1 \times X_2 \times \ldots \times X_n \ni (x_1, \ldots, x_n) \rightarrow \mathrm{pr}_k(x_1, \ldots, x_n) = x_k \in X_k.$$

Strengthening the hypotheses somewhat, we can rewrite this theorem in the form of a necessary and sufficient condition.

THEOREM VII.6.5. Let T be a mapping of an open subset V of the Cartesian product $X_1 \times X_2 \times \ldots \times X_n$ of n Banach spaces into a Banach space Y. In order that the mapping T be continuously differentiable on V, it is necessary and sufficient that T be continuously differentiable on V with respect to $X_k, k = 1, 2, \ldots, n$.

The proofs of Theorems VII.6.4 and VII.6.5 are patterned after those of Lemma VII.6.1 and Theorem VII.6.2.

7. MULTILINEAR MAPPINGS

DEFINITION. A mapping $X_1 \times X_2 \ni (x_1, x_2) \to F(x_1, x_2) \in Y$ of the Cartesian product of two vector spaces X_1 and X_2 into a vector space Y is said to be *bilinear* if the mappings $F(\cdot, x_2)$ and $F(x_1, \cdot)$ defined as

$$X_1 \ni x_1 \to F(x_1, x_2) \in Y,$$
$$X_2 \ni x_2 \to F(x_1, x_2) \in Y,$$

are linear, respectively, for x_1 and x_2.

In general, a mapping

$$X_1 \times X_2 \times \ldots \times X_n \ni (x_1, x_2, \ldots, x_n) \to F(x_1, x_2, \ldots, x_n) \in Y$$

of the product $X_1 \times X_2 \times \ldots \times X_n$ of n vector spaces into a vector space Y is called *n-linear* if the mappings

$$X_k \ni x_k \to F(x_1, \ldots, x_k, \ldots, x_n) \in Y, \quad k = 1, 2, \ldots, n,$$

are linear for the fixed variables $x_1, \ldots, x_{k-1}, x_{k+1}, \ldots, x_n$.

In a natural manner n-linear mappings form a vector space. The spaces of continuous n-linear mappings of the Cartesian product $X_1 \times X_2 \times \ldots \ldots \times X_n$ of n Banach spaces into a Banach space Y are denoted by the symbol $L(X_1, X_2, \ldots, X_n; Y)$. In the case $X_i = X$ we write $L_n(X, Y) := L(\underbrace{X \times \ldots \times X}_{n \text{ times}}, Y)$.

Remark. This should not be confused with the space of continuous linear mappings of the product $X_1 \times X_2 \times \ldots \times X_n$ into Y, denoted by the symbol $L(X_1 \times X_2 \times \ldots \times X_n, Y)$.

THEOREM VII.7.1. An n-linear mapping

$$X_1 \times X_2 \times \ldots \times X_n \ni (x_1, x_2, \ldots, x_n) \to F(x_1, x_2, \ldots, x_n) \in Y$$

of the Cartesian product of n Banach spaces into a Banach space Y is continuous at every point if and only if there exists an $A > 0$ such that for any $x_k \in X_k$, $k = 1, 2, \ldots, n$,

$$(*) \qquad \|F(x_1, \ldots, x_n)\| \leqslant A \cdot \|x_1\| \cdot \|x_2\| \cdot \ldots \cdot \|x_n\|.$$

PROOF. The theorem will be proved for the case $n = 2$; the proof is similar for other n.

Let $\|F(x_1, x_2)\| \leqslant A\|x_1\| \cdot \|x_2\|$, $x_1 \in X_1$, $x_2 \in X_2$. Then

$$\|F(x_1 + h_1, x_2 + h_2) - F(x_1, x_2)\|$$
$$= \|F(x_1, x_2) + F(x_1, h_2) + F(h_1, x_2) + F(h_1, h_2) - F(x_1, x_2)\|$$

$$\leqslant \|F(x_1, h_2)\| + \|F(h_1, x_2)\| + \|F(h_1, h_2)\|$$

$$\leqslant A(\|x_1\| \cdot \|h_2\| + \|h_1\| \cdot \|x_2\| + \|h_1\| \cdot \|h_2\|) < \varepsilon,$$

provided that

$$\|h_1\|, \|h_2\| < \min\left(\frac{1}{3A \cdot \|x_1\|}, \frac{1}{3A \cdot \|x_2\|}, \frac{1}{\sqrt{3A}}\right)\varepsilon.$$

The continuity of the mapping F has thus been demonstrated.

Conversely, if the mapping F is continuous, it is also continuous at the point $(x_1, x_2) = (0, 0)$. Thus,

$$(1) \qquad \|F(h_1, h_2)\| < \varepsilon \quad \text{for } \|(h_1, h_2)\| < \delta.$$

For any (x_1, x_2) such that $x_1 \neq 0$ and $x_2 \neq 0$, we have

$$\|F(x_1, x_2)\| = \left\|F\left(\frac{x_1 \cdot \frac{1}{3}\delta}{\|x_1\|}, \frac{x_2 \cdot \frac{1}{3}\delta}{\|x_2\|}\right)\right\| \cdot \frac{\|x_1\| \cdot \|x_2\|}{\left(\frac{1}{3}\delta\right)^2},$$

but since

$$\left\|\left(\frac{\frac{1}{3}\delta x_1}{\|x_1\|}, \frac{\frac{1}{3}\delta x_2}{\|x_2\|}\right)\right\| = \tfrac{1}{3}\delta + \tfrac{1}{3}\delta = \tfrac{2}{3}\delta < \delta,$$

we can apply formula (1). We obtain

$$\|F(x_1, x_2)\| \leqslant \frac{\varepsilon}{\left(\frac{1}{3}\delta\right)^2} \|x_1\| \cdot \|x_2\|.$$

On putting $A = \varepsilon / \left(\frac{1}{3}\delta\right)^2$, we obtain the result. $\square$

Let us note that for the inequality (*) the continuity of F at the point $(0, 0)$ is sufficient. Then by Theorem VII.7.1 we have the following

COROLLARY VII.7.2. An n-linear mapping which is continuous at one point (e.g. at zero) is continuous throughout the space.

A norm may be introduced in the space of n-linear mappings. Let $F \in L(X_1, X_2, \ldots, X_n; Y)$ and let

$$\|F\| := \inf\{A : \|F(x_1, \ldots, x_n)\| \leqslant A \cdot \|x_1\| \cdot \ldots \cdot \|x_n\|, \ x_k \in X_k\}.$$

Reasoning as in Section 2 for linear mappings, we have

$$\|F\| = \sup_{\|x_1\| \leqslant 1, \ldots, \|x_n\| \leqslant 1} \|F(x_1, \ldots, x_n)\|.$$

This formula shows that this is a well-defined norm.

The space $L(X_1, X_2, \ldots, X_n; Y)$ with the norm so introduced is a Banach space (provided Y is one). This is easily proved by analogy with the proof of Theorem VII.2.2, but we shall proceed in a different fashion;

namely, we shall demonstrate that this space is isomorphic to some Banach space.

THEOREM VII.7.3. The space $L(X_1, X_2; Y)$ is isometrically isomorphic to the space $L(X_1, L(X_2, Y))$.

PROOF. Let $T \in L(X_1, L(X_2, Y))$. In other words

$$X_1 \ni x_1 \to T(x_1) \in L(X_2, Y).$$

We define $F \in L(X_1, X_2; Y)$ as follows:

$$X_1 \times X_2 \ni (x_1, x_2) \to F(x_1, x_2) := T(x_1) \cdot x_2 \in Y.$$

The mapping

$$(2) \qquad L(X_1, L(X_2, Y)) \ni T \to F \in L(X_1, X_2; Y)$$

is obviously linear. We shall show that it is a mapping *onto* and a one-to-one mapping. Indeed, if $F \in L(X_1, X_2; Y)$, the mapping T defined by the formula $T(x_1) := F(x_1, \cdot)$ is an element of $L(X_1, L(X_2, Y))$. We thus have

$$X_1 \ni x_1 \to T(x_1) = F(x_1, \cdot) \in L(X_1, Y),$$

$$T(x_1) \cdot x_2 = F(x_1, x_2) \in Y.$$

This equation proves that T goes over onto F under the homomorphism (2). Hence it follows that this is a homomorphism *onto*.

If $F = 0$, then $T(x_1) \cdot x_2 = 0$ for every $x_1 \in X_1$ and every $x_2 \in X_2$. Hence $T = 0$. Thus this homomorphism is an isomorphism. We shall show it to be an isometry. We have

$$\|F\|_{L(X_1, X_2; Y)} = \sup_{\substack{\|x_1\| \leqslant 1 \\ \|x_2\| \leqslant 1}} \|F(x_1, x_2)\|_Y = \sup_{\|x_1\| \leqslant 1} \sup_{\|x_2\| \leqslant 1} \|T(x_1) \cdot x_2\|_Y$$

$$= \sup_{\|x_1\| \leqslant 1} \|T(x_1)\|_{L(X_2, Y)} = \|T\|_{L(X_1, L(X_2, Y))}. \quad \square$$

The foregoing theorem allows itself to be generalized immediately. We thus have:

THEOREM VII.7.4. A space $L(X_1, X_2, \ldots, X_n; Y)$ is isometrically isomorphic to the space $L(X_1, L(X_2, \ldots, L(X_n, Y)))$.

If a mapping $T \in L(X_1, L(X_2, \ldots, L(X_n, Y)))$, then there corresponds to it an $F \in L(X_1, X_2, \ldots, X_n; Y)$ such that

$$\Big(\big((T(x_1) x_2) \big) \ldots \Big) x_n = F(x_1, x_2, \ldots, x_n).$$

The proof of this theorem is analogous to that of Theorem VII.7.3.

A space $L(X_1, L(X_2, ..., L(X_n, Y)))$ is a Banach space (provided that Y is one). It thus follows from Theorem VII.7.4 that the space $L(X_1, X_2, ..., X_n, Y)$ is also a Banach space.

8. DERIVATIVES OF HIGHER ORDERS

DEFINITION. Let T be a differentiable mapping of an open subset V of a Banach space X into a Banach space Y. The mapping T is then said to be *twice differentiable at a point* $x_0 \in V \subset X$ if the derivative of T (that is, $X \supset V \ni x \to T'(x) \in L(X, Y)$) is differentiable at x_0. The derivative of the mapping $x \to T'(x)$ at x_0 is called the *second derivative of the mapping T at the point x_0* and is denoted by the symbol $T''(x_0)$.

Obviously, $T''(x_0) \in L(X, L(X, Y))$. We know from Theorem VII.7.3, however, that $L(X, L(X, Y)) \cong L(X, X; Y)$. Thus $T''(x_0) \in L(X, X; Y)$. Consequently we have

$$T''(x_0)(s, t) = (T''(x_0) \cdot s) \cdot t, \quad s, t \in X.$$

If T is twice-differentiable at every point of the set V, we say that it is *twice-differentiable on V*, and the mapping

$$X \supset V \ni x \to T''(x) \in L(X, X; Y)$$

is called the *second derivative of T*. If the mapping $x \to T''(x)$ is continuous on V, we say that T is *continuously twice-differentiable on V*. The latter definition is correct since the continuity of the first derivative on V follows from its differentiability on V.

We have the following formula for the second derivative.

THEOREM VII.8.1. Let T be a mapping of an open set V of a Banach space X into a Banach space V. If T is twice differentiable at a point $x \in V$, then

(1) $\quad T''(x)(k, h)$

$$= \lim_{t \to 0} \frac{1}{t^2} \{T(x+th+tk) - T(x+tk) - T(x+th) + T(x)\}.$$

PROOF. We define an auxiliary vector-valued function:

(2) $\quad [-1, 1] \ni a \to g(a) = T(x+ah+k) - T(x+ah).$

There exists an $r > 0$ such that for $||h|| < r, ||k|| < r$ the function g is continuously differentiable and

$$(3) \qquad g'(a) = T'(x+ah+k)h - T'(x+ah) \cdot h.$$

We can approximate the second derivative of T by g'. Indeed, for every $\varepsilon > 0$ there exists an $r_1 < r$ such that for every $||h|| < r_1, ||k|| < r_1$, $|a| < 1$ we have:

$$
\begin{aligned}
(4) \quad & ||g'(a) - T''(x)(k, h)|| \\
&= ||T'(x+ah+k)h - T'(x+ah)h - T''(x)(k, h)|| \\
&= ||T'(x+ah+k)h - T'(x)h - T''(x)(ah+k, h) - \\
&\quad - T'(x+ah)h + T'(x)h + T''(x)(ah, h)|| \\
&\leqslant ||R(x, ah+k) \cdot h|| + ||R(x, ah) \cdot h|| \\
&\leqslant (||R(x, ah+k)|| + ||R(x, ah)||)||h|| \\
&\leqslant \varepsilon(||ah+k|| + ||ah||)||h|| \\
&\leqslant \varepsilon(2||h|| + ||k||)||h||,
\end{aligned}
$$

where for every $u \in X$ the linear operator $R(x, u)$ is the remainder in the definition of the derivative of T'.

On the other hand we can approximate the derivative of g by the Mean-Value Theorem

$$
\begin{aligned}
(5) \quad & ||g(1) - g(0) - g'(0)|| \leqslant \sup_{a \in [0, 1]} ||g'(a) - g'(a)|| \\
&\leqslant \sup_{a \in [0, 1]} \{||g'(a) - T''(x)(k, h)|| + ||T''(x)(k, h) - g'(0)||\} \\
&\leqslant 2\varepsilon(2||h|| + ||k||)||h||.
\end{aligned}
$$

Combining (4) and (5) we obtain:

$$
\begin{aligned}
(6) \quad & ||g(1) - g(0) - T''(x)(k, h)|| \\
&\leqslant ||g(1) - g(0) - g'(0)|| + ||g'(0) - T''(x)(k, h)|| \\
&\leqslant 3\varepsilon(2||h|| + ||k||)||h||.
\end{aligned}
$$

Replacing in (6) h and k by th, tk we have

$$
\begin{aligned}
(7) \quad & ||T(x+th+tk) - T(x+th) - T(x+tk) - T(x) - T''(x)(tk, th)|| \\
&\leqslant 3\varepsilon t^2(2||h|| + ||k||)||h||.
\end{aligned}
$$

Thus dividing the both sides by t^2 and passing with t to zero we have for every $\varepsilon > 0$

(8) $\lim_{t \to 0}\left\{\dfrac{1}{t^2}\left(T(x+th+tk)-T(x+th)-T(x+tk)-T(x)\right)\right.$

$\left. -T''(x)(k,h)\right\}$

$\leqslant 3\varepsilon(2\|h\|+\|k\|)\|h\|.$ $\square$

The formula (1) of the Theorem VII.8.1 implies

COROLLARY VII.8.2. The second derivative is a blilinear symmetric mapping, i.e. $T''(x)(k,h) = T''(x)(h,k)$.

Derivatives of higher order are defined by induction.

DEFINITION. Let V be an open set in a Banach space X and let T be a mapping of V into a Banach space Y. The p-th derivative of T at the point x (denoted by $T^{(p)}(x)$) belongs to the space $L_p(X,Y)$.

Assume that the $(p-1)$-th derivative of T is defined in all points of some neighbourhood $U \subset V$ of x, and the mapping

(9) $V \ni x \to T^{(p-1)}(x) \in L_{p-1}(X,Y)$

is differentiable at x.

Then we put

(10) $T^{(p)}(x) := T^{(p-1)\prime}(x) \in L(X, L_{p-1}(X,Y)) \cong L_p(X,Y).$

The mapping T is then said to be p *times differentiable at the point* x.

When T is p times differentiable at every point of the set U we say that T is p *times differentiable on* U and we call the mapping

$U \ni x \to T^{(p)}(x)$

the p-*th derivative* of T.

The following theorem is a generalization of Corollary VII.8.2:

THEOREM VII.8.3. Let V be an open subset of a Banach space X and let T be a mapping of V into a Banach space Y. If T is p times differentiable at a point $x \in V$, then the p-linear mapping

$X \times X \times \ldots \times X \ni (h_1, h_2, \ldots, h_p) \to T^{(p)}(x)(h_1, h_2, \ldots, h_p) \in Y$

is symmetric. This means that

(11) $T^{(p)}(x)(h_1, h_2, \ldots, h_p) = T^{(p)}(x)(h_{j_1}, \ldots, h_{j_p}),$

where $j_1, j_2, \ldots, j_p$ is an arbitrary permutation of the numbers $1, 2, \ldots, p$.

189

PROOF. We carry out the proof by applying induction to p.

For $p = 1$, the theorem is obvious.

For $p = 2$, the theorem holds by virtue of Theorem VII.8.1.

Suppose that the theorem is valid for $p-1$. We take the mapping

$$X \ni x \to T^{(p-2)}(x)(h_3, h_4, \ldots, h_p) \in Y.$$

Its second derivative at the point x is a bilinear mapping

$$X \times X \ni (h_1, h_2) \to T^{(p)}(x)(h_1, h_2, \ldots, h_p) \in Y,$$

which is symmetric by Theorem VII.8.1. Thus we have

$$(12) \qquad T^{(p)}(x)(h_1, h_2, h_3, \ldots, h_p) = T^{(p)}(x)(h_2, h_1, \ldots, h_p).$$

By the inductive assumption

$$T^{(p-1)}(x)(h_2, h_3, \ldots, h_p) = T^{p-1}(x)(h_{i_2}, h_{i_3}, \ldots, h_{i_p}),$$

where $i_2, i_3, \ldots, i_p$ is an arbitrary permutation of the numbers from 2 to p.

Taking the derivative of both sides (with respect to x) at the point x, we get

$$(13) \qquad T^{(p)}(x)(h_1, h_2, \ldots, h_p) = T^{(p)}(x)(h_1, h_{i_2}, \ldots, h_{i_p}).$$

Since any permutation $j_1, j_2, \ldots, j_p$ of the numbers from 1 to p can be obtained by permuting the numbers $j_2, j_3, \ldots, j_p$ and transposing j_1 and j_2, it follows from (12) and (13) that

$$T^{(p)}(x)(h_1, h_2, \ldots, h_p) = T^{(p)}(x)(h_{j_1}, h_{j_2}, \ldots, h_{j_p}). \quad \square$$

Now we go on to define the second-order partial derivative.

DEFINITION. Let U be an open subset of the Cartesian product $X_1 \times X_2$ of two Banach spaces and let T be a mapping of U into a Banach space Y. Moreover, let T be differentiable on U with respect to X_1 and X_2. The mapping T is said to have a *second-order partial derivative* (also known as the *second partial derivative with respect to X_1*) at a point $(\mathring{x}_1, \mathring{x}_2) \in U$ if the mapping

$$X_1 \ni x_1 \to T'_{x_1}(x_1, \mathring{x}_2) \in L(X_1, Y)$$

is differentiable at the point $\mathring{x}_1$.

The derivative of this mapping at the point $\mathring{x}_1$ is called the *second partial derivative of the mapping T, with respect to X_1, at the point* $(\mathring{x}_1, \mathring{x}_2)$ and is denoted by the symbol $T''_{x_1 x_1}(\mathring{x}_1, \mathring{x}_2)$.

Plainly

$$T''_{X_1 X_1}(\mathring{x}_1, \mathring{x}_2) \in L(X_1, L(X_1, Y)) \cong L(X_1, X_1; Y).$$

The partial derivative of second order thus is a bilinear mapping.

If a mapping T has the second partial derivative with respect to X_1 at every point of the set U, then T is said to be *twice-differentiable with respect to X_1 on U*, and the mapping

$$X_1 \times X_2 \supset U \ni (x_1, x_2) \rightarrow T''_{X_1 X_1}(x_1, x_2) \in L(X_1, X_1; Y)$$

is called the *second partial derivative of the mapping T with respect to X_1*.

The second-order partial derivative with respect to X_2 can be defined in similar fashion.

A mapping T is said to have a *second partial derivative with respect to X_2* at a point $(\mathring{x}_1, \mathring{x}_2) \in U$ if the mapping

$$X_2 \ni x_2 \rightarrow T'_{X_2}(\mathring{x}_1, x_2) \in L(X_2, Y)$$

is differentiable at the point $\mathring{x}_2$.

The derivative of this mapping at the point $\mathring{x}_2$ is denoted by the symbol $T''_{X_2 X_2}(\mathring{x}_1, \mathring{x}_2)$ and is called the *second partial derivative of the mapping T with respect to X_2 at the point $(\mathring{x}_1, \mathring{x}_2)$*;

$$T''_{X_2 X_2}(\mathring{x}_1, \mathring{x}_2) \in L(X_2, L(X_2, Y)) \cong L(X_2, X_2; Y).$$

Similarly, if the mapping T has a second partial derivative with respect to X_2 at every point of the set U, then T is said to be *twice-differentiable with respect to X_2 on U*, and the mapping

$$X_1 \times X_2 \supset U \ni (x_1, x_2) \rightarrow T''_{X_2 X_2}(x_1, x_2) \in L(X_2, X_2; Y)$$

is called the *second partial derivative of the mapping T with respect to X_2*.

Now let us define mixed derivatives of the second order.

A mapping T is said to have a *second partial derivative with respect to X_1 and X_2* at a point $(\mathring{x}_1, \mathring{x}_2) \in U$ if the mapping

$$X_2 \ni x_2 \rightarrow T'_{X_1}(\mathring{x}_1, x_2) \in L(X_1, Y)$$

is differentiable at the point $\mathring{x}_2$.

The derivative of this mapping is denoted by the symbol $T''_{X_1 X_2}(\mathring{x}_1, \mathring{x}_2)$ and is called the *second-order partial derivative of the mapping T with respect to X_1 and X_2 at the point $(\mathring{x}_1, \mathring{x}_2)$*;

$$T''_{X_1 X_2}(\mathring{x}_1, \mathring{x}_2) \in L(X_2, L(X_1, Y)) \cong L(X_2, X_1; Y).$$

If T has a mixed partial derivative with respect to X_1 and X_2 at every point of U, the mapping

$$X_1 \times X_2 \supset U \ni (x_1, x_2) \to T''_{X_1X_2}(x_1, x_2) \in L(X_2, X_1; Y)$$

is called the *second-order partial derivative*, or *mixed partial derivative, of the mapping T with respect to X_1 and X_2*.

The mixed partial derivative with respect to X_2 and X_1 is defined analogously.

If the mapping

$$X_1 \ni x_1 \to T'_{X_2}(x_1, \mathring{x}_2) \in L(X_2, Y)$$

is differentiable at the point $\mathring{x}_1$, then its derivative at that point is denoted by the symbol $T''_{X_2X_1}(\mathring{x}_1, \mathring{x}_2)$ and is called the *second-order partial derivative of the mapping T with respect to X_2 and X_1 at the point* $(\mathring{x}_1, \mathring{x}_2)$.

If a mixed partial derivative of the mapping T with respect to X_2 and X_1 exists at every point of the set U, the mapping

$$X_1 \times X_2 \supset U \ni (x_1, x_2) \to T''_{X_2X_1}(x_1, x_2) \in L(X_1, X_2; Y)$$

is called the *second-order partial derivative of the mapping T with respect to X_2 and X_1* or the *mixed partial derivative* of the mapping T with respect to X_2 and X_1.

Let us emphasize that

$$T''_{X_2X_1}(x_1, x_2) \in L\big(X_1, L(X_2, Y)\big) \cong L(X_1, X_2; Y).$$

Now we shall show that the existence of partial derivatives follows from two-fold differentiability.

LEMMA VII.8.4. *Let U be an open subset of the Cartesian product $X_1 \times X_2$ of two Banach spaces and let T be a mapping of the set U into a Banach space Y. If T is twice-differentiable at a point (x_1, x_2) $\in U$, then all four second-order partial derivatives exist at that point and the following formulae hold:*

$$T''_{X_1X_1}(x_1, x_2)(h_1, t_1) = T''(x_1, x_2)\big((h_1, 0), (t_1, 0)\big),$$

$$T''_{X_2X_2}(x_1, x_2)(h_2, t_2) = T''(x_1, x_2)\big((0, h_2), (0, t_2)\big),$$

$$T''_{X_1X_2}(x_1, x_2)(h_2, t_1) = T''(x_1, x_2)\big((0, h_2), (t_1, 0)\big),$$

$$T''_{X_2X_1}(x_1, x_2)(h_1, t_2) = T''(x_1, x_2)\big((h_1, 0), (0, t_2)\big).$$

Furthermore, the mixed partial derivatives are equal, that is

$$T''_{X_1X_2}(x_1, x_2)(h_2, t_1) = T''_{X_2X_1}(x_1, x_2)(t_1, h_2).$$

In all of these formulae $t_1, h_1 \in X_1$, $t_2, h_2 \in X_2$.

PROOF. The twice-differentiability of T at the point (x_1, x_2) implies that the first derivative of T exists in a neighbourhood of that point. Hence, it follows from Lemma VII.6.1 that first-order partial derivatives exist in this neighbourhood and that

$$T'_{x_1}(x_1, x_2) \cdot t_1 = T'(x_1, x_2)(t_1, 0),$$
$$T'_{x_2}(x_1, x_2) \cdot t_2 = T'(x_1, x_2)(0, t_2),$$

and consequently we have

$$T'_{x_1}(x_1 + h_1, x_2) \cdot t_1 - T'_{x_1}(x_1, x_2) \cdot t_1$$
$$= T'(x_1 + h_1, x_2)(t_1, 0) - T'(x_1, x_2)(t_1, 0)$$
$$= T''(x_1, x_2)((h_1, 0), (t_1, 0)) + r(h_1) \cdot t_1,$$

whence

$$T''_{x_1 x_1}(x_1, x_2)(h_1, t_1) = T''(x_1, x_1)((h_1, 0), (t_1, 0)).$$

The proof for other derivatives is similar.

Since

$$T''_{x_1 x_2}(x_1, x_2)(h_2, t_1) = T''(x_1, x_2)((0, h_2), (t_1, 0)),$$
$$T''_{x_2 x_1}(x_1, x_2)(t_1, h_2) = T''(x_1, x_2)((t_1, 0), (0, h_2)),$$

and by the symmetry of the second derivative (cf. Theorem VII.8.1)

$$T''(x_1, x_2)((0, h_2), (t_1, 0)) = T''(x_1, x_2)((t_1, 0), (0, h_2)),$$

we obtain

$$T''_{x_1 x_2}(x_1, x_2)(h_2, t_1) = T''_{x_2 x_1}(x_1, x_2)(t_1, h_2). \quad \square$$

Introducing the injection mappings $i_{11}, i_{22}, i_{12}, i_{21}$:

$$X_1 \times X_1 \ni (h_1, t_1) \to i_{11}(h_1, t_1) = ((h_1, 0), (t_1, 0))$$
$$\in X_1 \times X_2 \times X_1 \times X_2,$$

$$X_1 \times X_2 \ni (h_1, t_2) \to i_{12}(h_1, t_2) = ((h_1, 0), (0, t_2))$$
$$\in X_1 \times X_2 \times X_1 \times X_2,$$

$$X_2 \times X_1 \ni (h_2, t_1) \to i_{21}(h_2, t_1) = ((0, h_2), (t_1, 0))$$
$$\in X_1 \times X_2 \times X_1 \times X_2,$$

$$X_2 \times X_2 \ni (h_2, t_2) \to i_{22}(h_2, t_2) = ((0, h_2), (0, t_2))$$
$$\in X_1 \times X_2 \times X_1 \times X_2,$$

we obtain

$$T''_{x_1 x_1}(x_1, x_2) = T''(x_1, x_2) \circ i_{11},$$
$$T''_{x_1 x_2}(x_1, x_2) = T''(x_1, x_2) \circ i_{21},$$

$$T''_{X_2X_1}(x_1, x_2) = T''(x_1, x_2) \circ i_{12},$$
$$T''_{X_2X_2}(x_1, x_2) = T''(x_1, x_2) \circ i_{22}.$$

Now we give the sufficient condition for the existence of the second derivative.

THEOREM VII.8.5. Let U be an open subset of the Cartesian product $X_1 \times X_2$ of two Banach spaces and let T be a mapping of the set U into a Banach space Y. Suppose now that in an open neighbourhood $V \subset U$ of a point $(\mathring{x}_1, \mathring{x}_2) \in U$ the mapping T is differentiable with respect to X_1 and X_2 and suppose that one of the partial derivatives is continuous on V.

Moreover, suppose that the second-order partial derivatives $T''_{X_1X_1}$ and $T''_{X_2X_1}$ exist at the point $(\mathring{x}_1, \mathring{x}_2)$ and suppose that the partial derivatives $T''_{X_1X_2}$ and $T''_{X_2X_2}$ exist at every point of the set V and are continuous on V or, conversely, suppose that the derivatives $T''_{X_1X_2}$ and $T''_{X_2X_2}$ exist at the point $(\mathring{x}_1, \mathring{x}_2)$ and the derivatives $T''_{X_1X_1}$ and $T''_{X_2X_1}$ exist and are continuous on V.

The mapping T then is twice-differentiable at the point $(\mathring{x}_1, \mathring{x}_2)$ and the following formula holds:

$$T'''(\mathring{x}_1, \mathring{x}_2)((h_1, h_2), (t_1, t_2)) = T''_{X_1X_1}(\mathring{x}_1, \mathring{x}_2)(h_1, t_1) +$$
$$+ T''_{X_1X_2}(\mathring{x}_1, \mathring{x}_2)(h_2, t_1) + T''_{X_2X_1}(\mathring{x}_1, \mathring{x}_2)(h_1, t_2) +$$
$$+ T''_{X_2X_2}(\mathring{x}_1, \mathring{x}_2)(h_2, t_2).$$

PROOF. The existence of both first-order partial derivatives at every point of the set V and the continuity of one of them on V imply (cf. Theorem VII.6.2, Corollary VII.6.3) that T is differentiable on V and

$$T'(x_1, x_2) = T'_{X_1}(x_1, x_2) \circ \mathrm{pr}_1 + T'_{X_2}(x_1, x_2) \circ \mathrm{pr}_2,$$

where $(x_1, x_2) \in V$ and pr_1 and pr_2 are the projections onto X_1 and X_2. We must now examine the differentiability of the mapping

$$X_1 \times X_2 \supset V \ni (x_1, x_2) \to T'(x_1, x_2) \in L(X_1 \times X_2, Y)$$

at the point $(\mathring{x}_1, \mathring{x}_2)$.

Its partial derivatives are expressed by

$$(T')'_{X_1}(\mathring{x}_1, \mathring{x}_2) \cdot h_1$$
$$= T''_{X_1X_1}(\mathring{x}_1, \mathring{x}_2) \cdot h_1 \circ \mathrm{pr}_1 + T''_{X_2X_1}(\mathring{x}_1, \mathring{x}_2) \cdot h_1 \circ \mathrm{pr}_2,$$
$$(T')'_{X_2}(\mathring{x}_1, \mathring{x}_2) \cdot h_2$$
$$= T''_{X_1X_2}(\mathring{x}_1, \mathring{x}_2) \cdot h_2 \circ \mathrm{pr}_1 + T''_{X_2X_2}(\mathring{x}_1, \mathring{x}_2) \cdot h_2 \circ \mathrm{pr}_2.$$

It follows from the hypotheses of the theorem that one of them exists at the point $(\mathring{x}_1, \mathring{x}_2)$ and the other exists at every point of the set V and is continuous on V. The mapping T thus has a second derivative at the point $(\mathring{x}_1, \mathring{x}_2)$. This derivative is expressed in terms of the second-order partial derivatives by means of the formula given above. $\square$

If we introduce the operations of projection pr_{11}, pr_{12}, pr_{21} and pr_{22}:

$$X_1 \times X_2 \times X_1 \times X_2 \ni ((h_1, h_2), (t_1, t_2))$$
$$\to \mathrm{pr}_{11}((h_1, h_2), (t_1, t_2)) = (h_1, t_1) \in X_1 \times X_1,$$
$$X_1 \times X_2 \times X_1 \times X_2 \ni ((h_1, h_2), (t_1, t_2))$$
$$\to \mathrm{pr}_{12}((h_1, h_2), (t_1, t_2)) = (h_1, t_2) \in X_1 \times X_2,$$
$$X_1 \times X_2 \times X_1 \times X_2 \ni ((h_1, h_2), (t_1, t_2))$$
$$\to \mathrm{pr}_{21}((h_1, h_2), (t_1, t_2)) = (h_2, t_1) \in X_2 \times X_1,$$
$$X_1 \times X_2 \times X_1 \times X_2 \ni ((h_1, h_2), (t_1, t_2))$$
$$\to \mathrm{pr}_{22}((h_1, h_2), (t_1, t_2)) = (h_2, t_2) \in X_2 \times X_2,$$

the formula for the second derivative takes on the form

$$T''(x_1, x_2) = T''_{X_1 X_1}(x_1, x_2) \circ \mathrm{pr}_{11} + T''_{X_1 X_2}(x_1, x_2) \circ \mathrm{pr}_{21} +$$
$$+ T''_{X_2 X_1}(x_1, x_2) \circ \mathrm{pr}_{12} + T''_{X_2 X_2}(x_1, x_2) \circ \mathrm{pr}_{22}.$$

If we strengthen the hypotheses somewhat, Theorem VII.8.5 assumes the form of a necessary and sufficient condition.

THEOREM VII.8.6. Let U be an open subset of the Cartesian product $X_1 \times X_2$ of two Banach spaces and let T be a mapping of U into a Banach space Y. The mapping T is twice-differentiable on U in a continuous manner if and only if all of its second-order partial derivatives exist and are continuous on U.

PROOF. If T is twice-differentiable on U, then Lemma VII.8.4 implies that second-order partial derivatives exist. Their continuity follows easily.

Conversely, if continuous second-order partial derivatives exist on U, the mappings

$$X_1 \times X_2 \ni (x_1, x_2) \to T'_{X_i}(x_1, x_2) \in L(X_i, Y), \quad i = 1, 2,$$

are continuously differentiable with respect to X_1 and X_2 on U. The continuity of the first-order partial derivatives follows from this. On applying Theorem VII.8.5 we get existence of the second derivative of T. Again its continuity is straightforward. $\square$

The concept of partial derivative can be extended by introducing partial derivatives of arbitrary order. Thus we have:

DEFINITION. Let U be an open subset of the Cartesian product $X_1 \times X_2 \times \ldots \times X_n$ of n Banach spaces and let T be a mapping of the set U into a Banach space Y. The mapping T is said to have a *partial derivative of order p with respect to* $X_{i_1}, X_{i_2}, \ldots, X_{i_p}, 1 \leqslant i_1, i_2, \ldots, i_p \leqslant n$, at a point $(\mathring{x}_1, \mathring{x}_2, \ldots, \mathring{x}_n) \in U$ if a partial derivative of order $p-1$ with respect to $X_{i_1}, X_{i_2}, \ldots, X_{i_{p-1}}$ exists in a neighbourhood of that point and if the mapping

$$X_{i_p} \ni x_{i_p} \to T^{(p-1)}_{X_{i_1} X_{i_2} \ldots X_{i_{p-1}}}(\mathring{x}_1, \ldots, x_{i_p}, \ldots, \mathring{x}_n) \in L(X_{i_{p-1}}, \ldots, X_{i_1}; Y)$$

is differentiable at the point $\mathring{x}_{i_p}$.

The derivative of this mapping at $\mathring{x}_{i_p}$ is called the *partial derivative of order p, at the point* $(\mathring{x}_1, \ldots, \mathring{x}_n) \in U$, *of the mapping T with respect to* $X_{i_1}, X_{i_2}, \ldots, X_i$ and is denoted by the symbol $T^{(p)}_{X_{i_1} X_{i_2} \ldots X_{i_{p-1}} X_{i_p}}(\mathring{x}_1, \ldots, \mathring{x}_n)$.

$$T^{(p)}_{X_{i_1} X_{i_2} \ldots X_{i_{p-1}} X_{i_p}}(x_1, \ldots, x_n) \in L\big(X_{i_p}, L(X_{i_{p-1}}, \ldots, X_1; Y)\big)$$
$$\cong L(X_{i_p}, X_{i_{p-1}}, \ldots, X_1; Y).$$

If a partial derivative of order p of the mapping T with respect to $X_{i_1}, \ldots, X_{i_p}$ exists at every point of the set U, the mapping

$$X_1 \times \ldots \times X_n \supset U \ni (x_1, \ldots, x_n) \to T^{(p)}_{X_{i_1} \ldots X_{i_p}}(x_1, \ldots, x_n)$$
$$\in L(X_{i_p}, \ldots, X_{i_1}; Y)$$

is called the *partial derivative, of order p, of the mapping T with respect to* $X_{i_1}, \ldots, X_{i_p}$.

Remark. The existence of the partial derivative of order $p-1$ in some neighbourhood of the point $(\mathring{x}_1, \ldots, \mathring{x}_n)$ need not necessarily be required in the definition of the partial derivative of order p. It is sufficient to require its existence on the set $\{\mathring{x}_1\} \times \{\mathring{x}_2\} \times \ldots \times \{\mathring{x}_{i_{p-1}}\} \times V \times \{\mathring{x}_{i_{p+1}}\} \times \ldots \times \{\mathring{x}_n\}$, where $V \subset X_{i_p}$ is a neighbourhood of the point $\mathring{x}_{i_p}$.

We shall now demonstrate that the p times differentiability of a mapping at a given point implies that all the partial derivatives of order up to and including p exist at the point.

LEMMA VII.8.7. Let U be an open subset of the Cartesian product $X_1 \times X_2 \times \ldots \times X_n$ of n Banach spaces and let T be a mapping of U into a Banach space Y. If T is p-fold differentiable at a point $(x_1, \ldots, x_n)$

$\in U$, then at that point T has all the partial derivatives of order up to and including p.

Furthermore,

$$T^{(m)}_{X_{k_1}\ldots X_{k_m}}(x_1, \ldots, x_n) = T^{(m)}(x_1, \ldots, x_n) \circ i_{\substack{k_m\ldots k_1 \\ (n)}},$$

where $i_{\substack{k_m\ldots k_1 \\ (n)}}$ is the injection operation and $1 \leqslant m \leqslant p$

$$i_{\substack{k_m\ldots k_1 \\ (n)}}: X_{k_m} \times X_{k_{m-1}} \times \ldots \times X_{k_1} \to (X_1 \times X_2 \times \ldots \times X_n)^m;$$

$$X_{k_m} \times X_{k_{m-1}} \times \ldots \times X_{k_1} \ni (x_{k_m}, x_{k_{m-1}}, \ldots, x_{k_1}) \to i_{\substack{k_m\ldots k_1 \\ (n)}}(x_{k_m}, \ldots, x_{k_1})$$

$$= \big((0, \ldots, x_{k_m}, \ldots, 0), (0, \ldots, x_{k_{m-1}}, \ldots, 0), \ldots$$

$$\ldots, (0, \ldots, x_{k_1}, \ldots, 0)\big)$$

$$\in \underbrace{(X_1 \times \ldots \times X_n) \times (X_1 \times \ldots \times X_n) \times \ldots \times (X_1 \times \ldots \times X_n)}_{m \text{ times}}.$$

This formula can be rewritten as

$$T^{(m)}_{X_{k_1}\ldots X_{k_m}}(x_1, \ldots, x_n)(h_{k_m}, \ldots, h_{k_1})$$

$$= T^{(m)}(x_1, \ldots, x_n)\big((0, \ldots, h_{k_m}, \ldots, 0), \ldots, (0, \ldots, h_{k_1}, \ldots, 0)\big).$$

PROOF. The proof is patterned after that of Lemma VII.8.4, and proceeds by induction. $\square$

By virtue of the relations given in Lemma VII.8.7 and by the symmetry of the m-th derivative (cf. Theorem VII.8.3), we have

$$T^{(m)}_{X_{k_1} X_{k_2}\ldots X_{k_m}}(x_1, \ldots, x_n)(h_{k_m}, \ldots, h_{k_1})$$

$$= T^{(m)}_{X_{k_{j_1}} X_{k_{j_2}}\ldots X_{k_{j_m}}}(x_1, \ldots, x_n)(h_{k_{j_m}}, h_{k_{j_{m-1}}}, \ldots, h_{k_{j_1}}),$$

where $k_{j_1}, k_{j_2}, \ldots, k_{j_m}$ is an arbitrary permutation of the numbers $k_1, k_2, \ldots, k_m$. These equations may also be written as

$$T^{(m)}_{X_{k_{j_1}} X_{k_{j_2}}\ldots X_{k_{j_m}}}(x_1, \ldots, x_n) \circ \sigma_j = T^{(m)}_{X_{k_1} X_{k_2}\ldots X_{k_m}}(x_1, \ldots, x_n),$$

where σ_j is a permutation operation:

$$\sigma_j: X_{k_m} \times X_{k_{m-1}} \times \ldots \times X_{k_1} \to X_{k_{j_m}} \times X_{k_{j_{m-1}}} \times \ldots \times X_{k_{j_1}},$$

$$X_{k_m} \times X_{k_{m-1}} \times \ldots \times X_{k_1} \ni (h_{k_m}, \ldots, h_{k_1}) \to \sigma_j(h_{k_m}, h_{k_{m-1}}, \ldots, h_{k_1})$$

$$:= (h_{k_{j_m}}, h_{k_{j_{m-1}}}, \ldots, h_{k_{j_1}}) \in X_{k_{j_m}} \times X_{k_{j_{m-1}}} \times \ldots \times X_{k_{j_1}}.$$

Thus, we have the following corollary:

COROLLARY VII.8.8. *If a mapping $T: X_1 \times \ldots \times X_n \supset U \to Y$ is p times differentiable at a point $(x_1, \ldots, x_n)$, then the partial derivatives of order $\leqslant p$ do not depend on the order of differentiation.*

A mapping $T: X_1 \times \ldots \times X_n \supset U \to Y$ which is differentiable at a point has n partial derivatives at that point, and a mapping which is p times differentiable has n^p partial derivatives of order p at a given point.

We now give a sufficient condition for a mapping to be p times differentiable.

THEOREM VII.8.9. *Let U be an open subset of the Cartesian product $X_1 \times X_2 \times \ldots \times X_n$ of n Banach spaces and let T be a mapping of the set U into a Banach space Y. Suppose that all the partial derivatives of T of order up to and including $p-2$ exist and are continuous in an open neighbourhood $V \subset U$ of a point $(x_1, \ldots, x_n) \in U$. Suppose that all the partial derivatives of order $p-1$ exist on V and that all, except perhaps n^{p-2} of them (of the form $T^{(p-1)}_{X_{k_1} \ldots X_{k_{p-2}} X_{l_0}}$, $1 \leqslant l_0 \leqslant n$), are continuous on V.*

If, furthermore, all the partial derivatives of order p exist and are continuous on V, with the exception perhaps of n^{p-1} of them (of the form $T^{(p)}_{X_{k_1} \ldots X_{k_{p-2}} X_{r_0}}$, $1 \leqslant r_0 \leqslant n$), which exist only at the point $(x_1, \ldots, x_n)$, then the mapping T is p times differentiable at the point $(x_1, \ldots, x_n)$ and

$$T^{(m)}(x_1, \ldots, x_n) = \sum_{k_1 = 1, \ldots, k_m = 1}^{n} T_{X_{k_1} \ldots X_{k_m}}(x_1, \ldots, x_n) \circ \mathrm{pr}_{\substack{k_m \ldots k_1 \\ (n)}},$$

where $1 \leqslant m \leqslant p$ and $\mathrm{pr}_{\substack{k_m \ldots k_1 \\ (n)}}$ is the *projection operation*

$$\mathrm{pr}_{\substack{k_m \ldots k_1 \\ (n)}} : (X_1 \times \ldots \times X_n)^m \to X_{k_m} \times \ldots \times X_{k_1},$$

$$(X_1 \times \ldots \times X_n)^m \ni \left((\underset{1}{x_1}, \ldots, \underset{1}{x_n}), (\underset{2}{x_1}, \ldots, \underset{2}{x_n}), \ldots, (\underset{m}{x_1}, \ldots, \underset{m}{x_n}) \right)$$

$$\to \mathrm{pr}_{\substack{k_m \ldots k_1 \\ (n)}} \left((\underset{1}{x_1}, \ldots, \underset{1}{x_n}), (\underset{2}{x_1}, \ldots, \underset{2}{x_n}), \ldots, (\underset{m}{x_1}, \ldots, \underset{m}{x_n}) \right)$$

$$= (\underset{1}{x_{k_m}}, \underset{2}{x_{k_{m-1}}}, \underset{3}{x_{k_{m-2}}}, \ldots, \underset{m}{x_{k_1}}) \in X_{k_m} \times \ldots \times X_{k_1}.$$

We may rewrite the foregoing as

$$T^{(m)}(x_1, \ldots, x_n)\big((h_1, \ldots, h_n)_1, (h_1, \ldots, h_n)_2, \ldots, (h_1, \ldots, h_n)_m\big)$$

$$= \sum_{k_1=1,\ldots,k_m=1}^{n} T^{(m)}_{x_{k_1} \ldots x_{k_m}}(x_1, \ldots, x_n)(h_{k_m 1}, h_{k_{m-1} 2}, \ldots, h_{k_1 m}).$$

PROOF. Repeated use of Theorem VII.6.5 shows that T is $(p-2)$ times differentiable continuously on V. By applying Theorem VII.6.4 to the derivative of order $p-2$ of the mapping T, we demonstrate that T is $p-1$ times differentiable on V. Application of Theorem VII.6.4 to the derivative of order $p-1$ of T demonstrates that T has the derivative of order p at the point $(x_1, \ldots, x_p)$. The formula given above follows from the construction of the proof. $\square$

When somewhat stronger assumptions are made, we can rewrite Theorem VII.8.9 as a necessary and sufficient condition.

THEOREM VII.8.10. Let U be an open subset of the Cartesian product $X_1 \times \ldots \times X_n$ of n Banach spaces and let T be a mapping of the set U into a Banach space Y. The mapping T is p times differentiable continuously on U if and only if all the partial derivatives of T of order up to and including p exist and are continuous on U.

PROOF. If T is p-fold differentiable on U, then the existence of partial derivatives and the relation

$$T^{(m)}_{x_{k_1} \ldots x_{k_m}}(x_1, \ldots, x_n) = T^{(m)}(x_1, \ldots, x_n) \circ i_{k_m \ldots k_1 \atop (n)}$$

follow from Lemma VII.8.7. The continuity of partial derivatives follows easily.

Conversely, if continuous partial derivatives of order up to and including p exist, then by Theorem VII.8.9 all the derivatives of order up to and including p exist and the following formula holds:

$$T^{(m)}(x_1, \ldots, x_n) = \sum_{k_1=1,\ldots,k_m=1}^{n} T^{(m)}_{x_{k_1} \ldots x_{k_m}}(x_1, \ldots, x_n) \circ \mathrm{pr}_{k_m \ldots k_1 \atop (n)}.$$

The continuity of the derivatives again follows easily. $\square$

Now we shall consider the example of mappings of the space R^n into an arbitrary Banach space Y in order to discuss the results obtained above.

For this purpose, note first of all that the space of continuous p-linear mappings of the space R^n into a Banach space Y is isometrically isomorphic to the space Y. This isomorphism is given by

$$F(t_1, t_2, \ldots, t_p) = t_1 t_2 \ldots t_p y,$$

$$F \in L_p(R^1; Y), \quad y \in Y, \ t_1 \in R^1, \ \ldots, \ t_p \in R^1.$$

We write

$$L_p(R^1; Y) \cong Y.$$

Let T be a mapping of an open subset U of the space R^n into a Banach space Y. The partial derivative of the mapping T with respect to the i-th variable at a point $(x_1, \ldots, x_n) \in U$ thus is an element of the space Y and may be evaluated by the following formula:

$$\frac{\partial T(x_1, \ldots, x_n)}{\partial x_i} = \lim_{h \to 0} \frac{T(x_1, \ldots, x_i + h, \ldots, x_n) - T(x_1, \ldots, x_n)}{h}.$$

We have here introduced the notation used earlier,

$$T_i'(x_1, \ldots, x_n) := \frac{\partial T(x_1, \ldots, x_n)}{\partial x_i}.$$

The partial derivatives of higher orders are also elements of Y. They can be evaluated by means of the limits of difference quotients. If we bring in the notation

$$\frac{\partial^m T(x_1, \ldots, x_n)}{\partial x_{i_m} \ldots \partial x_{i_1}} := T_{i_1 \ldots i_m}^{(m)}(x_1, \ldots, x_n),$$

we obtain

$$\frac{\partial^m T(x_1, \ldots, x_n)}{\partial x_{i_m} \ldots \partial x_{i_1}} = \lim_{h \to 0} \frac{1}{h} \times$$

$$\times \left[\frac{\partial^{m-1} T(x_1, \ldots, x_{i_m} + h, \ldots, x_n)}{\partial x_{i_{m-1}} \ldots \partial x_{i_1}} - \frac{\partial^{m-1} T(x_1, \ldots, x_n)}{\partial x_{i_{m-1}} \ldots \partial x_{i_1}} \right].$$

Theorem VII.8.9 tells us that for a mapping which is p times differentiable at a point $(x_1, \ldots, x_n)$ we have the formula

$$T^{(p)}(x_1, \ldots, x_n) \left(\underset{1}{(h_1, \ldots, h_n)}, \ldots, \underset{p}{(h_1, \ldots, h_n)} \right)$$

$$= \sum_{k_1 = 1, \ldots, k_p = 1}^{n} \frac{\partial^p T(x_1, \ldots, x_n)}{\partial x_{k_p} \partial x_{k_{p-1}} \ldots \partial x_{k_1}} \cdot \underset{1}{h_{k_p}} \underset{2}{h_{k_{p-1}}} \ldots \underset{p}{h_{k_1}}.$$

On the other hand, the Corollary to Lemma VII.8.7 speaks of the symmetry of the partial derivatives:

$$\frac{\partial^m T(x_1, \ldots, x_n)}{\partial x_{k_m} \partial x_{k_{m-1}} \ldots \partial x_{k_1}} = \frac{\partial^m T(x_1, \ldots, x_n)}{\partial x_{k_{j_m}} \partial x_{k_{j_{m-1}}} \ldots \partial x_{k_{j_1}}}$$

for $1 \leqslant m \leqslant p$ and any permutation $j_1, j_2, \ldots, j_m$ of the numbers $1, 2, \ldots, m$.

It should be remarked that mixed partial derivatives are not always equal. Here is an example.

Let $f: R^2 \to R^1$

$$R^2 \ni (x, y) \to f(x, y) := \begin{cases} \dfrac{x \cdot y \cdot (x^2 - y^2)}{x^2 + y^2} & \text{for } x, y \neq 0, \\ 0 & \text{for } x, y = 0. \end{cases}$$

This is undoubtedly a function continuous on R^2 which has derivatives of all orders everywhere apart from the point $(0, 0)$. Thus, we have

$$\frac{\partial f(0, y)}{\partial x} = \lim_{x \to 0} \frac{f(x, y) - f(0, y)}{x} = \lim_{x \to 0} \frac{y \cdot (x^2 - y^2)}{x^2 + y^2} = -y,$$

$$\frac{\partial f(0, 0)}{\partial x} = \lim_{x \to 0} \frac{f(x, 0) - f(0, 0)}{x} = 0,$$

$$\frac{\partial f(x, 0)}{\partial y} = \lim_{y \to 0} \frac{f(x, y) - f(x, 0)}{y} = \lim_{y \to 0} \frac{x \cdot (x^2 - y^2)}{x^2 + y^2} = x,$$

$$\frac{\partial f(0, 0)}{\partial y} = \lim_{y \to 0} \frac{f(0, y) - f(0, 0)}{y} = 0.$$

Now let us evaluate the mixed partial derivatives of the second order:

$$\frac{\partial^2 f(0, 0)}{\partial y \, \partial x} = \lim_{y \to 0} \frac{1}{y} \left(\frac{\partial f(0, y)}{\partial x} - \frac{\partial f(0, 0)}{\partial x} \right) = -1,$$

$$\frac{\partial^2 f(0, 0)}{\partial x \, \partial y} = \lim_{x \to 0} \frac{1}{x} \left(\frac{\partial f(x, 0)}{\partial y} - \frac{\partial f(0, 0)}{\partial y} \right) = 1.$$

We thus see that

$$\frac{\partial^2 f(0, 0)}{\partial y \cdot \partial x} \neq \frac{\partial^2 f(0, 0)}{\partial x \cdot \partial y} .$$

To end this section we shall consider directional derivatives of higher orders.

Let T be a mapping of an open set U in a Banach space X into a Banach space Y. We recall that the *directional derivative of a mapping T at a point x in the direction e* is the limit of the following expression (provided that limit exists):

$$\nabla_e T(x) = \lim_{t \to 0} \frac{T(x + t \cdot e) - T(x)}{t}, \qquad t \in R^1.$$

If the mapping T is differentiable (strongly) at a point x, then

$$T'(x) \cdot e = \nabla_e T(x).$$

We consider the mapping

$$X \supset U \ni x \to \nabla_{e_1} T(x) \in Y.$$

The directional derivative of this mapping in the direction e_2 (provided such does exist) is denoted by the symbol $\nabla_{e_2} \nabla_{e_1} T(x)$ and is called the *second-order directional derivative of the mapping T in the directions e_1 and e_2*. Accordingly we have

$$\nabla_{e_2} \nabla_{e_1} T(x) = \lim_{t \to 0} \frac{\nabla_{e_1} T(x + t \cdot e_2) - \nabla_{e_1} T(x)}{t}.$$

If the mapping T is twice-differentiable at the point x, then second-order directional derivatives in all directions exist at that point and

$$\nabla_{e_2} \nabla_{e_1} T(x) = T''(x)(e_2, e_1).$$

For we have

$$\nabla_{e_1} T(x + t \cdot e_2) - \nabla_{e_1} T(x) = T'(x + t \cdot e_2) \cdot e_1 - T'(x) \cdot e_1$$
$$= T''(x)(t \cdot e_2) \cdot e_1 + r(t \cdot e_2) \cdot e_1.$$

Dividing both sides by t and going to the limit, we obtain the formula above.

Moreover, because of the symmetry of the second derivatives, the second-order directional derivative in this case does not depend on the order of differentiation. Thus we have

$$\nabla_{e_2} \nabla_{e_1} T(x) = \nabla_{e_1} \nabla_{e_2} T(x).$$

Directional derivatives of higher orders can be introduced by induction:

$$\nabla_{e_p} \nabla_{e_{p-1}} \cdots \nabla_{e_1} T(x)$$

$$= \lim_{t \to 0} \frac{\nabla_{e_{p-1}} \cdots \nabla_{e_1} T(x + t \cdot e_p) - \nabla_{e_{p-1}} \cdots \nabla_{e_1} T(x)}{t}.$$

A mapping T which is p-differentiable (strongly) at a point x has a directional derivative of order p in every direction at that point and

$$\nabla_{e_p} \ldots \nabla_{e_1} T(x) = T^{(p)}(x)(e_p, \ldots, e_1).$$

Furthermore, by virtue of the symmetry of the p-th derivative, the directional derivative does not depend on the order of differentiation:

$$\nabla_{e_p} \ldots \nabla_{e_1} T(x) = \nabla_{e_{i_p}} \ldots \nabla_{e_{i_1}} T(x),$$

where $i_1, \ldots, i_p$ is an arbitrary permutation of the numbers $1, 2, \ldots, p$.

Now, let $X = R^n$, and $e_1, e_2, \ldots, e_n$ be a canonical basis in R^n. We already know that

$$\nabla_{e_i} T(x) = \frac{\partial T(x)}{\partial x_i}.$$

Similarly, it is readily seen that

$$\nabla_{e_{k_m}} \ldots \nabla_{e_{k_1}} T(x) = \frac{\partial^m T(x)}{\partial x_{k_m} \ldots \partial x_{k_1}}.$$

9. THE TAYLOR FORMULA

LEMMA VII.9.1. Let g be a real-valued function which is $p+1$ times differentiable on the open interval $]a, b[$. If f is a $p+1$ times differentiable mapping of the interval $]a, b[$ into a Banach space X then

$$f \cdot g^{(p+1)} - (-1)^{p+1} \cdot f^{(p+1)} \cdot g$$

$$= \frac{d}{dt} \left[\sum_{k=1}^{p+1} (-1)^{k-1} f^{(k-1)} \cdot g^{(p+1-k)} \right].$$

PROOF. We have

$$\frac{d}{dt} \left[\sum_{k=1}^{p+1} (-1)^{k-1} f^{(k-1)} \cdot g^{(p+1-k)} \right]$$

$$= \sum_{k=1}^{p+1} (-1)^{k-1} f^{(k)} \cdot g^{(p+1-k)} + \sum_{k=1}^{p+1} (-1)^{k-1} f^{(k-1)} \cdot g^{(p+2-k)}$$

$$= \sum_{k=1}^{p+1} (-1)^{k-1} f^{(k)} \cdot g^{(p+1-k)} + \sum_{k=0}^{p} (-1)^{k} \cdot f^{(k)} \cdot g^{(p+1-k)}$$

$$= (-1)^{p} f^{(p+1)} \cdot g + f \cdot g^{(p+1)} = f \cdot g^{(p+1)} - (-1)^{p+1} f^{(p+1)} \cdot g. \quad \square$$

We now give the Taylor formula for vector-valued functions.

THEOREM VII.9.2. Let f be a mapping of the open interval $]a, b[$ into a Banach space X. If f is continuously $p+1$ times differentiable on $]a, b[$, then for $t, t_0 \in]a, b[$ we have

$$f(t) = \sum_{k=0}^{p} f^{(k)}(t_0) \frac{(t-t_0)^k}{k!} + \int_{t_0}^{t} f^{(p+1)}(s) \frac{(t-s)^p}{p!} ds.$$

PROOF. Suppose that

$$g(s) := \frac{(t-s)^p}{p!}.$$

We invoke Lemma VII.9.1. Since $g^{(p+1)}(s) \equiv 0$, we obtain
$$s$$

$$f^{(p+1)}(s) \cdot \frac{(t-s)^p}{p!} = \frac{d}{ds} \left(\sum_{k=0}^{p} f^{(k)}(s) \frac{(t-s)^k}{k!} \right).$$

Integration of both sides of this equation between t_0 and t yields

$$f(t) = \sum_{k=0}^{p} f^{(k)}(t_0) \frac{(t-t_0)^k}{k!} + \int_{t_0}^{t} f^{(p+1)}(s) \frac{(t-s)^p}{p!} ds. \quad \square$$

Remark. In this section we integrate vector-valued functions. This topic will be discussed at greater length in Chapter IX. Here we note only that if $Y = R^n$ and $k: [a, b] \ni t \rightarrow k(t) = (k_1(t), ..., k_n(t)) \in R^n$, then

$$\int_{a}^{b} k(t)dt = \left(\int_{a}^{b} k_1(t)dt, ..., \int_{a}^{b} k_n(t)dt \right) \in R^n.$$

THEOREM VII.9.3 (The Taylor Formula). Let T be a mapping of an open set U of a Banach space X into a Banach space Y. Let T be continuously $p+1$ times differentiable on U. If the interval $[x, x+h]$ is contained in the set U, the following formula holds:

$$T(x+h) = T(x) + \frac{1}{1!} T'(x) \cdot h + \frac{1}{2!} T''(x) \cdot (h, h) +$$

$$+ ... + \frac{1}{p!} T^{(p)}(x) \cdot h^{(p)} + R_{p+1}(x, h),$$

where $h^{(m)} = (h, ..., h)$ (m times) and

$$R_{p+1}(x, h) = \left(\int_{0}^{1} \frac{(1-s)^p}{p!} T^{(p+1)}(x+s \cdot h)ds \right) \cdot h^{(p+1)}.$$

Furthermore, for every $\varepsilon > 0$ there exists an r such that for $\|h\| < r$ we have

$$\left\| T(x+h) - T(x) - \frac{1}{1!} T'(x) \cdot h - \ldots - \right.$$
$$\left. - \frac{1}{(p+1)!} T^{(p+1)}(x) \cdot h^{(p+1)} \right\| \leqslant \varepsilon \|h\|^{p+1}.$$

PROOF. Since U is an open set, there exists a $\delta > 0$ such that U also contains the interval $]x - \delta h, x + (1+\delta)h[$ along with the interval $[x, x+h]$.

We introduce the auxiliary function:
$$]-\delta, 1+\delta[\ni t \rightarrow f(t) := T(x+t \cdot h) \in Y.$$
This function is continuously $p+1$ times differentiable and

$$f'(t) = T'(x+t \cdot h) \cdot h,$$
$$f''(t) = T''(x+t \cdot h) \cdot (h, h),$$
$$\cdots\cdots\cdots\cdots\cdots\cdots\cdots\cdots\cdots$$
$$f^{(p+1)}(t) = T^{(p+1)}(x+t \cdot h) \cdot h^{(p+1)}.$$

Application of Theorem VII.9.2 to f gives us

$$f(t) = f(0) + \frac{t}{1!} f'(0) + \frac{t^2}{2!} f''(0) + \ldots + \frac{t^p}{p!} f^{(p)}(0) +$$
$$+ \int_0^t \frac{(t-s)^p}{p!} f^{(p+1)}(s) \, ds.$$

On putting $t = 1$, we get

$$T(x+h) = T(x) + \frac{1}{1!} T'(x) \cdot h + \frac{1}{2!} T''(x) \cdot (h, h) +$$
$$+ \ldots + \frac{1}{p!} T^{(p)}(x) \cdot h^{(p)} + \int_0^1 \frac{(1-s)^p}{p!} T^{(p+1)}(x+s \cdot h) \cdot h^{(p+1)} ds.$$

However, since (cf. Chapter IX)

$$\int_0^1 \frac{(1-s)^p}{p!} T^{(p+1)}(x+s \cdot h) \cdot h^{(p+1)} ds$$
$$= \left(\int_0^1 \frac{(1-s)^p}{p!} T^{(p+1)}(x+s \cdot h) \, ds \right) \cdot h^{(p+1)},$$

the first part of the theorem has been proved.

205

We now go on to prove the second part of the theorem.

$$\left\| T(x+h) - T(x) - \frac{1}{1!} T'(x) \cdot h - \ldots - \frac{1}{p!} T^{(p)}(x) \cdot h^{(p)} - \right.$$

$$\left. - \frac{1}{(p+1)!} T^{(p+1)}(x) \cdot h^{(p+1)} \right\|$$

$$= \left\| \int_0^1 \frac{(1-s)^p}{p!} T^{(p+1)}(x+s \cdot h) \cdot h^{(p+1)} ds - \right.$$

$$\left. - \frac{1}{(p+1)!} T^{(p+1)}(x) \cdot h^{(p+1)} \right\|$$

$$= \left\| \int_0^1 \frac{(1-s)^p}{p!} \left(T^{(p+1)}(x+s \cdot h) \cdot h^{(p+1)} - T^{(p+1)}(x) \cdot h^{(p+1)} \right) ds \right\|$$

$$\leqslant \int_0^1 \frac{(1-s)^p}{p!} ds \sup_{0 \leqslant s \leqslant 1} \| T^{(p+1)}(x+s \cdot h) - T^{(p+1)}(x) \| \cdot \| h \|^{p+1}$$

$$= \frac{1}{(p+1)!} \sup_{0 \leqslant s \leqslant 1} \| T^{(p+1)}(x+s \cdot h) - T^{(p+1)}(x) \| \cdot \| h \|^{p+1}.$$

The continuity of $T^{(p+1)}$ on U implies that there exists an $r > 0$ such that

$$\| T^{(p+1)}(x+s \cdot h) - T^{(p+1)}(x) \| \leqslant (p+1)! \cdot \varepsilon$$

for $0 \leqslant s \leqslant 1$, $\| h \| < r$.

Finally, we obtain

$$\left\| T(x+h) - T(x) - \frac{1}{1!} T'(x) \cdot h - \right.$$

$$\left. - \ldots - \frac{1}{(p+1)!} T^{(p+1)}(x) \cdot h^{(p+1)} \right\| \leqslant \varepsilon \| h \|^{p+1}$$

for $\| h \| < r$. $\square$

For the remainder in the Taylor formula we have

$$\| R_{p+1}(x, h) \| = \left\| \left(\int_0^1 \frac{(1-s)^p}{p!} T^{(p+1)}(x+s \cdot h) ds \right) \cdot h^{(p+1)} \right\|$$

$$\leqslant \int_0^1 \frac{(1-s)^p}{p!} ds \cdot \sup_{0 \leqslant s \leqslant 1} \|T^{(p+1)}(x+s \cdot h)\| \cdot \|h\|^{p+1}.$$

The fact that the derivative of order $p+1$ is continuous yields

$$\sup_{0 \leqslant s \leqslant 1} \|T^{(p+1)}(x+s \cdot h)\| = C < \infty.$$

Finally, therefore,

$$\|R_{p+1}(x, h)\| \leqslant C/(p+1)! \, \|h\|^{p+1}.$$

In other words, the remainder after $p+1$ terms is an infinitesimal of order $p+1$.

The value of the k-th derivative of the mapping T for an increment h is called the *k-th differential of the mapping T*:

$$d^k T(x, h) := T^{(k)}(x) \underbrace{(h, \ldots, h)}_{k \text{ times}}.$$

Let us evaluate the k-th differential for the mapping $T: R^n \to Y$. Let $e_1, \ldots, e_n$ be a canonical basis in R^n. For $i = 1, 2, \ldots, k$ we have $h = \sum_{j_i=1}^n h_{j_i} \cdot e_{j_i}$ and

$$T^{(k)}(x) \underbrace{(h, \ldots, h)}_{k \text{ times}} = \sum_{j_1, \ldots, j_k=1}^n h_{j_k} h_{j_{k-1}} \ldots h_{j_1} T^{(k)}(x)(e_{j_k}, \ldots, e_{j_1}).$$

We know, however, that

$$T^{(k)}(x)(e_k, \ldots, e_1) = \frac{\partial^k T(x)}{\partial x_k \ldots \partial x_1}.$$

Consequently we have

$$d^k T(x, h) = \sum_{j_1, \ldots, j_k=1}^n h_{j_k} \ldots h_{j_1} \frac{\partial^k T(x)}{\partial x_{j_1} \ldots \partial x_{j_k}}.$$

Example. If $n = k = 2$, then

$$d^2 T(x, h) = \frac{\partial^2 T(x)}{\partial x_1^2} h_1^2 + 2 \frac{\partial^2 T(x)}{\partial x_1 \partial x_2} h_1 \cdot h_2 + \frac{\partial^2 T(x)}{\partial x_2^2} h_2^2.$$

In general, when $n = 2$,

$$d^k T(x, h) = \sum_{r=0}^k \binom{k}{r} \frac{\partial^k T(x)}{\partial x_1^r \partial x_2^{k-r}} h_1^r \cdot h_2^{k-r}.$$

It is sometimes convenient to use a symbolic formula of the form

$$d^k T(x_0, h) = \left(h_1 \frac{\partial}{\partial x_1} + h_2 \frac{\partial}{\partial x_2} + \ldots + h_n \frac{\partial}{\partial x_n} \right)^k T(x_0),$$

which signifies that we expand the expression in parentheses as if it were an algebraic sum, and then apply the resultant differential operators to T, taking the derivatives at the point x_0.

10. WEAK DERIVATIVES (GATEAUX DERIVATIVES)

In Section 3 we introduced the concept of a directional derivative of a mapping. If T is a mapping of an open subset U of a Banach space X into a Banach space Y, a *directional derivative* (*in the direction e*) *of the mapping T at a point* $x_0 \in U$ is what we call the expression $\nabla_e T(x_0)$:

$$\nabla_e T(x_0) := \lim_{t \to 0} \frac{T(x_0 + t \cdot e) - T(x_0)}{t} .$$

This may be rewritten as

$$\nabla_e T(x_0) = \frac{d}{dt} T(x_0 + t \cdot e)|_{t=0} .$$

We also know that if the mapping T has a directional derivative in every direction at the point x_0, then the mapping

$$X \ni e \to \nabla_e T(x_0) \in Y$$

is not, in general, linear. An interesting case is that when this mapping is linear and continuous. We then have the following definition:

DEFINITION. A mapping T is said to be *weakly differentiable at a point* x_0 if it has the directional derivative in every direction at that point and if the mapping

$$X \ni e \to \nabla_e T(x_0) \in Y$$

is linear and continuous on X. This mapping is denoted by the symbol $\nabla T(x_0)$ and is called a *weak derivative of the mapping T at the point* x_0. Clearly, $\nabla T(x_0) \in L(X, Y)$. We also have the formula

$$\nabla_e T(x_0) = \nabla T(x_0) \cdot e.$$

The weak derivative of a mapping is also called a *Gateaux derivative*, and a weakly differentiable mapping is referred to as a *Gateaux differentiable mapping*.

208

LEMMA VII.10.1. A mapping T which is strongly differentiable at a point x_0 is also weakly differentiable at that point and $\nabla T(x_0) = T'(x_0)$.

PROOF. Theorem VII.3.2 implies the existence of the directional derivative in every direction and the formula $\nabla_e T(x_0) = T'(x_0) \cdot e$. Hence $\nabla T(x_0) = T'(x_0)$. $\square$

Example. Now we give an example of a function which is weakly differentiable but does not have the Fréchet derivative. Let $f: R^2 \to R^1$,

$$f(x_1, x_2) := \begin{cases} 0 & \text{for } (x_1, x_2) = (0, 0), \\ x_1 + x_2 + \dfrac{x_1^3 \cdot x_2}{x_1^4 + x_2^2} & \text{for } (x_1, x_2) \neq (0, 0). \end{cases}$$

Let $h = (h_1, h_2)$; then

$$\nabla_h f(0) = \lim_{t \to 0} \frac{f(t \cdot h_1, t \cdot h_2) - f(0, 0)}{t}$$

$$= \lim_{t \to 0} \left(\frac{t \cdot h_1 + t \cdot h_2}{t} + \frac{t^4 \cdot h_1^3 \cdot h_2}{t^4 \cdot h_1^4 + t^2 \cdot h_2^2} \right) = h_1 + h_2.$$

A directional derivative thus exists in every direction and is continuous with respect to h. The function f consequently is weakly differentiable at the point $x_0 = (0, 0)$ and $\nabla f(0) = (1, 1)$.

We shall demonstrate that the function f is not strongly differentiable at the point $(0, 0)$. For if the strong derivative existed, then $f'(0) = \nabla f(0)$. We evaluate the remainder:

$$\frac{\|r(h_1, h_2)\|}{\|(h_1, h_2)\|} = \frac{\|f(h_1, h_2) - f(0, 0) - h_1 - h_2\|}{\|(h_1, h_2)\|}$$

$$= \left| \frac{h_1^3 h_2}{(h_1^4 + h_2^2)(h_1^2 + h_2^2)^{1/2}} \right|.$$

Running along the parabola $h_2 = h_1^2$, we have

$$\frac{\|r(h_1, h_2)\|}{\|(h_1, h_2)\|} = \frac{|h_1^5|}{2h_1^4(h_1^2 + h_1^4)^{1/2}} = \frac{1}{2(1 + h_1^2)^{1/2}} \xrightarrow[h_1 \to 0]{} \frac{1}{2}.$$

The function f thus is not strongly differentiable at the point $x_0 = (0, 0)$.

Now we give several theorems concerning the weak differentiability.

LEMMA VII.10.2. Let X be a Banach space and let T be a mapping of an open subset U of the space X into another Banach space Y. If at every point of U the mapping T has a directional derivative in an arbitrary direction and if for every $h \in X$ the mapping $X \supset U \ni x \to \nabla_h T(x) \in Y$

is continuous on U, then for every $x \in U$ the mapping $X \ni h \to \nabla_h T(x)$ $\in Y$ is linear.

PROOF. We fix $x_0 \in U$ and we then prove:

(a) Homogeneity:

$$\nabla_{a \cdot h} T(x_0) = \lim_{t \to 0} \frac{T(x_0 + ta \cdot h) - T(x_0)}{t} .$$

Setting $at = s$, we get

$$\nabla_{a \cdot h} T(x_0) = \lim_{s \to 0} \frac{T(x_0 + s \cdot h) - T(x_0)}{s/a} = a \cdot \nabla_h T(x_0).$$

(b) Additivity:

Remark. In the sequel we shall use some formulae which will be proved in Section 1 of Chapter IX.

By virtue of the continuity of the directional derivative we may write

(1)
$$\left\| \frac{T((x_0 + t \cdot (h_1 + h_2)) - T(x_0 + t \cdot h_1)}{t} - \nabla_{h_2} T(x_0) \right\|$$

$$= \left\| \frac{1}{t} \int_0^t (\nabla_{h_2} T(x_0 + t \cdot h_1 + s \cdot h_2) - \nabla_{h_2} T(x_0)) ds \right\|$$

$$\leqslant \frac{1}{|t|} |t| \sup_{\|u\| \leqslant |t| \|h_2\|} \|\nabla_{h_2} T(x_0 + t \cdot h_1 + u) - \nabla_{h_2} T(x_0)\| < \varepsilon$$

for $|t|(\|h_1\| + \|h_2\|) < \delta$.

On the other hand,

(2)
$$\left\| \frac{T(x_0 + t \cdot h_1) - T(x_0)}{t} - \nabla_{h_1} T(x_0) \right\|$$

$$= \frac{1}{|t|} \left\| \int_0^t (\nabla_{h_1} T(x_0 + s \cdot h_1) - \nabla_{h_1} T(x_0)) ds \right\|$$

$$\leqslant \sup_{\|u\| \leqslant |t| \|h_1\|} \|\nabla_{h_1} T(x_0 + u) - \nabla_{h_1} T(x_0)\| < \varepsilon$$

for $|t| \|h_1\| < \delta$.

210

Addition of both sides of (1) and (2) yields

$$\left\| \frac{T(x_0 + t \cdot (h_1 + h_2)) - T(x_0)}{t} - \nabla_{h_1} T(x_0) - \nabla_{h_2} T(x_0) \right\| < 2\varepsilon$$

for $|t| < \delta/(\|h_1\| + \|h_2\|)$. $\square$

THEOREM VII.10.3. Let T be a mapping of an open subset U of a Banach space X into another Banach space Y. If T is weakly differentiable on U and if the mapping

$$X \supset U \ni x \to \nabla T(x) \in L(X, Y)$$

is continuous on U, then T is strongly differentiable on U and

$$\nabla T(x) = T'(x) \quad \text{for } x \in U.$$

PROOF. We invoke the formula (cf. Section IX.1)

$$(3) \qquad \int_0^1 \nabla_h T(x + s \cdot h) ds = T(x + h) - T(x).$$

The continuity of the weak derivative on U implies that for a fixed $x_0 \in U$

$$(4) \qquad \|\nabla T(x_0 + u) - \nabla T(x_0)\| < \varepsilon, \quad \text{provided that } \|u\| < \delta.$$

Since the weak derivative is a continuous linear mapping, it is sufficient to inspect the remainder in order to prove that T is strongly differentiable.

When we make use of (3) and (4) the result is

$$\|r(x_0; h)\| = \|T(x_0 + h) - T(x_0) - \nabla T(x_0) \cdot h\|$$
$$= \|T(x_0 + h) - T(x_0) - \nabla_h T(x_0)\|$$
$$= \left\| \int_0^1 (\nabla_h T(x_0 + s \cdot h) - \nabla_h T(x_0)) ds \right\|$$
$$\leqslant \int_0^1 \|\nabla_h T(x_0 + s \cdot h) - \nabla_h T(x_0)\| ds$$
$$\leqslant \sup_{0 \leqslant s \leqslant 1} \|\nabla T(x_0 + s \cdot h) - \nabla T(x_0)\| \cdot \|h\| \leqslant \varepsilon \|h\|$$

for $\|h\| < \delta$. $\square$

In the case of a finite-dimensional space, the hypotheses of Theorem VII.10.3 may be weakened significantly.

THEOREM VII.10.4. Let X be a finite-dimensional Banach space and let T be a mapping of an open subset U of a space X into a Banach space Y. If at every point of U the mapping T has a directional derivative in an arbitrary direction and if for every $h \in X$ the mapping

$$X \supset U \ni x \to \nabla_h T(x) \in Y$$

is continuous on U, then:

(i) the mapping T is weakly differentiable on U (i.e. the Gateaux derivative exists);

(ii) the mapping $X \supset U \ni x \to \nabla T(x) \in L(X, Y)$ is continuous on U;

(iii) the mapping T is strongly differentiable on U and

$$T'(x) = \nabla T(x) \quad \text{for } x \in U.$$

PROOF. (i) By Lemma VII.10.2 we know that the mapping $X \ni h \to \nabla_h T(x)$ is linear. But every linear mapping in a finite-dimensional Banach space is continuous (cf. Section 2). Therefore the weak derivative exists and $\nabla T(x) \cdot h = \nabla_h T(x)$.

(ii) Let $e_1, e_2, \ldots, e_n$ be a basis in X. For any $h \in X$ we have $h = \sum_{i=1}^{n} h^i e_i$. Accordingly,

$$\|\nabla T(x_1) \cdot h - \nabla T(x_0) \cdot h\| \leqslant \sum_{i=1}^{n} |h^i| \cdot \|\nabla T(x_1) \cdot e_i - \nabla T(x_0) \cdot e_i\|.$$

By virtue of the continuity of the mapping $x \to \nabla T(x) \cdot e_i \in Y$, $i = 1, 2, \ldots, n$, there exist $\delta_i > 0$ such that

(5) $\qquad \|\nabla T(x_1) \cdot e_i - \nabla T(x_0) \cdot e_i\| < \varepsilon \quad \text{for } \|x_1 - x_0\| < \delta_i.$

On putting $\delta = \min_i \delta_i$, we have

(6) $\qquad \|\nabla T(x_1) \cdot h - \nabla T(x_0) \cdot h\| \leqslant \sum_{i=1}^{n} |h_i| \cdot \varepsilon \quad \text{for } \|x_1 - x_0\| < \delta.$

The formula $X \ni h \to \sum_{i=1}^{n} |h_i| \in R^1$ defines in X a norm which is equivalent to the original norm (cf. Section 2). Hence, there exist $a > 0$ such that

(7) $\qquad \sum_{i=1}^{n} |h_i| \leqslant a\|h\|.$

212

Substituting (7) into (6), we get

$$||\nabla T(x_1) \cdot h - \nabla T(x_0) \cdot h|| \leqslant a\varepsilon ||h|| \quad \text{for } ||x_1 - x_0|| < \delta.$$

Next we take the upper bounds of both sides with respect to $||h|| \leqslant 1$ and we have

$$||\nabla T(x_1) - \nabla T(x_0)||_{L(X,Y)} = \sup_{||h|| \leqslant 1} ||\nabla T(x_1) \cdot h - \nabla T(x_0) \cdot h||_Y \leqslant a\varepsilon$$

for $||x_1 - x_0|| < \delta$.

Item (iii) follows from (ii) and Theorem VII.10.3. $\square$

The concept of a Gateaux derivative of the second, third, and higher order may also be introduced.

DEFINITION. Let T be a mapping of an open subset U of a Banach space X into a Banach space Y. Suppose that T has a Gateaux derivative in a neighbourhood $V \subset U$ of a point $x_0 \in U$. The mapping T is said to be *twice weakly differentiable at* x_0 if the mapping $X \supset V \ni x \rightarrow \nabla T(x) \in L(X, Y)$ has the Gateaux derivative at x_0. The Gateaux derivative of this mapping at x_0 is called the *second Gateaux derivative of the mapping T at the point* x_0 and is denoted by the symbol $\nabla^2 T(x_0)$. Plainly, $\nabla^2 T(x_0) \in L\big((X, L(X, Y)\big) \cong L(X, X; Y) = L_2(X; Y)$.

If T has a second Gateaux derivative at every point of the set V, then the mapping

$$X \supset V \ni x \rightarrow \nabla T(x) \in L(X, Y)$$

is called the *Gateaux derivative* (or *weak derivative*) *of the mapping T,* and the mapping

$$X \supset V \ni x \rightarrow \nabla^2 T(x) \in L_2(X; Y)$$

is called the *second Gateaux derivative of the mapping T.*

A Gateaux derivative of order p can be defined in similar fashion. A mapping T is said to have a *Gateaux derivative of order p* at a point $x_0 \in U$ if it has the Gateaux derivative of order $p-1$ in some neighbourhood V of the point x_0 and if the mapping $X \supset V \ni x \rightarrow \nabla^{p-1} T(x) \in L_{p-1}(X; Y)$ is weakly differentiable at x_0. The Gateaux derivative of this mapping at x_0 is called the *Gateaux derivative of order p of the mapping T at the point* x_0. It is denoted by the symbol $\nabla^p T(x_0)$; $\nabla^p T(x_0) \in L_p(X; Y)$.

If at every point of the set V the mapping T has a Gateaux derivative of order p, then T is said to be *p times weakly differentiable on V,* and the

mapping $X \supset V \ni x \to \nabla^p T(x) \in L_p(X; Y)$ is called a *Gateaux deriv-*
ative of order p of the mapping T.

In Section 8, as we recall, we defined a directional derivative of
order p,

$$\nabla_{h_p} \nabla_{h_{p-1}} \cdots \nabla_{h_1} T(x) := \frac{\partial^p}{\partial t_p \cdots \partial t_1} T\left(x + \sum_{i=1}^{p} t_i \cdot h_i\right)\bigg|_{t_1 = \ldots = t_p = 0}.$$

LEMMA VII.10.5. If a mapping T has a weak derivative of order p at
a point x_0, then a directional derivative of order p exists in arbitrary
directions and

$$\nabla_{h_p} \nabla_{h_{p-1}} \cdots \nabla_{h_1} T(x_0) = \nabla^p T(x_0)(h_p, h_{p-1}, \ldots, h_1).$$

PROOF. For the proof we apply induction with respect to p. The
lemma is valid for $p = 1$. Suppose that the thesis holds for $p-1$. We
then have

$$\nabla_{h_p} \cdots \nabla_{h_1} T(x_0)$$

$$= \lim_{t \to 0} \frac{\nabla_{h_{p-1}} \cdots \nabla_{h_1} T(x_0 + t \cdot h_p) - \nabla_{h_{p-1}} \cdots \nabla_{h_1} T(x_0)}{t}$$

$$= \lim_{t \to 0} \frac{1}{t} [\nabla^{p-1} T(x_0 + t \cdot h_p)(h_{p-1}, \ldots, h_1) -$$

$$- \nabla^{p-1} T(x_0)(h_{p-1}, \ldots, h_1)]$$

$$= \left(\lim_{t \to 0} \frac{\nabla^{p-1} T(x_0 + t \cdot h_p) - \nabla^{p-1} T(x_0)}{t}\right)(h_{p-1}, \ldots, h_1)$$

$$= (\nabla \cdot \nabla^{p-1} T(x_0) \cdot h_p)(h_{p-1}, \ldots, h_1)$$

$$= \nabla^p T(x_0)(h_p, h_{p-1}, \ldots, h_1). \quad \square$$

The analogue of Theorem VII.10.3 follows:

THEOREM VII.10.6. Let T be a mapping of an open subset U of a Banach space X into a Banach space Y. If T is p times weakly differentiable
on U and if the mapping $X \supset U \ni x \to \nabla^p T(x) \in L_p(X; Y)$ is continuous
on U, then the mapping T is strongly differentiable on U and $\nabla^p T(x)$
$= T^{(p)}(x)$, $x \in U$.

PROOF. Since the mapping

$$x \to \nabla^p T(x) \in L_p(X; Y)$$

is continuous on U, the mapping

$$x \to \nabla(\nabla^{p-1}T)(x) \in L(X, L_{p-1}(X; Y))$$

is also continuous on U. Therefore, by virtue of Theorem VII.10.3, the mapping

$$U \ni x \to \nabla^{p-1}T(x) \in L_{p-1}(X; Y)$$

is strongly differentiable on U and hence is continuous on U. We can therefore apply the induction method. For we have shown that if the Gateaux derivative of order p is continuous on U, the derivative of order $p-1$ is also continuous on U and strongly differentiable. For $p = 1$ the theorem is true by virtue of Theorem VII.10.3. Employing induction, we obtain the thesis. $\square$

The counterpart of Theorem VII.10.4 can also be given:

THEOREM VII.10.7. Let X be a finite-dimensional Banach space and let T be a mapping of an open subset U of the space X into a Banach space Y. If at every point of the set U the mapping T has a directional derivative of order p in arbitrary directions and if for every set $(h_1, h_2, ..., h_p) \subset X$ the mapping

$$X \supset U \ni x \to \nabla_{h_p} ... \nabla_{h_1}T(x) \in Y$$

is continuous on U, then

(i) the mapping T is p times weakly differentiable on U,

(ii) the mappings

$$X \supset U \ni x \to \nabla^k T(x) \in L_k(X; Y), \quad k = 1, 2, ..., p,$$

are continuous on U,

(iii) the mapping T is p times strongly differentiable on U and

$$\nabla^k T(x) = T^{(k)}(x), \quad k = 1, 2, ..., p.$$

PROOF. For $p = 1$ the thesis follows from Theorem VII.10.4. For higher p, we carry out the proof by using the induction method. $\square$

11. EXTREMA OF FUNCTIONS AND FUNCTIONALS

DEFINITION. Let X be a metric space. A function $f: X \to R^1$ is said to have a *local minimum (maximum)* at a point $x_0 \in X$ if there exists an $r > 0$ such that for every $x \in K(x_0, r)$

$$f(x) \geqslant f(x_0) \quad (f(x) \leqslant f(x_0)).$$

THEOREM VII.11.1. Let f be a function differentiable on an open set U of a Banach space X. If f has a local minimum (maximum) at a point $x_0 \in U$, then $f'(x_0) = 0$.

PROOF. We introduce an auxiliary function

$$[-1, 1] \ni t \to g_h(t) := f(x_0 + t \cdot h).$$

It is easily seen that

$$g_h'(t) = f'(x_0 + t \cdot h) \cdot h,$$

whereby

(1) $\qquad g_h'(0) = f'(x_0) \cdot h.$

The function g_h has a local minimum (maximum) at the point $t = 0$. Thus, by virtue of Theorem V.3.1

(2) $\qquad g_h'(0) = 0.$

From (1) and (2) we obtain

$$f'(x_0) \cdot h = 0.$$

Taking the family of functions (g_h), $h \in V$, where V is an appropiately small neighbourhood of zero at X, we have

$$f'(x_0) \cdot h = 0, \quad h \in V, \quad \text{and hence } f'(x_0) = 0. \quad \square$$

The theorem above is a necessary condition for the existence of a local extremum. A sufficient condition can also be given.

THEOREM VII.11.2. Let f be a function defined on an open subset U of a Banach space, $2k$ times continuously differentiable on U, and for a certain $x_0 \in U$ let

$$f^{(j)}(x_0) = 0, \quad j = 1, 2, \ldots, 2k-1.$$

If there exists a $c > 0$ such that

$$d^{2k}f(x_0, h) \geqslant c\|h\|^{2k} \quad (\leqslant -c\|h\|^{2k}),$$

then the function f has a local minimum (maximum) at the point x_0.

PROOF (we carry it out for a minimum). Since $2k-1$ first derivatives vanish, the Taylor formula for the function f takes on the form

(3) $\qquad f(x_0 + h) - f(x_0) = \dfrac{1}{(2k)!} d^{2k}f(x_0, h) + r(x_0; h),$

where

(4) $\qquad |r(x_0; h)| \leqslant \|h\|^{2k}\varepsilon \quad \text{for } \|h\| < \delta.$

216

On the other hand, we have

(5) $\qquad d^{2k}f(x_0, h) \geqslant c||h||^{2k}.$

Addition of both sides of (3) and (5) yields

$$f(x_0 + h) - f(x_0) \geqslant \frac{1}{(2k)!} c||h||^{2k} + r(x_0; h).$$

When we put $\varepsilon = \frac{1}{2}c/(2k)!$ and use (4), then for $||h|| \leqslant \delta_1$ we obtain

$$f(x_0 + h) - f(x_0) \geqslant \left(c - \frac{1}{2}c\right) \cdot \frac{1}{(2k)!} \cdot ||h||^{2k}$$

$$= \frac{1}{2} \cdot \frac{c}{(2k)!} ||h||^{2k} \geqslant 0.$$

This inequality shows that we have a local minimum at the point x_0. $\square$

DEFINITION. A $2k$-linear mapping of the Cartesian product $X \times X \times \ldots \times X$ into R^1 is said to be *positive* (*negative*) *definite* if there exists a $c > 0$ such that

$$Q(h, h, \ldots, h) \geqslant c||h||^{2k} \quad (\leqslant -c||h||^{2k}) \quad \text{for every } h \in X.$$

The mapping Q is called *positive* (*negative*) if

$$Q(h, h, \ldots, h) > 0 \quad (< 0) \quad \text{for } h \neq 0.$$

A positive (negative) definite mapping clearly is positive (negative), but not conversely. The converse theorem is, however, valid for $X = R^n$.

LEMMA VII.11.3. If $X = R^n$, then the $2k$-linear positive (negative) mapping Q is positive (negative) definite.

PROOF. Let

$$Q(h, h, \ldots, h) > 0 \quad \text{for } h \neq 0.$$

Then

$$Q\left(\frac{h}{||h||}, \ldots, \frac{h}{||h||}\right) > 0.$$

In R^n every multilinear mapping is continuous and hence the function $R^n \ni h \to Q(h, h, \ldots, h) \in R^1$ is continuous.

The unit sphere in R^n (i.e. $\{x: ||x|| = 1\}$) is compact and thus the function $h \to Q(h, h, \ldots, h)$ attains its lower bound on it. Hence, there exists an $h_0 \in R^n$, $||h_0|| = 1$, such that

$$Q(h_0, h_0, \ldots, h_0) = c = \inf_{||h|| = 1} Q(h, h, \ldots, h).$$

Therefore $c > 0$ and consequently

$$Q\left(\frac{h}{||h||}, \frac{h}{||h||}, ..., \frac{h}{||h||}\right) \geqslant c > 0 \quad \text{for every } h \in R^n.$$

Finally, we have

$$Q(h, h, ..., h) \geqslant c||h||^{2k}. \quad \square$$

We most frequently investigate the positive definiteness of a second differential of a function in R^n. We know that a second differential is represented by the matrix which elements are

$$a_{ij} = \frac{\partial^2 f}{\partial x_i \partial x_j}.$$

It is known from algebra, however, that for the quadratic form represented by the matrix

$$[a_{ij}] = \begin{bmatrix} a_{11} \cdots a_{1n} \\ \cdots \cdots \cdots \\ a_{n1} \cdots a_{nn} \end{bmatrix}$$

to be positive, it is necessary and sufficient that all determinants d_k $(k = 1, 2, ..., n)$,

$$d_k := \begin{vmatrix} a_{11} \cdots a_{1k} \\ \cdots \cdots \cdots \\ a_{k1} \cdots a_{kk} \end{vmatrix}$$

be positive. For the second differential to be negative it is necessary and sufficient that $(-1)^k d_k > 0$, $k = 1, 2, ..., n$. On the other hand, if we have only $d_k \geqslant 0$ or $(-1)^k d_k \geqslant 0$ $(k = 1, 2, ..., n)$, that is, if there exists a k, $1 \leqslant k \leqslant n$, such that $d = 0$, then the quadratic form is only semidefinite, that is, $Q(h, h) \geqslant 0$ or $Q(h, h) \leqslant 0$, $h \in R$, and there exists an $h_0 \neq 0$ such that $Q(h_0, h_0) = 0$. In this case we cannot determine whether we have a maximum or a minimum, or whether there is no extremum point at all. All that has to be done then is to examine the function directly in a neighbourhood of the given point.

In a case where neither of the two possibilities

$$d_k \geqslant 0 \quad \text{or} \quad (-1)^k d_k \geqslant 0, \quad k = 1, 2, ..., n,$$

occurs, the quadratic form is indefinite, i.e. there exists a vector $h_1 \in R^n$ such that $Q(h_1, h_1) > 0$, there exists a vector $h_0 \in R^n$ such that $Q(h_0, h_0) = 0$, and there exists a vector $h_2 \in R^n$ such that $Q(h_2, h_2) < 0$. The proof of Theorem VII.11.2 shows that we have no extremum point (saddle point) in this case.

Examples. 1. Inspect the extrema of the functions $f: R^2 - \{0\} \to R^1$;

$$f(x, y) := xy\ln(x^2 + y^2).$$

We calculate the partial derivatives:

$$\frac{\partial f}{\partial x}(x, y) = y\ln(x^2 + y^2) + \frac{2x^2 y}{x^2 + y^2},$$

$$\frac{\partial f}{\partial y}(x, y) = x\ln(x^2 + y^2) + \frac{2xy^2}{x^2 + y^2}.$$

From the equations

$$\frac{\partial f}{\partial x}(x, y) = 0, \qquad \frac{\partial f}{\partial y}(x, y) = 0$$

we obtain the following points which are "suspected" of being extrema:

$$(0, 1), (0, -1), (1, 0), (-1, 0), \left(\frac{1}{\sqrt{2e}}, \frac{1}{\sqrt{2e}}\right),$$

$$\left(-\frac{1}{\sqrt{2e}}, \frac{1}{\sqrt{2e}}\right), \left(\frac{1}{\sqrt{2e}}, -\frac{1}{\sqrt{2e}}\right),$$

$$\left(-\frac{1}{\sqrt{2e}}, -\frac{1}{\sqrt{2e}}\right).$$

Next we evaluate the second-order partial derivatives

$$\frac{\partial^2 f}{\partial x^2}(x, y) = \frac{2x^3 y + 6xy^3}{(x^2 + y^2)^2},$$

$$\frac{\partial^2 f}{\partial y^2}(x, y) = \frac{2xy^3 + 6x^3 y}{(x^2 + y^2)^2},$$

$$\frac{\partial^2 f}{\partial x \partial y}(x, y) = \frac{\partial^2 f}{\partial y \partial x}(x, y) = \ln(x^2 + y^2) + \frac{2x^4 + 2y^4}{(x^2 + y^2)^2}.$$

The matrix of the second differential is of the form

(a) for the points $\left(\frac{1}{\sqrt{2e}}, \frac{1}{\sqrt{2e}}\right), \left(-\frac{1}{\sqrt{2e}}, -\frac{1}{\sqrt{2e}}\right)$:

$$\begin{bmatrix} 2 & 0 \\ 0 & 2 \end{bmatrix}; \quad \text{hence } d_1 = 2, d_2 = 4.$$

Thus, we have a local minimum at these points.

(b) for the points $\left(\dfrac{1}{\sqrt{2e}}, -\dfrac{1}{\sqrt{2e}}\right), \left(-\dfrac{1}{\sqrt{2e}}, \dfrac{1}{\sqrt{2e}}\right)$:

$$\begin{bmatrix} -2 & 0 \\ 0 & -2 \end{bmatrix}; \quad \text{hence } d_1 = -2, d_2 = 4.$$

Therefore, we have a local maximum here.

(c) for the points $(0, 1)$, $(0, -1)$, $(1, 0)$, $(-1, 0)$:

$$\begin{bmatrix} 0 & 2 \\ 2 & 0 \end{bmatrix}; \quad \text{hence } d_1 = 0, d_2 = -4.$$

One can check that there is no extremum at these points.

12. EULER–LAGRANGE EQUATIONS

In the calculus of variations one looks for extremal points, critical points, maximal and minimal points of functionals given in integral form. It is one of the oldest and the most difficult parts of analysis—it were mainly attempts to incorporate the calculus of variations into the body of "general analysis" which led to creation of functional analysis.

Let I denote an interval $[a, b]$ and let a funtion $L: I \times R^n \times R^n \to R^1$ of class C^2 be given: we shall denote points of $I \times R^n \times R^n$ by (t, x, p). Consider the Banach space $C^1(I, R^n)$ consisting of continuously differentiable vector fields on I with values in R^n, equipped with the norm

$$\|u\| := \sum_{j=1}^{n} \sup |u_j(I)| + \sum_{j=1}^{n} \sup \left| \frac{du_j}{dt}(I) \right|.$$

We shall look for a necessary condition for an extremum of the functional J on $C^1(I, R^n)$ defined as

$$C^1(I, R^n) \ni u \to J(u)$$

$$:= \int_a^b L\left(t, u_1(t), \ldots, u_n(t), \frac{du_1}{dt}(t), \ldots, \frac{du_n}{dt}(t)\right) dt.$$

THEOREM VII.12.1. Let

$$u \to J(u) = \int_a^b L\left((t, u(t), \frac{du}{dt}(t)\right) dt$$

be defined on $C^1(I, R^n)$. If a function $u \in C^2(I, R^n)$ is an extremal point of J then it satisfies a system of equations, known as *Euler–Lagrange equations*

$$\frac{\partial L}{\partial x_j}\left(t, u(t), \frac{du}{dt}(t)\right) - \frac{d}{dt}\frac{\partial L}{\partial p_j}\left(t, u(t), \frac{du}{dt}(t)\right) = 0$$

for $j = 1, \dots, n$.

PROOF. As we know, a necessary condition for an extremum is that the directional derivative vanishes for all directions. Hence for $\varphi \in C_0^1(I, R^n)$ we have

$$\nabla_\varphi J(u) = \frac{df}{ds}\bigg|_{s=0} = 0,$$

where the function $s \to f(s)$ stands for $J(u+s\varphi)$, i.e.

$$f(s) = J(u+s\varphi) = \int_a^b L(t, u+s\varphi, u'+s\varphi')dt.$$

One checks without difficulty that the hypotheses on differentiation under the integral sign are satisfied and thus

$$\int_a^b \left(\sum_{j=1}^n \frac{\partial L}{\partial x_j}\varphi_j + \sum_{j=1}^n \frac{\partial L}{\partial p_j}\frac{d\varphi_j}{dt}\right)dt = 0.$$

Integrating by parts the second summand and taking into account that φ vanishes in a neighbourhood of the end-points of $[a, b]$ we get

$$\int_a^b \sum_{j=1}^n \left(\frac{\partial L}{\partial x_j} - \frac{d}{dt}\frac{\partial L}{\partial p_j}\right)\varphi_j dt = 0.$$

Now, since φ is arbitrary, the Mean-Value Theorem for integrals gives us the assertion of the theorem. $\square$

13. DIFFERENTIATION ON NON-OPEN SETS

In preceding chapters, in defining the concept of derivative we took an open set U of a Banach space X and decomposed the increment $T(x_0+h) - T(x_0)$ (where $x_0+h \in U$ and $x_0 \subset U$) into a linear part $L_{x_0}h$ and a remainder which was an infinitesimal of order higher than h. It is necessary to take an open set U because the linear mapping L_{x_0} is determined uniquely if its values are known on a neighbourhood of zero. Accordingly, the fact that U is an open set ensures the uniqueness of the derivative (Lemma VII.3.1).

At times, however, it becomes necessary to differentiate on sets which are not open. One might try to define a derivative in that event by taking an arbitrary (in general not open) set V in a Banach space and considering only increments of h such that $x_0 + h \in V$. In accordance with what has been said above, the linear mapping L_{x_0} will then, in general, not be uniquely determined.

Example. Let $V \subset R^2$, $V := \{(x, y) \in R^2 : x = y\}$. Consider the function

$$f(x, y) = x + y.$$

The derivative of this function at a point (x_0, y_0) is a linear mapping of R^2 into R^1, and hence is in the form of a row matrix. We have

$$f(x_0 + h_1, y_0 + h_2) - f(x_0, y_0) = h_1 + h_2,$$

but in order that $(x_0 + h_1, y_0 + h_2) \in V$ and $(x_0, y_0) \in V$, we must have $h_1 = h_2$. It is thus seen that the derivative at the point (x_0, y_0) has the form $(1 + a, 1 - a)$, where a is an arbitrary number. In defining the derivative, therefore, we must choose a particular a.

In the general case, we have the following definition:

DEFINITION. A mapping T of a subset V (in general, not open) of a Banach space X into a Banach space Y is said to be *differentiable on V* if there exists an open set U containing V and a mapping T_1 which is differentiable on U and such that $T_1|V = T$. The derivative of the mapping T_1 at a point $x_0 \in V$ is called the *derivative of the mapping T* at that point.

The foregoing definition does not uniquely determine the derivative of a mapping defined on a non-open set. For in general there are many extensions of a given mapping. The derivative, however, is determined uniquely for the interior of the set V (provided it is not the empty set). The definition above is immediately transferable to derivatives of higher orders.

DEFINITION. A mapping T of a subset V of a Banach space X into a Banach space Y is said to be *p times continuously differentiable on V* if there exists an open set U containing V and a mapping T_1 which is p times continuously differentiable on U and such that $T_1|V = T$. By the *p-th derivative of the mapping T* we mean the p-th derivative of the mapping T_1.

13. DIFFERENTIATION ON NON-OPEN SETS

Let us consider an important example. Suppose that $X = R^1$, and $V = [a, b]$. For $U =]a-\varepsilon, b+\varepsilon[$ we take

$$T_1(x) := \begin{cases} T(x) & \text{for } x \in [a, b], \\ \lim_{h \to +0} \dfrac{T(a+h)-T(a)}{h}(x-a)+T(a) & \\ & \text{for } x \in]a-\varepsilon, a], \\ \lim_{h \to -0} \dfrac{T(b+h)-T(b)}{h}(x-b)+T(b) & \\ & \text{for } x \in [b, b+\varepsilon[. \end{cases}$$

The mapping T_1 thus is differentiable on U if T is differentiable on $]a, b[$ and if the limits written above exist for the difference quotients, the right-hand limit at the point a and the left-hand limit at the point b. These limits are called, respectively, the *right-hand derivative of the mapping T at the point a* and the *left-hand derivative of the mapping T at the point b*. Henceforth in the book, whenever we say that a mapping T is differentiable on $[a, b]$ we shall take it to mean that it is differentiable on $]a, b[$, and that the right- and left-hand derivatives, respectively, exist at the points a and b.

These results may be generalized to derivatives of higher order by defining the suitable one-sided derivatives and may also be generalized to mappings of n-dimensional cubes into R^n.

VIII. THE METHOD OF SUCCESSIVE APPROXIMATIONS
THE LOCAL INVERTIBILITY OF MAPPINGS
CONSTRAINED EXTREMA

1. THE METHOD OF SUCCESSIVE APPROXIMATIONS. THE BANACH PRINCIPLE

Let X be a metric space and let P be a (continuous) mapping of X into itself: $X \ni x \to P(x) \in X$.

DEFINITION. A *fixed point of a mapping* $P: X \to X$ is a point $\tilde{x} \in X$ such that $P(\tilde{x}) = \tilde{x}$.

Clearly, a mapping may have one fixed point, several fixed points, or none at all.

The method of successive approximations is useful for finding the fixed points of a continuous mapping. Take an arbitrary point $x_0 \in X$ and form the sequence

$$x_0, \quad x_1 := P(x_0), \quad x_2 := P(x_1) = P(P(x_0)) = P^2(x_0),$$
$$\ldots, \quad x_n := P(P^{n-1}(x_0)) = P^n(x_0), \quad \ldots$$

It may happen that the sequence (x_n) converges to the element $\tilde{x}$. We shall show that $\tilde{x}$ is then a fixed point of the mapping P. Since $x_n \to \tilde{x}$, and since the mapping P is continuous, we have

$$d(\tilde{x}, x_n) < \varepsilon \quad \text{for } n > N,$$
$$d(P(x_{n-1}), P(\tilde{x})) < \varepsilon \quad \text{for } n > M,$$

whereby

$$d(\tilde{x}, P(\tilde{x})) \leqslant d(\tilde{x}, x_n) + d(x_n, P(\tilde{x}))$$
$$\leqslant d(\tilde{x}, x_n) + d(P(x_{n-1}), P(\tilde{x})) \leqslant 2\varepsilon \quad \text{for } n > N, M,$$

and consequently

$$d(\tilde{x}, P(\tilde{x})) < 2\varepsilon.$$

The arbitrariness of ε implies

$$P(\tilde{x}) = \tilde{x}.$$

It may be, of course, that the sequence (x_n) is not convergent, regardless of how the point x_0 is chosen. The method of successive approximations breaks down in that case. We shall demonstrate, however, that for a class of mappings called contractions this method yields a fixed point which is the only fixed point of that mapping.

DEFINITION. Let (X, d) be a metric space. A mapping $P: X \to X$ is called a *contraction* if there exists a positive number $q < 1$ such that for every pair $x, y \in X$ the following inequality holds

$$d(P(x), P(y)) \leqslant q \cdot d(x, y).$$

THEOREM VIII.1.1 (Banach Principle). Let P be a contraction of a complete metric space X. Then:

(i) for every $x_0 \in X$ the sequence $x_n := P^n(x_0)$ converges to $\tilde{x} \in X$ and $\tilde{x}$ is a fixed point of the mapping P;

(ii) the point $\tilde{x}$ is the only fixed point of the mapping P, that is, if $P(x) = x$ for some $x \in X$, then $x = \tilde{x}$;

(iii) if q is as in the definition, then

$$d(\tilde{x}, x_m) \leqslant \frac{q^m}{1-q} d(x_1, x_0).$$

PROOF. (i) We define the sequence $x_n := P^n(x_0)$. Since P is a contraction, we have

$$d(x_{m+1}, x_m) = d\big(P(x_m), P(x_{m-1})\big) \leqslant q \cdot d(x_m, x_{m-1}).$$

Applying this inequality m times, we get

(1) $\qquad d(x_{m+1}, x_m) \leqslant q^m \cdot d(x_1, x_0).$

Moreover, we have

$$d(x_{m+k}, x_m) \leqslant d(x_{m+k}, x_{m+k-1}) + d(x_{m+k-1}, x_m)$$
$$\leqslant d(x_{m+k}, x_{m+k-1}) + d(x_{m+k-1}, x_{m+k-2}) + d(x_{m+k-2}, x_m).$$

Repeated application of the triangle inequality to the right-hand side of this inequality yields

(2) $\qquad d(x_{m+k}, x_m) \leqslant \sum_{j=0}^{k-1} d(x_{m+j+1}, x_{m+j}).$

Using (1) and (2) and the formula for the sum of terms of a geometric progression, we get

$$d(x_{m+k}, x_m) \leqslant \sum_{j=0}^{k-1} q^{m+j} d(x_1, x_0) = q^m \frac{1-q^k}{1-q} d(x_1, x_0)$$

$$\leqslant \frac{q^m}{1-q} d(x_1, x_0).$$

Therefore (x_n) is a Cauchy sequence and because of the completeness of the space X there exists an $\tilde{x} \in X$ such that $x_n \to \tilde{x}$. Being a contraction as it is, P is at the same time a continuous mapping. By virtue of the foregoing arguments, we have $P(\tilde{x}) = \tilde{x}$.

(ii) If there exists an $x \neq \tilde{x}$ such that $P(x) = x$, then

$$d(x, \tilde{x}) = d\big(P(x), P(\tilde{x})\big) \leqslant q \cdot d(x, \tilde{x}),$$

that is,

$$d(x, \tilde{x}) \leqslant q \cdot d(x, \tilde{x}),$$

whereby either $q \geqslant 1$, which contradicts the hypotheses of the theorem, or $d(x, \tilde{x}) = 0$, which contradicts the assumption that $x \neq \tilde{x}$.

(iii) We already obtained the inequality

$$d(x_{m+k}, x_m) \leqslant q^m \frac{1-q^k}{1-q} d(x_1, x_0).$$

Passing to the limit $k \to \infty$, we obtain

$$d(\tilde{x}, x_m) \leqslant \frac{q^m}{1-q} d(x_1, x_0). \quad \square$$

Example. Let $\overline{K(0, r)}$ be a closed ball in a Banach space X. Let P be a differentiable mapping of that ball into itself and let $\|P'(x)\| \leqslant q$, $q < 1$, $x \in \overline{K(0, r)}$. By the Mean-Value Theorem we have

$$\|P(x) - P(x')\| \leqslant \sup_{z \in [x, x']} \|P'(z)\| \, \|x - x'\| \leqslant q \|x - x'\|.$$

Therefore P is a contraction and has exactly one fixed point.

LEMMA VIII.1.2. Let X be a complete metric space and let A be a metric space. Suppose that $(P_a)_{a \in A}$ is an equicontracting family of mappings of the space X into itself, i.e. there exists a q, $0 \leqslant q < 1$, such that

$$d(P_a(x), P_a(x')) \leqslant q \cdot d(x, x')$$

for every $a \in A$ and any $x, x' \in X$. Suppose, moreover, that for every $x \in X$ the mapping $A \ni a \to P_a(x) \in X$ is continuous on A.

Then for every $a \in A$ there exists a (unique) $\tilde{x}(a) \in X$, such that $P_a(\tilde{x}(a)) = \tilde{x}(a)$ and the mapping $A \ni a \to \tilde{x}(a) \in X$ is continuous on A.

PROOF. The existence and uniqueness of fixed points $\tilde{x}(a)$ follows directly from Theorem VIII.1.1. It is therefore sufficient to prove that the mapping $a \to \tilde{x}(a)$ is continuous.

Take a point a_0 and the corresponding point $\tilde{x}(a_0)$. Let us set up the sequence

$$x_0 := \tilde{x}(a_0), \quad x_1 := P_a(\tilde{x}(a_0)), \quad \ldots, \quad x_n := P_a^n(\tilde{x}(a_0)).$$

It is easily seen that

$$\lim_{n \to \infty} x_n = \tilde{x}(a).$$

On employing item (iii) of Theorem VIII.1.1, we have

$$d\big(\tilde{x}(a), \tilde{x}(a_0)\big) = d\big(\tilde{x}(a), x_0\big) \leqslant \frac{1}{1-q} d(x_1, x_0)$$

$$= \frac{1}{1-q} d\big(P_a(x_0), P_{a_0}(x_0)\big).$$

If use is made of the continuity of the mapping $a \to P_a(x)$ for $x = x_0$, the result is

$$d\big(\tilde{x}(a), \tilde{x}(a_0)\big) \leqslant \frac{1}{1-q}\varepsilon \quad \text{for } d(a, a_0) < \delta. \;\square$$

Now we give a theorem on the solution of a certain equation in a Banach space. This theorem will be used in the next section in proving the theorem on local invertibility.

THEOREM VIII.1.3. Let $\overline{K(0, r)}$ be a closed ball (with centre at 0 and radius r) in a Banach space X. Let S be a mapping of this ball into X such that

(a) $S(0) = 0$,

(b) there exists a $q, 0 \leqslant q < 1$, such that

$$\|S(x) - S(x')\| \leqslant q \cdot \|x - x'\|, \quad x \in K, \; x' \in K.$$

Then the equation

$$S(x) + y = x$$

has exactly one solution for $y \in \overline{K(0, (1-q) \cdot r)}$. Let us denote this solution by the symbol $R(y)$. The solution has the following properties:

(i) $R(0) = 0$,

(ii) $\|R(y) - R(y')\| \leqslant \dfrac{1}{1-q} \|y - y'\|$ for $y, y' \in \overline{K(0, (1-q) \cdot r)}$.

PROOF. Let us define a family of mappings $(P_y), y \in \overline{K(0, (1-q) \cdot r)}$,

$$\overline{K(0, r)} \ni x \to P_y(x) := \big(y + S(x)\big) \in X.$$

We shall show that every P_y maps the ball $\overline{K(0, r)}$ into itself. For we have

$$\|P_y(x)\| = \|y + S(x)\| = \|y + S(x) - S(0)\|$$
$$\leqslant \|y\| + \|S(x) - S(0)\| \leqslant \|y\| + q\|x\|$$
$$\leqslant (1-q) \cdot r + q \cdot r = r.$$

Next we shall demonstrate that the family (P_y), $y \in \overline{K(0, (1-q) \cdot r)}$, satisfies the hypotheses of Lemma VIII.1.2, i.e. is an equicontracting family.

$$\|P_y(x) + P_y(x')\| = \|y + S(x) - y - S(x')\|$$
$$= \|S(x) - S(x')\|$$
$$\leqslant q \cdot \|x - x'\|.$$

By Lemma VIII.1.2, there exists for every P_y a unique fixed point $\tilde{x}$ (y). Denoting $R(y) := \tilde{x}(y)$, we have

$$R(y) = y + S(R(y)).$$

Since $S(0) = 0$, therefore $R(0)$ must be zero. We thus have $R(0) = 0$.

Now we go on to prove property (ii). We have

$$\|R(y) - R(y')\| = \|y + S(R(y)) - y' - S(R(y'))\|$$
$$\leqslant \|y - y'\| + \|S(R(y)) - S(R(y'))\|$$
$$\leqslant \|y - y'\| + q\|R(y) - R(y')\|,$$

whence

$$(1 - q)\|R(y) - R(y')\| \leqslant \|y - y'\|,$$

and consequently

$$\|R(y) - R(y')\| \leqslant \frac{1}{1-q}\|y - y'\|. \quad \square$$

Example. Let A be a continuous linear mapping of a Banach space X into itself. In other words, $A \in L(X, X)$. Suppose, moreover, that $\|A\| < 1$. We seek the inverse of the mapping $I - A$. Therefore we must solve the equation $(I - A)x = y$ for every $y \in X$, which could be written in the form $Ax + y = x$, considered in Theorem VIII.1.3. Note that the mapping A satisfies the hypotheses of Theorem VIII.1.3. For we have

(a) $A(0) = 0$ (by virtue of linearity),

(b) $\|A(x) - A(x')\| \leqslant \|A\| \cdot \|x - x'\|$.

Thus there exists one (only one) solution

$$x = (I - A)^{-1} \cdot y.$$

The linearity of the mapping $(I - A)^{-1}$ is obvious. Its continuity follows from item (ii) of Theorem VIII.1.3. We have

$$\|(I-A)^{-1} \cdot y - (I-A)^{-1} \cdot y'\| = \|x - x'\| \leqslant \frac{\|y - y'\|}{1 - \|A\|}.$$

Now we shall show that

$$(3) \qquad (I-A)^{-1} = \sum_{k=0}^{\infty} A^k.$$

By the inequality

$$\left\| \sum_{k=0}^{\infty} A^k \right\| \leqslant \sum_{k=0}^{\infty} \|A\|^k = \frac{1}{1 - \|A\|},$$

the right-hand side of equation (3) is well-defined. To verify this equation, it is sufficient to multiply both sides by $I - A$. The evaluation is immediate.

The series on the right-hand side of formula (3) is often called a *Neumann series*.

2. THE LOCAL INVERTIBILITY OF MAPPINGS. THE RANK THEOREM

LEMMA VIII.2.1. Let T be a mapping of an open set U of a Banach space X into a Banach space Y which is continuously differentiable in $K(x, r) \subset U$. If

$$r(x, h) := T(x+h) - T(x) - T'(x) \cdot h,$$

then for every $\varepsilon > 0$ there exists a $\delta > 0$ such that

$$\|r(x, h_1) - r(x, h_2)\| \leqslant \varepsilon \|h_1 - h_2\| \quad \text{for } \|h_1\| < \delta, \|h_2\| < \delta.$$

PROOF. We have

$$r(x, h_1) - r(x, h_2) = T(x+h_1) - T(x+h_2) - T'(x)(h_1 - h_2).$$

The expression $T'(x+h_2)(h_1 - h_2)$ is added to, and subtracted from, the right-hand side of the equation. This gives us

$$r(x, h_1) - r(x, h_2) = T(x+h_2 + (h_1 - h_2)) - T(x+h_2) - \\ - T'(x+h_2)(h_1 - h_2) + \big(T'(x+h_2) - T'(x)\big)(h_1 - h_2),$$

whence

$$\|r(x, h_1) - r(x, h_2)\| \leqslant \|T'(x+h_2) - T'(x)\| \, \|h_1 - h_2\| + \\ + \|T(x+h_2 + (h_1 - h_2)) - T(x+h_2) - T'(x+h_2)(h_1 - h_2)\|.$$

Invoking Lemma VII.5.4, we have

$$\|T(x+h_2+(h_1-h_2))-T(x+h_2)-T'(x+h_2)(h_1-h_2)\|$$
$$\leqslant \|h_1-h_2\| \sup_{v\in[x+h_1,x+h_2]} \|T'(v)-T'(x)\|.$$

The continuity of the derivative in $K(x,r)$ implies that

$$(1) \qquad \sup_{v\in[x+h_1,x+h_2]} \|T'(v)-T'(x)\| < \frac{1}{2}\varepsilon$$

for $\|h_1\| < \delta < r$, $\|h_2\| < \delta < r$. Inequality (1) leads to

$$(2) \qquad \|T'(x+h_2)-T'(x)\| < \frac{1}{2}\varepsilon.$$

Using (1) and (2), we obtain

$$\|r(x,h_1)-r(x,h_2)\| < \varepsilon\|h_1-h_2\|$$

for $\|h_1\| < \delta < r$, $\|h_2\| < \delta < r$. $\square$

We now give a generalization of Theorem III.2.6 to the case of a Banach space.

THEOREM VIII.2.2 (The Inverse Mapping Theorem). Let T be a differentiable mapping of an open set U of a Banach space X into a Banach space Y. If the derivative of T is continuous in a ball with centre at the point $x_0 \in U$ and radius r and if the mapping $T'(x_0)$ has a continuous inverse, then

(i) there exists a neighbourhood U_0 of the point x_0 on which the mapping T is one-to-one with an open image; $(T/U_0)^{-1}$ is continuous on $V = T(U_0)$

$$Y \supset V \ni y \to (T|U_0)^{-1}(y) \in U_0 \subset X,$$

(ii) the mapping $(T|U_0)^{-1}$ is differentiable at the point $y_0 = T(x_0)$ and

$$((T|U_0)^{-1})'(T(x_0)) = (T'(x_0))^{-1}.$$

PROOF. We denote

$$\Delta y := T(x_0+\Delta x)-T(x_0), \qquad \Delta x := x-x_0.$$

From the definition of derivative we have

$$\Delta y = T'(x_0)\cdot\Delta x+r(\Delta x),$$

hence

$$(T'(x_0))^{-1}(\Delta y) = \Delta x+(T'(x_0))^{-1}((r(\Delta x)).$$

233

The problem therefore reduces to one of solving the equation

(3) $S(\Delta x) + z = \Delta x,$

where

$$S(\Delta x) := -\big(T'(x_0)\big)^{-1}\big(r(\Delta x)\big), \quad z := \big(T'(x_0)\big)^{-1}(\Delta y).$$

Let us now take $0 < \varepsilon < \|T'(x_0)^{-1}\|^{-1}$ and the corresponding $\delta > 0$ such that in accordance with Lemma VIII.2.1

(4) $\|r(\Delta x_1) - r(\Delta x_2)\| \leqslant \varepsilon\|\Delta x_1 - \Delta x_2\|$

for $\|\Delta x_1\| \leqslant \delta$, $\|\Delta x_2\| \leqslant \delta$.

We verify that S satisfies the hypotheses of Theorem VIII.1.3. Let $\Delta x \in \overline{K(0, \delta)}$; then

(a) $S(0) = \big(T'(x_0)\big)^{-1}\big(r(0)\big) = 0,$ because $r(0) = 0,$

(b) $\|S(\Delta x_1) - S(\Delta x_2)\| = \|\big(T'(x_0)\big)^{-1}\big(r(\Delta x_1) - r(\Delta x_2)\big)\|$

$\leqslant \|\big(T'(x_0)\big)^{-1}\| \cdot \|r(\Delta x_1) - r(\Delta x_2)\|.$

Using (4), we get

$$\|S(\Delta x_1) - S(\Delta x_2)\| \leqslant \|\big(T'(x_0)\big)^{-1}\| \cdot \|r(\Delta x_1) - r(\Delta x_2)\|$$
$$\leqslant q\|\Delta x_1 - \Delta x_2\|,$$
$$q := \varepsilon\|\big(T'(x_0)\big)^{-1}\| < 1.$$

By virtue of Theorem VIII.1.3 equation (3) thus has exactly one solution which we write in the form

$$\Delta x = R\big(\big(T'(x_0)\big)^{-1}\Delta y\big).$$

This solution exists provided that $\|\big(T'(x_0)\big)^{-1}\Delta y\| \leqslant (1-q)\delta$, that is,

$$\|\Delta y\| \leqslant (1-q)\,\delta\|\big(T'(x_0)\big)^{-1}\|^{-1} =: p.$$

Since $\Delta x = x - x_0$, $\Delta y = y - y_0$, we can write

(5) $x = x_0 + R\big(\big(T'(x_0)\big)^{-1}(y - y_0)\big).$

The mapping $y \rightarrow x$ defined by the foregoing formula is the inverse to a restriction $T|K(x_0, \eta)$ of the mapping T $(\eta = \delta \cdot (1-q)/(1+q))$. Henceforth we shall write $T^{-1} \overset{\text{df}}{=} \big(T|K(x_0, \eta)\big)^{-1}$. Equation (5) and item (ii) of Theorem VIII.1.3 demonstrate that the inverse mapping is continuous on $K(y_0, p)$. We thus have

$$Y \supset K(y_0, p) \ni y \rightarrow T^{-1}(y) = x_0 + R\big(\big(T'(x_0)\big)^{-1}(y - y_0)\big).$$

We shall now show that the inverse mapping is differentiable. We have

(6) $T^{-1}(y_0+\Delta y)-T^{-1}(y_0) = T^{-1}(y)-T^{-1}(y_0) = x-x_0 = \Delta x.$

On the other hand,

(7) $\Delta x = (T'(x_0))^{-1}\Delta y - (T'(x_0))^{-1}(r(\Delta x)).$

From (6) and (7) we have

$$T^{-1}(y_0+\Delta y)-T^{-1}(y_0)$$
$$= (T'(x_0))^{-1}\Delta y - (T'(x_0))^{-1}(r(\Delta x)).$$

We evaluate the remainder

(8) $\|(T'(x_0))^{-1}(r(\Delta x))\| \leqslant \|(T'(x_0))^{-1}\| \cdot \|r(\Delta x)\|.$

Item (ii) of Theorem VIII.1.3 implies that

(9) $\|\Delta x\| = \|R((T'(x_0))^{-1}\Delta y)\| \leqslant (1-q)^{-1}\|(T'(x_0))^{-1}\Delta y\|$

$$\leqslant \frac{1}{1-q}\|(T'(x_0))^{-1}\| \cdot \|\Delta y\|$$

for $\|\Delta y\| \leqslant p$. Using (8) and (9), we obtain

(10) $$\frac{\|(T'(x_0))^{-1}(r(\Delta x))\|}{\|\Delta y\|} \leqslant \frac{\|(T'(x_0))^{-1}\| \cdot \|r(\Delta x)\| \cdot \|\Delta x\|}{\|\Delta y\| \cdot \|\Delta x\|}$$

$$\leqslant \frac{1}{1-q}\|(T'(x_0))^{-1}\|^2 \frac{\|r(\Delta x)\| \cdot \|\Delta y\|}{\|\Delta x\| \cdot \|\Delta y\|}$$

$$= \frac{1}{1-q}\|(T'(x_0))^{-1}\|^2 \frac{\|r(\Delta x)\|}{\|\Delta x\|}.$$

It follows from (9) that $\Delta x \to 0$ as $\Delta y \to 0$, and since $\frac{\|r(\Delta x)\|}{\|\Delta x\|} \xrightarrow[\Delta x \to 0]{} 0$,
therefore by virtue of (10) we have

(11) $$\frac{\|(T'(x_0))^{-1}(r(\Delta x))\|}{\|\Delta y\|} \xrightarrow[\Delta y \to 0]{} 0.$$

Equations (7) and (11) imply that the mapping T^{-1} is differentiable at
the point $y_0 = T(x_0)$ and also imply the formula

$$(T^{-1})'(T(x_0)) = (T'(x_0))^{-1}. \quad \square$$

In the special case when $X = R^n$ and $Y = R^n$, a system of n functions
of n variables

$$T(x) = \sum_{i=1}^{n} t^i(x) \cdot e_i$$

corresponds to the mapping; the vectors $e_1, e_2, ..., e_n$ form a canonical basis of the space R^n. The Jacobi matrix

$$T_j'^i(x) = \frac{\partial t^i(x)}{\partial x^j}, \quad i, j = 1, 2, ..., n, \quad x = (x^1, x^2, ..., x^n)$$

corresponds to the derivative of the mapping T.

Now let us invoke Theorem VIII.2.2. The continuity of the derivative in some neighbourhood of a point x_0 is clearly equivalent to the continuity of all the partial derivatives of the functions t^i in that neighbourhood. On the other hand, the existence of the mapping $(T'(x_0))^{-1}$ is equivalent to the nonvanishing of the determinant of the Jacobi matrix. Thus, if all the (first-order) partial derivatives of the functions t^i, $i = 1, 2, ..., n$, are continuous in some neighbourhood of the point $x_0 = (x_0^1, x_0^2, ..., x_0^n)$ and $\det\left[\dfrac{\partial t^i(x_0)}{\partial x^j}\right] \neq 0$, then the inverse mapping exists in some neighbourhood of the point $y_0 = T(x_0)$.

The matrix of the derivative of the inverse mapping is determined by means of the familiar rules for evaluating an inverse matrix. We thus have

$$(T^{-1})_l'^k(y_0) = \frac{D_k^l(x_0)}{\det\left[\dfrac{\partial t^i(x_0)}{\partial x^j}\right]}, \quad k, l = 1, 2, ..., n,$$

where $D_k^l(x_0)$ is the cofactor of the element $\dfrac{\partial t^i(x_0)}{\partial x^k}$ of the matrix $\left[\dfrac{\partial t^i(x_0)}{\partial x^j}\right]$.

LEMMA VIII.2.3. Let A be a linear mapping of a Banach space X onto a Banach space Y. If there exists an $m > 0$ such that $m||x|| \leqslant ||A \cdot x||$ for every $x \in X$, then there exist $A^{-1} \in L(Y, X)$ and $||A^{-1}|| \leqslant 1/m$.

PROOF. First of all, we shall show that A is one-to-one. Let $x_1, x_2 \in X$ and $A \cdot x_1 = A \cdot x_2$. We then have

$$0 = ||A \cdot (x_1 - x_2)|| \geqslant m||x_1 - x_2||,$$

whereby $||x_1 - x_2|| = 0$, and hence $x_1 = x_2$. We have thus proved the existence of the mapping $A^{-1} \colon Y \to X$. The linearity of this mapping is obvious; we shall demonstrate its continuity. If $y = A \cdot x$ then $x = A^{-1} \cdot y$ and

$$m \cdot ||A^{-1} \cdot y|| = m \cdot ||x|| \leqslant ||A \cdot x|| = ||y||,$$

and therefore

$$\|A^{-1} \cdot y\| \leqslant \frac{1}{m} \|y\|.$$

The mapping A thus is bounded and $\|A^{-1}\| \leqslant 1/m$. $\square$

Now we give without proof an important theorem of functional analysis, the Banach theorem on isomorphism.

THEOREM VIII.2.4 (Banach). Let A be a continuous linear mapping of a Banach space X onto a Banach space Y. If A is one-to-one (i.e. if a linear mapping A^{-1} exists), then A^{-1} is a continuous mapping.

The theorem may also be reformulated as follows:

A continuous algebraic isomorphism of two Banach spaces is at the same time also a topological isomorphism (homeomorphism).

The proof of this theorem is not simple and is based on, among other things, Baire's category theorem. A generalization of the Banach theorem is proved at the end of Part III of this book.

Let us now present some facts concerning the differentiability of mappings of a Banach space into a space of continuous linear mappings.

LEMMA VIII.2.5. Let X, Y, Z, and Q be Banach spaces, let T be a mapping of Q into the space $L(X, Y)$, and let F be a mapping of Q into the space $L(Y, Z)$. If T and F are differentiable at a point $q_0 \in Q$, the mapping G defined by the formula

$$Q \ni q \to G(q) := F(q) \circ T(q) \in L(X, Z)$$

is differentiable at q_0 and the following formula holds:

$$G'(q_0) \cdot h = F'(q_0) \cdot h \circ T(q_0) + F(q_0) \circ T'(q_0) \cdot h, \quad h \in Q.$$

PROOF. We have

$$G(q_0+h) - G(q_0) = F(q_0+h) \circ T(q_0+h) - F(q_0) \circ T(q_0)$$
$$= \big(F(q_0) + F'(q_0) \cdot h + r_1(h)\big) \circ \big(T(q_0) + T'(q_0) \cdot h + r_2(h)\big) -$$
$$- F(q_0) \circ T(q_0)$$
$$= F'(q_0) \cdot h \circ T(q_0) + F(q_0) \circ T'(q_0) \cdot h + \text{remainder}.$$

It is easily shown that the remainder is an infinitesimal of a higher order. $\square$

THEOREM VIII.2.6. Let X, Y, and Z be Banach spaces, and let $H \subset L(X, Y)$ be a set of linear homeomorphisms of the spaces X and Y. If T

237

is a mapping of Z into H, differentiable at a point $z_0 \in Z$ ($Z \ni z \to T(z)$ $\in H \subset L(X, Y)$), the mapping F of Z into the set $H^{-1} := \{L \in L(Y, X):$ $L^{-1} \in H\}$ defined by the formula $Z \ni z \to F(z) := (T(z))^{-1} \in H^{-1}$ $\subset L(Y, X)$ is differentiable at the point z_0 and for $h \in Z$ we have

$$F'(z_0) \cdot h = -(T(z_0))^{-1} \circ T'(z_0) \cdot h \circ (T(z_0))^{-1} \in L(Y, X).$$

PROOF. We know (cf. example to Theorem VIII.1.3) that for $A \in L(X, X)$ such that $\|A\| < 1$ the mapping $(I+A)^{-1}$ exists and we have

$$(I+A)^{-1} = \sum_{k=0}^{\infty} (-1)^k A^k, \quad \|(I+A)^{-1}\| \leqslant \frac{1}{1-\|A\|}.$$

Moreover

$$(I+A)^{-1} - I + A = \sum_{k=2}^{\infty} (-1)^k A^k.$$

Consequently we have

$$(12) \qquad \|(I+A)^{-1} - I + A\| = \left\|\sum_{k=2}^{\infty} (-1)^k A^k\right\| \leqslant \sum_{k=2}^{\infty} \|A\|^k = \frac{\|A\|^2}{1-\|A\|}.$$

We evaluate the derivative of the mapping F at the point z_0:

$$(13) \qquad F(z_0+h) - F(z_0) = (T(z_0+h))^{-1} - (T(z_0))^{-1}$$
$$= (T(z_0) + T'(z_0) \cdot h + r(h))^{-1} - (T(z_0))^{-1}$$
$$= \Big(T(z_0) \circ \big(I + (T(z_0))^{-1} \circ (T'(z_0) \cdot h) +$$
$$+ (T(z_0))^{-1} \circ r(h)\big)\Big)^{-1} - (T(z_0))^{-1}$$
$$= \Big(I + (T(z_0))^{-1} \circ (T'(z_0) \cdot h) +$$
$$+ (T(z_0))^{-1} \circ r(h)\Big)^{-1} \circ (T(z_0))^{-1} - (T(z_0))^{-1}.$$

We denote

$$(14) \qquad A := (T(z_0))^{-1} \circ T'(z_0) \cdot h + (T(z_0))^{-1} \circ r(h).$$

We have

$$(15) \qquad \|A\| \leqslant \|(T(z_0))^{-1}\| \left(\|T'(z_0)\| \cdot \|h\| + \|r(h)\|\right).$$

It follows from (15) that, on taking suitably small δ_1, we have $\|A\| < 1$ for $\|h\| < \delta_1$. By virtue of (13) and (14),

$$(16) \qquad F(z_0+h) - F(z_0) = (I+A)^{-1} \circ (T(z_0))^{-1} - (T(z_0))^{-1}.$$

When we use (16), the result is

(17) $F(z_0+h)-F(z_0)+(T(z_0))^{-1}\circ T'(z_0)\cdot h\cdot (T(z_0))^{-1}$
$= (I+A)^{-1}\circ (T(z_0))^{-1}-(T(z_0))^{-1}+$
$+(T(z_0))^{-1}\circ T'(z_0)\cdot h\cdot (T(z_0))^{-1}$
$= (I-A)\circ (T(z_0))^{-1}-(T(z_0))^{-1}+$
$+((I+A)^{-1}-I+A)\circ (T(z_0))^{-1}+$
$+(T(z_0))^{-1}\circ T'(z_0)\cdot h\cdot (T(z_0))^{-1}$
$= -(T(z_0))^{-1}\circ r(h)\circ (T(z_0))^{-1}+$
$+((I+A)^{-1}-I+A)\cdot (T(z_0))^{-1}.$

From (12) and (17) we have

$\|F(z_0+h)-F(z_0)+(T(z_0))^{-1}\circ T'(z_0)\cdot h\circ (T(z_0))^{-1}\|$
$\leqslant \|(T(z_0))^{-1}\|\cdot \|r(h)\|\cdot \|(T(z_0))^{-1}\|+$
$+\|(T(z_0))^{-1}\|\dfrac{\|A\|^2}{1-\|A\|}.$

Taking suitably small $\delta_1 \geqslant \delta_2 > 0$, we have for $\|h\| < \delta_2$

$\|F(z_0+h)-F(z_0)+(T(z_0))^{-1}\circ T'(z_0)\cdot h\circ (T(z_0))^{-1}\|$
$\leqslant \|h\|\varepsilon.$ $\square$

If the mapping T is p times differentiable at a point $z_0 \in Z$, then Lemma VIII.2.5, the theorem on the derivative of compositions of mappings, and the formula

$$F'(z_0)\cdot h = -(T(z_0))^{-1}\circ T'(z_0)\cdot h\circ (T(z_0))^{-1}$$

imply that F is also p times differentiable at that point.

We shall now demonstrate that if there is at least one linear homeomorphism of the Banach spaces X and Y, the set of such homeomorphisms is an open set in the space $L(X, Y)$.

Suppose that L is that homeomorphism and take the ball

$$K\left(0, \frac{1}{2\|L^{-1}\|}\right) \subset L(X, Y).$$

If $P \in K(0, 1/2\|L^{-1}\|)$, then there is a continuous linear mapping

$$(L+P)^{-1} = (I+L^{-1}\circ P)^{-1}\circ L^{-1}.$$

Thus, a ball with centre at L and radius $1/2\|L^{-1}\|$ in the space $L(X, Y)$ contains only homeomorphisms.

In the special case of Theorem VIII.2.6, when $Z = L(X, Y)$, and the mapping T is the restriction of the identity mapping to the (open) set of homeomorphisms $H \subset L(X, Y)$, the mapping F,

$$L(X, Y) \supset H \ni L \to F(L) = L^{-1} \in L(Y, X),$$

is continuously differentiable infinitely many times on H.

The theorems given above enable us to formulate a modified version of the Inverse Mapping Theorem.

THEOREM VIII.2.7. Let U be an open set of a Banach space X and let T be a mapping of U into a Banach space Y. If T is p times continuously differentiable on U and if $(T'(x))^{-1} \in L(Y, X)$ exists at a point x of the set U, then T is invertible on a neighbourhood U_0 of the point x, and the inverse mapping is p times continuously differentiable on $T(U_0)$.

PROOF. The invertibility of the mapping T and the formula $(T^{-1})'(y) = (T'(T^{-1}(y)))^{-1}$ follow from Theorem VIII.2.2. This formula, the theorem on the derivative of a composition of mappings, Theorem VIII.2.6, and the openness of the set of isomorphisms $L(X, Y)$ imply the existence of all derivatives of order up to and including p. $\square$

DEFINITION. A mapping $T: U \to V \subset Y, U \subset X$ will be called a *diffeomorphism of class C^k* (or *C^k-diffeomorphism*) if it is invertible and if both T and T^{-1} are k times continuously differentiable.

THEOREM VIII.2.8. (The Rank Theorem). If a set $O \subset R^n$ is open, $f \in C^p(O, R^m)$, the rank of the derivative $f'(x)$ being independent of x and equal to $r, r > 0$, then for every $a \in O$ there exist neighbourhoods $U \ni a$, $V \ni b = f(a)$, open cubes $P \subset R^n$, $Q \subset R^m$ and C^k-diffeomorphisms $u: P \to U, w: V \to Q$, such that $g := w \circ f \circ u$ has the form

$$g(x_1, \ldots, x_n) = (x_1, \ldots, x_r, 0, \ldots, 0).$$

PROOF. On carrying out the affine automorphisms, we can assume that $a = 0, b = 0$, and that

$$f'(0): (y_1, \ldots, y_n) \to (y_1, \ldots, y_r, 0, \ldots, 0).$$

Let $v: O \to R^n$ be given by the formula

$$v(x) := (f_1(x), \ldots, f_r(x), x_{r+1}, \ldots, x_n), \quad \text{where } f = (f_i)_{i=1}^m.$$

Then $v'(0) = I_{R^n}$, and consequently by the Theorem VIII.2.7 we know that there exist a $U \subset O$ and an open cube P such that $v|U \to P$ is a diffeomorphism of class C^k. Now we set $u := (v|U)^{-1}$. Plainly

$$f \circ u(t) = (t_1, \ldots, t_r, g_{r+1}(t), \ldots, g_m(t)), \quad \text{where } g_i \in C^k(P).$$

The rank of the matrix $(f \circ u)'(t)$ is equal r for $t \in P$, and hence $\partial g_l / \partial t_s = 0$ for $l, s > r$, that is, the g_j are independent of $t_{r+1}, \ldots, t_m$.

Suppose that $P = P^r \times P^{r-n}$, where $P^r \subset R^r$, $P^{n-r} \subset R^{n-r}$ are open cubes. Now let us define the diffeomorphism $h \in C^k(P^r \times R^{m-r}, P^r \times R^{m-r})$ by the formula

$$h(t_1, \ldots, t_m)$$

$$= (t_1, \ldots, t_r; t_{r+1} - g_{r+1}(t_1, \ldots, t_r), \ldots, t_m - g_m(t_1, \ldots, t_r)).$$

Finally, suppose that Q is a cube in R^m such that

$$h \circ f \circ u(P) \subset Q \subset P^r \times R^{m-r}, \quad V := h^{-1}(Q).$$

On setting $w = h|V$, we get the required relation

$$w \circ f \circ u(x_1, \ldots, x_n) = (x_1, \ldots, x_r, 0, \ldots, 0). \quad \square$$

3. IMPLICIT MAPPINGS

Let us recall (cf. Chapter I) that we use the term *relation* to denote a subset of the Cartesian product $X \times Y$ of any two sets X and Y. A relation $\mathcal{R} \subset X \times Y$ is called a *mapping of X into Y* if $(x, y_1) \in \mathcal{R}$ and $(x, y_2) \in \mathcal{R}$ implies that $y_1 = y_2$.

Suppose that H is a mapping of an open subset U of the Cartesian product $X \times Y$ of normed spaces X and Y into Y; $X \times Y \ni (x, y) \to H(x, y) \in Y$. We define the relation

$$\mathcal{R} := \{(x, y) \in X \times Y : H(x, y) = 0\}.$$

The mapping H is said to *generate* an implicit mapping of X into Y in a neighbourhood of a point $(x_0, y_0) \in \mathcal{R}$ if there exist $r_1, r_2 > 0$ such that the relation $\mathcal{F}$ defined by the formula

$$\mathcal{F} := \mathcal{R} \cap (K(x_0, r_1) \times K(y_0, r_2))$$

is a mapping of $K(x_0, r_1)$ into $K(y_0, r_2)$. We then have $H(x, \mathcal{F}(x)) = 0$.

We now give theorems concerning the existence of implicit mappings.

THEOREM VIII.3.1 (Graves). Let X and Y be Banach spaces and let H be a mapping of an open subset V of the Cartesian product $X \times Y$ into Y:

$$X \times Y \supset V \ni (x, y) \to H(x, y) \in Y,$$

Furthermore, suppose that there exists a point $(x_0, y_0) \in V$ such that $H(x_0, y_0) = 0$ and let H be continuously differentiable in a neighbourhood $U \subset V$ of the point (x_0, y_0).

If $\big(H'_Y(x_0, y_0)\big)^{-1} \in L(Y, Y)$ exists, then H generates an implicit mapping continuous on $K(x_0, r_1)$;

$$X \supset K(x_0, r_1) \ni x \to \mathscr{F}(x) \in K(y_0, r_2) \subset Y.$$

This mapping is continuously differentiable and

$$\mathscr{F}'(x) = -\big(H'_Y(x, y)\big)^{-1} \circ H'_X(x, y),$$

where $y = \mathscr{F}(x)$.

PROOF. We have the relation $\mathscr{R} := \{(x, y) \in X \times Y : H(x, y) = 0\}$. This relation is not an empty set since the point $(x_0, y_0) \in \mathscr{R}$. We seek r_1 and r_2 such that $\mathscr{R} \cap \big(K(x_0, r_1) \times K(y_0, r_2)\big)$ be a mapping.

Let us introduce auxiliary mapping

$$(1) \qquad X \times Y \ni (x, y) \to T(x, y) := \big(x, H(x, y)\big) \in X \times Y.$$

We shall show that the mapping T is invertible in a neighbourhood of the point (x_0, y_0). First of all, we evaluate the derivative

$$T'(x_0, y_0)\,(\Delta x, \Delta y)$$
$$= \big(\Delta x, H'_X(x_0, y_0) \cdot \Delta x + H'_Y(x_0, y_0) \cdot \Delta y\big).$$

An elementary computation shows that $T'(x_0, y_0)$ has an inverse, given as

$$X \times Y \ni (x', y') \to \Big(x', \big(H'_Y(x_0, y_0)\big)^{-1} \cdot \big(y' - H'_X(x_0, y_0)x'\big)\Big)$$
$$\in X \times Y.$$

The estimate

$$\|\big(T'(x_0, y_0)\big)^{-1} \cdot (x', y')\|$$
$$\leqslant \|x'\| + \|\big(H'_Y(x_0, y_0)\big)^{-1}\| \big(\|y'\| + \|H'_X(x_0, y_0)\| \cdot \|x'\|\big)$$
$$\leqslant \big(1 + \|\big(H'_Y(x_0, y_0)\big)^{-1}\|\big(1 + \|H'_X(x_0, y_0)\|\big)\big) \cdot \|(x', y')\|$$

gives us the continuity of $\big(T'(x_0, y_0)\big)^{-1}$. Theorem VIII.2.2 can thus be applied to T assuring that the restriction of T to a suitable neighbourhood W of the point (x_0, y_0) has a continuously differentiable inverse. We abuse notation letting T^{-1} denote this inverse. If follows that for a given $r_2 > 0$ such that $K\big((x_0, y_0), r_2\big) \subset W$ there exists $r_1 > 0$ such that

$$(2) \qquad T^{-1}\big(K(x_0, 0), r_1\big) \subset K\big((x_0, y_0), r_2\big)$$

(note that $T(x_0, y_0) = (x_0, 0)$).

Let us define the relation

$$X \times Y \supset W \supset F := \{T^{-1}(x, 0): x \in K(x_0, r_1)\}.$$

Denote by pr_X (pr_Y) the canonical projection of $X \times Y$ onto X (Y).

$$\mathrm{pr}_X \circ T^{-1}(x, 0) = x \in K(x_0, r_1),$$
$$\mathrm{pr}_Y \circ T^{-1}(x, 0) \subset K(y_0, r_2),$$

thus

(3) $\mathscr{F} \subset K(x_0, r_1) \times K(y_0, r_2).$

Now it is easily to check that $\mathscr{F}$ is a mapping.

In fact, if $(x, y_1) \in \mathscr{F}$ and $(x, y_2) \in \mathscr{F}$ then $T(x, y_1) = T(x, y_2) = (x, 0)$.

But since the mapping T is one-to-one for points of the set $\mathscr{F}$ it implies $y_1 = y_2$.

The mapping (defined by) $\mathscr{F}$ is given by the formula

$$X \supset K(x_0, r_1) \ni x \to \mathscr{F}(x) = \mathrm{pr}_Y \circ T^{-1}(x, 0) \in Y$$

which displays also its continuous differentiability.

Observe now that $\mathscr{F}$ satisfies the equation

$$H(x, \mathscr{F}(x)) = 0.$$

By differentiation we get

$$H'_X(x, \mathscr{F}(x)) + H'_Y(x, \mathscr{F}(x)) \circ \mathscr{F}'(x) = 0$$

whence

$$\mathscr{F}'(x) = -\left(H'_Y(x, \mathscr{F}(x))\right)^{-1} \circ H'_X(x, \mathscr{F}(x)). \quad \square$$

Theorem VIII.3.1 applies to the case when H is a mapping of the product $X \times Y$ into Y. One would like to generalize this theorem to the case of three different spaces.

THEOREM VIII.3.2. Let X, Y, and Z be Banach spaces and let H be a mapping of an open subset V of the Cartesian product $X \times Y$ into Z. Let there be a point $(x_0, y_0) \in V$ such that $H(x_0, y_0) = 0$ and let H be continuously differentiable in some neighbourhood $U \subset V$ of the point (x_0, y_0).

If $\left(H'_Y(x_0, y_0)\right)^{-1} \in L(Z, Y)$ exists, then in a neighbourhood of the point x_0 the mapping H generates an implicit mapping which is continuous on that neighbourhood:

$$X \supset K(x_0, r_1) \ni x \to F(x) \in K(y_0, r_2) \subset Y.$$

The mapping F is differentiable at x_0 and

$$F'(x_0) = -(H'_Y(x_0, y_0))^{-1} \circ H'_X(x_0, y_0).$$

PROOF. Since the continuous linear mapping $(H'_Y(x_0, y_0))^{-1}$ exists, the mapping $H'_Y(x_0, y_0)$ is a topological isomorphism of the spaces Y and Z. We introduce the mapping $(H'_Y(x_0, y_0))^{-1} \circ H: X \times Y \to Y$. It satisfies the hypotheses of Theorem VIII.3.1 since its partial derivative with respect to Y at the point (x_0, y_0) is an identity mapping. Hence, by Theorem VIII.3.1 there exists an implicit mapping $x \to F(x)$ such that

$$(H'_Y(x_0, y_0) \circ H)(x, F(x)) = 0, \quad x \in K(x_0, r_1),$$

whereby

$$H(x, F(x)) = 0, \quad x \in K(x_0, r_1).$$

The formula for the derivative is also easily justified. $\square$

If the mapping H is p times differentiable in some neighbourhood of the point (x_0, y_0) and if the mapping $(H'_Y(x, y))^{-1} \in L(Z, Y)$ exists at every point (x, y) of this neighbourhood, the implicit mapping generated by H is p times continuously differentiable in that neighbourhood. The proof of this follows immediately from the equation

$$F'(x) = -(H'_Y(x, F(x)))^{-1} \circ H'_X(x, F(x)).$$

Now let us consider the case of finite-dimensional spaces. Suppose that $X = R^n$, $Y = Z = R^m$. A system of m functions of $n+m$ variables corresponds to the mapping H. If $e_1, e_2, \ldots, e_m$ is a canonical basis in the space R^m, then

$$H(x, y) = \sum_{i=1}^{m} h^i(x_1, \ldots, x_n, y_1, \ldots, y_m) \cdot e_i.$$

The condition $H(x_0, y_0) = 0$ is equivalent to the system of equations

$$h^1(\mathring{x}_1, \ldots, \mathring{x}_n, \mathring{y}_1, \ldots, \mathring{y}_m) = 0,$$
$$h^2(\mathring{x}_1, \ldots, \mathring{x}_n, \mathring{y}_1, \ldots, y_m) = 0,$$
$$\cdots \cdots \cdots \cdots \cdots \cdots \cdots$$
$$h^m(\mathring{x}_1, \ldots, \mathring{x}_n, \mathring{y}_1, \ldots, y_m) = 0,$$

where $x_0 = (\mathring{x}_1, \ldots, \mathring{x}_n)$, $y_0 = (\mathring{y}_1, \ldots, \mathring{y}_m)$.

The differentiability of the mapping H in a neighbourhood of the point (x_0, y_0) is clearly equivalent to the differentiability of the functions h^i. On the other hand, the existence of the mapping $(H'_Y(x_0, y_0))^{-1}$ is equivalent to the inequality

$$\frac{\partial(h^1, \ldots, h^m)}{\partial(y_1, \ldots, y_m)}\bigg|_{\substack{x=x_0 \\ y=y_0}} \neq 0.$$

The implicit mapping F is given by a system of m functions of n variables,

$$F(x) = \sum_{j=1}^{m} f^j(x_1, \ldots, x_n) e_j.$$

The matrix of the derivative of F is expressed by

$$\left[\frac{\partial f^j}{\partial x_s}\right] = -\left[\frac{\partial h^i}{\partial y_k}\right]^{-1}\left[\frac{\partial h^l}{\partial x^p}\right].$$

All of the partial derivatives in this formula are taken at the point $(x, y) = (x_0, y_0)$.

Special Cases.

1. $X = Y = R^1$,

$$H(x, y) = h(x, y) \in R^1:$$

$$\frac{dF(x_0)}{dx} = -\frac{\dfrac{\partial h(x_0, y_0)}{\partial x}}{\dfrac{\partial h(x_0, y_0)}{\partial y}}.$$

2. $X = R^n$, $Y = R^1$,

$$H(x, y) = h(x_1, \ldots, x_n, y) \in R^1;$$

$$F'(x_0) = -\frac{\left(\dfrac{\partial h(x_0, y_0)}{\partial x_1}, \ldots, \dfrac{\partial h(x_0, y_0)}{\partial x_n}\right)}{\dfrac{\partial h(x_0, y_0)}{\partial y}}.$$

Let $X = Y = R^1$ and $H(x, y) := \dfrac{x^2}{a^2} + \dfrac{y^2}{b^2} - 1$; then

$$H(x, y) = 0 = \frac{x^2}{a^2} + \frac{y^2}{b^2} - 1,$$

and hence $\dfrac{x^2}{a^2} + \dfrac{y^2}{b^2} = 1$. This is the equation of an ellipse. Furthermore, we have

$$\frac{\partial H(x, y)}{\partial y} = 2\frac{y}{b^2} \neq 0, \quad \text{when } y \neq 0.$$

It is thus seen that an implicit function exists in a neighbourhood of an arbitrary point lying on an ellipse, with the exception of the points $(-a, 0)$ and $(a, 0)$. This is related to the fact that

$$y = \pm b \cdot \left(1 - \frac{x^2}{a^2}\right)^{1/2}$$

and for every point on the ellipse, with the exception of the two mentioned, there exists a neighbourhood in which the sign is determined uniquely.

4. CONSTRAINED EXTREMA

In the preceding chapter we were engaged in examining extrema of functions and functionals defined on a Banach space. In mathematics and physics it frequently becomes necessary to inspect the extrema of functions under given auxiliary conditions.

We may want to find, for example, the distance of a point $(\mathring{x}_1, \mathring{x}_2, \mathring{x}_3)$ $\in R^3$ from a surface given by the equation $g(x_1, x_2, x_3) = 0$. The problem here reduces to one of finding the minimum of the function f defined as $f(x) := (x_1 - \mathring{x}_1)^2 + (x_2 - \mathring{x}_2)^2 + (x_3 - \mathring{x}_3)^2$ under the constraint $g(x_1, x_2, x_3) = 0$.

Now we shall consider how the constrained extrema problem is solved. We shall first discuss the concept of the direct sum of two vector spaces.

Let X_1 and X_2 be two vector spaces. We introduce the structure of vector space into the Cartesian product $X_1 \times X_2$. If $(x_1, x_2) \in X_1 \times X_2$, $(x_1', x_2') \in X_1 \times X_2$, then

$$(x_1, x_2) + (x_1', x_2') := (x_1 + x_1', x_2 + x_2'),$$
$$\alpha(x_1, x_2) := (\alpha x_1, \alpha x_2).$$

The Cartesian product $X_1 \times X_2$ with a vector structure so introduced is called the (algebraic) direct sum of the spaces X_1 and X_2 and is denoted by $X = X_1 \oplus X_2$.

If X_1 and X_2 are normed spaces, the direct sum is also a normed space. For $(x_1, x_2) \in X_1 \times X_2$,

$$\|(x_1, x_2)\| = \|x_1\| + \|x_2\|.$$

We now consider a special case of direct sum. Let X be a vector space and let $X_1 \subset X$ and $X_2 \subset X$ be subspaces of X such that $X_1 \cap X_2 = \{0\}$. Next we form the direct sum $X_1 \oplus X_2 = Y$. The elements of the space $X_1 \oplus X_2$ are pairs (x_1, x_2), and they may be written in the form $x_1 + x_2$. We then have $Y \subset X$. This inclusion means that we identify the space $Y = X_1 \oplus X_2$ with the subspace Y of the space X spanned by vectors of the form $x_1 + x_2$, $x_1 \in X_1$, $x_2 \in X_2$. Since X_1 and X_2 have only the null vector in common, every vector $x \in X_1 \oplus X_2$ is uniquely representable in the form $x = x_1 + x_2$ where $x_1 \in X_1$, $x_2 \in X_2$.

Now, let X be a normed space and let X_1 and X_2 be subspaces of X, and furthermore let $X_1 \cap X_2 = \{0\}$. Two norms are defined in $X_1 \oplus X_2$:

(i) the norm of the Cartesian product $X_1 \times X_2$,

(ii) the norm of the space X.

Norm (i) is usually stronger than norm (ii). If these norms are equivalent, the space $X_1 \oplus X_2$ is called the *topological direct sum* of the subspaces X_1 and X_2. In other words, it may be said that: the algebraic direct sum $X_1 \oplus X_2$ of the subspaces $X_1 \subset X$ and $X_2 \subset X$ is a topological direct sum if and only if the projections pr_1 and pr_2,

$$X_1 \oplus X_2 \ni x \to \mathrm{pr}_1(x) := x_1 \in X_1,$$
$$X_1 \oplus X_2 \ni x \to \mathrm{pr}_2(x) := x_2 \in X_2,$$

are continuous in the topology of the space X.

It is readily seen that when $X_1 \oplus X_2$ is a topological direct sum, the subspaces X_1 and X_2 are closed in $X_1 \oplus X_2$.

It is known from algebra that for every subspace X_1 of a vector space X there is a subspace X_2 such that $X = X_1 \oplus X_2$ (algebraic direct sum). The subspace X_2 is called the *algebraic complement* of the subspace X_1 in X. On the other hand, if we have a closed subspace X_1 in a normed space X, then in general it is not possible to select for it a subspace X_2 such that $X = X_1 \oplus X_2$ (topological direct sum). This can always be done, however, if X is a Hilbert space. For there is an *orthogonal decomposition theorem* which states that an orthogonal complement exists for every closed subspace of a Hilbert space. The proof of this theorem is given in Chapter XIII.

247

DEFINITION. Let G be a mapping of a Banach space X into a Banach space Y differentiable at $x_0 \in X$ such that $G(x_0) = 0$. The point x_0 is called *regular point* of the set $M := \{x \in X: G(x) = 0\} = G^{-1}(0)$ if the derivative of G at x_0 maps X onto Y, that is, $G'(x_0) \cdot X = Y$.

Examples. 1. If $Y = R^1$, then the point $x_0 \in M \subset X$ is regular if and only if $G'(x_0) \neq 0$.

2. If $X = R^n$, $Y = R^m$, $m \leqslant n$, then the point $x_0 \in M \subset X$ is regular precisely when the rank of the matrix $\left[\dfrac{\partial g_i}{\partial x_j} (x_0) \right]$ is m.

DEFINITION. Let $X_1 := \{x \in X: G'(x_0) \cdot x = 0\}$ or $X_1 = (G'(x_0))^{-1}(0)$. The set $T_{x_0} := \{x_0 + x_1: x_1 \in X_1\}$ is called the *tangent plane to the set* $M = G^{-1}(0)$ *at the point* $x_0 \in M$.

Note that the set X_1 is a linear subspace of the space X whereas the tangent plane T_{x_0} is not in general a linear subspace.

THEOREM VIII.4.1 (First Theorem of Lusternik). Let G be a mapping of a Banach space X into a Banach space Y which is continuously differentiable in some neighbourhood U of the point $x_0 \in M = G^{-1}(0)$. If the point x_0 is a regular point of the set M and if X decomposes into the topological direct sum of the subspaces X_1 and X_2 (that is, $X = X_1 \oplus X_2$, where $X_1 = (G'(x_0))^{-1}(0)$ and X_2 is its topological complement), then:

(i) there exists a neighbourhood $V_1 \subset T_{x_0}$ of the point x_0 in the plane tangent to M at x_0, this neighbourhood being homeomorphic to some neighbourhood $V_2 \subset M$ of the point x_0 in M,

(ii) there exists a neighbourhood of zero $W \subset X_1$, and a mapping $F: W \to X_2$ such that $x_0 + x_1 + F(x_1) \in M$ and $F(x_1) = o(||x_1||)$, that is

$$\lim_{x_1 \to 0} \frac{||F(x_1)||}{||x_1||} = 0.$$

A certain lemma will be needed for the proof of this theorem.

LEMMA VIII.4.2. Let G be a mapping of a Banach space X into a Banach space Y which is differentiable at a point $x_0 \in M = G^{-1}(0)$. If x_0 is a regular point of the set M, and the space X is the topological direct sum of spaces X_1 and X_2, where $X_1 = (G'(x_0))^{-1}(0)$, then the mapping $G'(x_0)|_{X_2}$ is an isomorphism of X_2 to Y.

PROOF. The restriction of the mapping $G'(x_0)$ to the subspace X_2 is a one-to-one mapping. For if $G'(x_0) \cdot h = 0$ for $h \in X_2$, then the definition of the subspace X_1 implies that $h \in X_1$; since $X_1 \cap X_2 = \{0\}$, we have $h = 0$.

The point x_0 is a regular point, whence $G'(x_0) \cdot X = Y$, and thus $G'(x_0)|_{X_2} \cdot X_2 = Y$. The mapping $G'(x_0)|_{X_2}$ thus is a continuous algebraic isomorphism of the spaces X_2 and Y. The subspace X_2 is a component of the topological direct sum and hence is a closed subspace of the Banach space X. It is therefore a Banach space. The Banach isomorphism theorem (Theorem VIII.2.5) can thus be applied to the mapping $G'(x_0)|_{X_2}$. $\square$

We now proceed with the proof of the first Lusternik theorem. To begin with we define the mapping

$$X_1 \times X_2 \ni (x_1, x_2) \rightarrow H(x_1, x_2) := G(x_0 + x_1 + x_2) \in Y.$$

We verify that the mapping H satisfies the hypotheses of the theorem on implicit mappings (Theorem VIII.3.2):

(i) $H(0, 0) = 0$.

(ii) The mapping H is continuously differentiable in a neighbourhood of the point $(0, 0)$ since it is a composition of the differentiable mapping $(x_1, x_2) \rightarrow (x_0 + x_1 + x_2)$ and the mapping G. Moreover, $H'_{x_2}(0, 0) = G'(x_0)|_{X_2}$.

It follows from Lemma VIII.4.2 that there is a mapping $(G'(x_0)|_{X_2})^{-1} = (H'_{x_2}(0, 0))^{-1} \in L(Y, X_2)$.

The mapping H thus satisfies the hypotheses of Theorem VIII.3.2. Hence, there exists an implicit mapping F generated by H,

$$X_1 \supset K(0, r) \ni x_1 \rightarrow F(x_1) \in X_2,$$
$$H(x_1, F(x_1)) = 0 \quad \text{for } x_1 \in K(0, r) \subset X_1.$$

But $H(x_1, F(x_1)) = G(x_0 + x_1 + F(x_1)) = 0$, and hence $x_0 + x_1 + F(x_1) \in M$.

The mapping F is differentiable at the point $x_1 = 0$ and

$$F'(0) = -(G'(x_0)|_{X_2})^{-1} \circ G'(x_0)|_{X_1},$$

whence $F'(0) \cdot x_1 = 0$ for $x_1 \in X_1$.

Note further that $F(0) = 0$. Accordingly, we have

$$F(x_1) = F(0) + F'(0) \cdot x_1 + o(\|x_1\|) = o(\|x_1\|).$$

Item (ii) of the theorem has thus been proved.

Since every point $x \in T_{x_0}$ is of the form $x = x_0 + x_1$, where $x_1 \in X_1$, the mapping

$$T_{x_0} \ni x = (x_0 + x_1) \rightarrow (x_0 + x_1 + F(x_1)) \in M$$

is a continuous mapping of a neighbourhood $V_1 \subset T_{x_0}$ of the point x_0 into a neighbourhood of the point x_0 in M.

Conversely, every point $x \in M$ may be uniquely represented in the form $x = x_0 + x_1 + x_2$, where $x_1 \in X_1$, $x_2 \in X_2$. Since the space X is the topological direct sum of the subspaces X_1 and X_2, the mapping

$$M \ni x = (x_0 + x_1 + x_2) \to (x_1 + x_0) \in T_{x_0}$$

is continuous on M. If it is restricted to a neighbourhood $V_2 \subset M$ of the point x_0 such that its image is contained in V_1, the mapping is continuous. Item (i) of the theorem has thus been proved. □

We now introduce the concept of constrained extremum.

DEFINITION. Let X be a metric space, let Y be vector space, and let G be a mapping of X into Y. Let $M := \{x \in X : G(x) = 0\}$. A function $f \colon X \to R^1$ is said to have a *local constrained minimum* (*maximum*) (constrained by the set M) at a point $x_0 \in M$ if there exist $r > 0$ such that

$$f(x_0) \leqslant f(x) \quad (f(x_0) \geqslant f(x)) \quad \text{for } x \in K(x_0, r) \cap M.$$

Now let us find the necessary condition for the existence of a local constrained extremum. Henceforth we shall assume that the hypotheses of the first Lusternik theorem are satisfied, i.e.:

(i) X and Y are Banach spaces;

(ii) G is a mapping of X into Y continuously differentiable in some neighbourhood of the point $x_0 \in X$;

(iii) $x_0 \in X$ is a regular point of the set $M := G^{-1}(0)$;

(iv) $X_1 := \big(G'(x_0)\big)^{-1}(0)$;

(v) $X = X_1 \oplus X_2$ (topological direct sum).

LEMMA VIII.4.3. Let f be a function defined on an open subset U of a Banach space X with values in R^1. Let $x_0 \in X$ be a regular point of the set $M = G^{-1}(0)$. If f is differentiable at x_0 and has a local constrained minimum (maximum) at that point, then $f'(x_0) \cdot x_1 = 0$ for every $x_1 \in X_1$.

PROOF. (a.a.) We shall prove the theorem for the minimum. The proof for the maximum is analogous.

Suppose that there exists a point $x_1 \in X_1$ such that $f'(x_0) \cdot x_1 = a \neq 0$. At the point x_0 we have a local minimum and thus

(1) $f(x_0 + h) - f(x_0) \geqslant 0 \quad \text{for } x_0 + h \in M.$

By the first theorem of Lusternik we have

(2) $\qquad x_0 + t \cdot x_1 + F(t \cdot x_1) \in M \quad$ for $|t| < \delta$.

Formulae (1) and (2) yield

(3) $\qquad 0 \leqslant f(x_0 + t \cdot x_1 + F(t \cdot x_1)) - f(x_0)$

$\qquad = f'(x_0) \cdot (t \cdot x_1) + f'(x_0) \cdot (F(t \cdot x_1)) + r((t \cdot x_1) + F(t \cdot x_1))$

$\qquad = t \cdot a + f'(x_0) \cdot (F(t \cdot x_1)) + r(t \cdot x_1 + F(t \cdot x_1))$.

The first Lusternik theorem implies

(4) $\qquad \|F(t \cdot x_1)\| < \varepsilon |t| \quad$ for $|t| < \delta_1$.

In addition, we have

(5) $\qquad \|r(t \cdot x_1 + F(t \cdot x_1))\| \leqslant \|t \cdot x_1 + F(t \cdot x_1)\| \cdot \varepsilon$

$\qquad$ for $\|t \cdot x_1 + F(t \cdot x_1)\| < \delta_2$.

However, from inequality (4) it follows that

(6) $\qquad \|t \cdot x_1 + F(t \cdot x_1)\| \leqslant |t| \cdot (\|x_1\| + \varepsilon) \quad$ for $|t| < \delta_1$.

When we introduce the notation

$\qquad g(t) := f'(x_0) \cdot (F(t \cdot x_1)) + r(t \cdot x_1 + F(t \cdot x_1))$

and make use of (4), (5), and (6), we obtain

(7) $\qquad |g(t)| \leqslant |t| \big((\|f'(x_0)\| \cdot \varepsilon) + \varepsilon \cdot (\|x_1\| + \varepsilon) \big)$

$\qquad$ for $|t| < \min(\delta_1, \delta_2/\|x_1\|)$.

By (3) we have

$\qquad t \cdot a + g(t) \geqslant 0$,

but this contradicts (7), inasmuch as for sufficiently small t the second term on the left-hand side is much smaller than the first and hence their sum cannot be always positive. The sum changes sign as t changes to $-t$. $\square$

THEOREM VIII.4.4 (Second Theorem of Lusternik). Suppose that the hypotheses of the first Lusternik theorem are satisfied. Let f be a function defined on an open set of the space X with values in R^1, and let that function be differentiable at the point $x_0 \in M$. If f has a local constrained extremum at x_0, there is a linear functional $\Lambda \in Y^*$ such that

$\qquad f'(x_0) = \Lambda \circ G'(x_0)$.

The functional Λ is known as the *Lagrange functional*.

PROOF. Let $h \in X = X_1 \oplus X_2$. Hence $h = h_1 + h_2$, where $h_1 \in X_1$, and $h_2 \in X_2$. By Lemma VIII.4.3 we have

$$(8) \qquad f'(x_0) \cdot h_1 = 0.$$

The definition of the subspace X_1 yields

$$(9) \qquad G'(x_0) \cdot h_1 = 0.$$

Moreover, Lemma VIII.4.2 implies the existence of the mapping

$$(10) \qquad \big(G'(x_0)|_{X_2}\big)^{-1}.$$

Using (8), (9), and (10), we can write

$$
\begin{aligned}
f'(x_0) \cdot h &= f'(x_0) \cdot (h_1 + h_2) = f'(x_0) \cdot h_2 \\
&= f'(x_0) \circ \big(G'(x_0)|_{X_2}\big)^{-1} \circ \big(G'(x_0)|_{X_2}\big) \cdot h_2 \\
&= f'(x_0) \circ \big(G'(x_0)|_{X_2}\big)^{-1} \circ G'(x_0)(h_1 + h_2) \\
&= f'(x_0) \circ \big(G'(x_0)|_{X_2}\big)^{-1} \circ G'(x_0) \cdot h.
\end{aligned}
$$

On denoting

$$\Lambda = f'(x_0) \circ \big(G'(x_0)|_{X_2}\big)^{-1},$$

we have

$$f'(x_0) = \Lambda \circ G'(x_0).$$

It is easily seen that $\Lambda \in Y^*$. $\square$

The second Lusternik theorem speaks of the necessary condition for the existence of a constrained extremum. We shall now give the sufficient condition.

Theorem VIII.4.5.[1] Suppose that the hypotheses of the first Lusternik theorem are satisfied and, moreover, let the mapping G be two times continuously differentiable in some neighbourhood of the point x_0. Let f be a function defined on an open subset of the space X and let it be two times continuously differentiable in some neighbourhood of the point x_0.

If there exists a $\Lambda \in Y^*$ such that $f'(x_0) = \Lambda \circ G'(x_0)$, and $c > 0$ such that

$$\big(f''(x_0) - \Lambda \circ G''(x_0)\big)(h, h) \geqslant c\|h\|^2 \qquad (\leqslant -c\|h\|^2)$$

for $h \in X_1$,

---

[1] The proof of this theorem is due to Wiktor Szczyrba.

the function f has a local constrained minimum (maximum) at the point x_0.

PROOF. We introduce the auxiliary function

$$X \ni x \rightarrow g(x) := f(x) - \Lambda \circ G(x).$$

This function is two times continuously differentiable in a neighbourhood of the point x_0. Its derivatives are of the form

(11) $g'(x) = f'(x) - \Lambda \circ G'(x), \quad g''(x) = f''(x) - \Lambda \circ G''(x).$

We expand the function g by the Taylor formula about the point x_0. We then have:

$$g(x_0 + h) - g(x_0) = g'(x_0) \cdot h + \frac{1}{2} g''(x_0)(h, h) + o(||h||^2).$$

By (11) we obtain

(12) $f(x_0 + h) - f(x_0) - \Lambda \circ G(x_0 + h) + \Lambda \circ G(x_0)$
$$= (f'(x_0) - \Lambda \circ G'(x_0)) \cdot h + \frac{1}{2}(f''(x_0) - \Lambda \circ G''(x_0))(h, h) +$$
$$+ o(||h||^2).$$

However, we recall that $G(x_0) = 0$ and $f'(x_0) - \Lambda \circ G'(x_0) = 0$. Moreover, we are interested in the case where $(x_0 + h) \in M = G^{-1}(0)$. Therefore $G(x_0 + h) = 0$. Formula (12) thus takes on the form:

(13) $f(x_0 + h) - f(x_0) = \frac{1}{2}(f''(x_0) - \Lambda \circ G''(x_0))(h, h) + o(||h||^2)$
     for $(x_0 + h) \in M.$

By virtue of the first Lusternik theorem for $||h|| < r$ we have the decomposition $h = h_1 + F(h_1)$, where $h_1 \in X_1, F(h_1) \in X_2$ and

(14) $||F(h_1)|| \leqslant ||h_1|| \varepsilon \quad$ for $||h_1|| < \delta_1.$

By (13) and (14) we obtain

(14') $f(x_0 + h) - f(x_0) = \frac{1}{2}\big((f''(x_0) - \Lambda \circ G''(x_0))(h_1, h_1) +$
$$+ (f''(x_0) - \Lambda \circ G''(x_0))(h_1, F(h_1)) +$$
$$+ (f''(x_0) - \Lambda \circ G''(x_0))(F(h_1), h_1) +$$
$$+ (f''(x_0) - \Lambda \circ G''(x_0))(F(h_1), F(h_1))\big) + o(||h||^2).$$

But

(15) $|(f''(x_0) - \Lambda \circ G''(x_0))(h_1, F(h_1))|$
$$\leqslant ||(f''(x_0) - \Lambda \circ G''(x_0))|| \cdot ||h_1|| \cdot ||F(h_1)|| \leqslant A \cdot \varepsilon ||h_1||^2$$
     for $||h_1|| < \delta_1.$

Similarly, we get

(16) $\quad |(f''(x_0)-\Lambda \circ G''(x_0))\,(F(h_1),\,h_1)| \leqslant A \cdot \varepsilon ||h_1||^2,$

(17) $\quad |(f''(x_0)-\Lambda \circ G''(x_0))\,(F(h_1),\,F(h_1))| \leqslant A \cdot \varepsilon^2 ||h_1||^2,$

$\quad\quad |o(||h||^2)| \leqslant \varepsilon ||h||^2 \quad$ for $||h|| < \delta_2,$

and thus

(18) $\quad |o(||h||^2)| \leqslant \varepsilon ||h||^2 = \varepsilon ||h_1 + F(h_1)||^2 \leqslant \varepsilon (||h_1|| + \varepsilon ||h_1||)^2$

$\quad\quad = \varepsilon (1+\varepsilon)^2 ||h_1||^2 \quad$ for $||h_1|| < \delta_1.$

The hypotheses of the theorem yield

(19) $\quad (f''(x_0)-\Lambda \circ G''(x_0))(h_1,\,h_1) \geqslant c||h_1||^2.$

On adding both sides of (14′) and (19) and using (15) to (18), we arrive at

$$f(x_0+h)-f(x_0) \geqslant \tfrac{1}{2}\,c||h_1||^2 - b||h_1||^2 \quad \text{for } x_0+h \in M,$$

where $b := \tfrac{1}{2}\varepsilon(2A+A \cdot \varepsilon+(1+\varepsilon)^2).$

When we take ε so small that $\tfrac{1}{2}c > b$, we get

$$f(x_0+h)-f(x_0) \geqslant (c-b)||h_1||^2 > 0 \quad \text{for } x_0+h \in M.$$

Thus, we have a local constrained minimum at the point x_0. $\square$

The foregoing theorem may be generalized to the case of derivatives of higher orders.

THEOREM VIII.4.6. Suppose that the hypotheses of the first Lusternik theorem are satisfied and furthermore let G be $2n$ times continuously differentiable in some neigbourhood of the point x_0. Let f be a function defined on an open set of the space X and let it be $2n$ times continuously differentiable in some neighbourhood of the point x_0. If there exists a $\Lambda \in Y^*$ such that

$$f^{(k)}(x_0)-\Lambda \circ G^{(k)}(x_0) = 0 \quad (k = 1, 2, ..., 2n-1)$$

and a $c > 0$ such that

$$(f^{(2n)}(x_0)-\Lambda \circ G^{(2n)}(x_0))\,(h, ..., h) \geqslant c||h||^{2n} \;(\leqslant -c||h||^{2n})$$
$$\text{for } h \in X_1,$$

the function f has a local constrained minimum (maximum) at the point x_0.

Special Case. Constrained Extrema in R^n. Let $X = R^n$, $Y = R^m$, $m \leqslant n$,

$G: R^n \to R^m$. The mapping G is represented by a system of m functions of n variables

$$G = \left(G_i(x_1, \ldots, x_n)\right)_{i=1}^m.$$

Let $f: R^n \to R^1$. We seek constrained extremal points which are also regular points. The problem reduces to that of solving a system of two operator equations:

(20) $\qquad f'(x) = \Lambda \circ G'(x), \qquad \text{where } \Lambda \in (R^m)^*,$

$\qquad\qquad G(x) = 0.$

However, we know that every covector $\Lambda \in (R^m)^*$ is represented by a system of m real numbers $\lambda_1, \ldots, \lambda_m$, and the derivative $G'(x)$ is a rectangular $m \times n$ matrix of rank m. The system of operator equations (20) reduces to a system of $n+m$ scalar equations

$$\frac{\partial f(x)}{\partial x^j} = \sum_{i=1}^m \lambda_i \frac{\partial G_i(x)}{\partial x^j}, \qquad j = 1, 2, \ldots, n,$$

$$G_k(x_1, \ldots, x_n) = 0, \qquad k = 1, 2, \ldots, m,$$

where $x = (x_1, \ldots, x_n)$,

with $n+m$ variables $(\lambda_i)_{i=1, 2, \ldots, m}$, $(x_k)_{k=1, 2, \ldots, n}$. On solving this system of equations, we obtain the coordinates of the point "suspected" of being a local constrained extremum. The numbers λ_i in the final result are of no interest to us as they play a merely auxiliary role. They are frequently referred to as *Lagrange multipliers*.

Having points "suspected" of being an extremum, we must resort to the sufficient condition. We examine the positive (negative) definiteness of the second derivative for $h \in X_1 = \{h \in R^n \colon G'(x_0) \cdot h = 0\}$. By virtue of Lemma VII.11.3, it is sufficient to investigate whether

$$\left(f''(x) - \Lambda \circ G''(x)\right)(h, h) > 0 \ (< 0) \qquad \text{for } h \in X_1.$$

The problem reduces to one of inspecting the quadratic form

(21) $\qquad \displaystyle\sum_{i,j=1}^n \left(\frac{\partial^2 f(x)}{\partial x_i \partial x_j} - \sum_{k=1}^m \lambda_k \frac{\partial^2 G_k(x)}{\partial x_i \partial x_j} \right) h_i h_j,$

where $h \in X_1, h = (h_1, h_2, \ldots, h_n)$. The condition $h \in X_1$ is equivalent to the equation $G'(x) \cdot h = 0$, which in matrix form becomes

(22) $\qquad \displaystyle\sum_{i=1}^n \frac{\partial G_k(x)}{\partial x_i} h_i = 0, \qquad k = 1, 2, \ldots, m.$

The rank of the matrix $\left[\dfrac{\partial G_k(x)}{\partial x_i}\right]$ is equal to m, and thus $h_{i_1}, h_{i_2}, \ldots, h_{i_m}$ are expressed linearly by $h_{i_{m+1}}, h_{i_{m+2}}, \ldots, h_{i_n}$, where $i_1, i_2, \ldots, i_n$ is a permutation of the numbers $1, 2, \ldots, n$. Thus we have

$$(23) \qquad h_{i_l} = \sum_{k=1}^{n-m} a_l^k h_{i_{m+k}}, \qquad l = 1, 2, \ldots, m.$$

Substitution of (23) into (21) yields a quadratic form in $n-m$ variables which we shall investigate by the determinantal method (cf. Section 11, Chapter VII).

In general, when we inspect the $2k$-th differential we proceed in the same way and obtain a form in $n-m$ variables of order $2k$, and we must examine the positive (negative) definiteness of this form.

Examples of constrained extrema. 1. Let us examine the extrema of the functions $f\colon \boldsymbol{R}^2 \to \boldsymbol{R}^1$,

$$f(x, y) := \frac{x}{a} + \frac{y}{b},$$

with the constraint

$$G(x, y) := x^2 + y^2 - 1 = 0.$$

We must solve the system of three equations

$$\frac{\partial f(x, y)}{\partial x} - \lambda \frac{\partial G(x, y)}{\partial x} = 0,$$

$$\frac{\partial f(x, y)}{\partial y} - \lambda \frac{\partial G(x, y)}{\partial y} = 0,$$

$$G(x, y) = 0$$

for the variables x, y, and λ. These equations are of the form

$$\frac{1}{a} - 2\lambda x = 0,$$

$$\frac{1}{b} - 2\lambda y = 0,$$

$$x^2 + y^2 = 1.$$

This system has two solutions

$$x_1 = \frac{|ab|}{a(a^2+b^2)^{1/2}}, \quad y_1 = \frac{|ab|}{b(a^2+b^2)^{1/2}}, \quad \lambda_1 = \frac{a^2+b^2}{2|ab|};$$

$$x_2 = -\frac{|ab|}{a(a^2+b^2)^{1/2}}, \quad y_2 = -\frac{|ab|}{b(a^2+b^2)^{1/2}},$$

$$\lambda_2 = -\frac{a^2+b^2}{2|ab|}.$$

Thus we have obtained two points suspected of being extrema. The subspace $X_1 = \{h \in R^2: G'(x) \cdot h = 0\}$ is the same for both points, $X_1 = \{(h_1, h_2) \in R^2: h_2 = -b \cdot h_1/a\}$; moreover

$$f''(x, y) = 0,$$

$$G''(x, y)(h, h)$$

$$= \frac{\partial^2 G(x, y)}{\partial x \, \partial x} h_1 h_1 + 2 \frac{\partial^2 G(x, y)}{\partial x \, \partial y} h_1 h_2 + \frac{\partial^2 G(x, y)}{\partial y \, \partial y} h_2 h_2,$$

$$(f''(x, y) - \lambda G''(x, y))(h, h) = -\lambda \cdot 2(h_1^2 + h_2^2)$$

$$= -\lambda \cdot 2\big(1 + (b/a)^2\big) h_1^2 > 0 \quad \text{for } \lambda = \lambda_2, \quad (<0 \text{ for } \lambda = \lambda_1).$$

Therefore, we have a local constrained maximum at the point (x_1, y_1) and a local constrained minimum at the point (x_2, y_2).

2. Let X be a real Hilbert space. Let $A \in L(X, X)$ be a linear, continuous, and symmetric operator, that is, $(A \cdot x \,|\, y) = (x \,|\, A \cdot y)$ for every pair of elements of the space X. Let

$$f(x) := (A \cdot x \,|\, x).$$

We are looking for the extremum of the function f on the unit sphere. The constraint is of the form $G(x) := ||x||^2 - 1 = 0$. Therefore

$$f(x+h) - f(x) = \big(A \cdot (x+h) \,|\, (x+h)\big) - (A \cdot x \,|\, x)$$

$$= 2(A \cdot x \,|\, h) + (A \cdot h \,|\, h).$$

Hence $f'(x) = 2A \cdot x$. Similarly, $G'(x) = 2x$. Thus we have

$$f'(x) - \Lambda \circ G'(x) = 0, \quad ||x|| = 1.$$

But

$$(R^1)^* \cong R^1, \quad \text{that is,} \quad \Lambda \cdot x = \lambda \cdot x, \ \lambda \in R^1.$$

Finally therefore

$$A \cdot x = \lambda \cdot x, \quad ||x|| = 1,$$

that is, the extremal points of the quadratic form generated by the symmetric operator on the unit sphere are normed eigenvectors of that operator.[1]

---

[1] *Remark*. An operator A in infinite-dimensional Hilbert space does not in general possess eigenvalues: it can have only a continuous spectrum. On the other hand, when $A = A^*$ is compact (i.e. when it maps a ball into a conditionally compact set), A has "many" eigenvalues: its spectrum is purely discrete. Problems of this kind led to the extremely elegant theory known as spectral theory; it is the subject of a monograph by the present author: *Methods of Hilbert Spaces*, Warszawa 1967.

IX. ORDINARY DIFFERENTIAL EQUATIONS

In this section we shall generalize the concept of the Riemann integral to vector-valued functions.

Let f be a function defined on the interval $[a, b] \subset R^1$ with values in a Banach space X

$$R^1 \supset [a, b] \ni t \to f(t) \in X.$$

We recall that a *partition π of the interval* $[a, b]$ is a set of points $(t_i)_{i=1,\ldots,n}$ of that interval $(a = t_1 < t_2 < \ldots < t_n = b)$. We define the sum corresponding to the function f and the partition π:

$$S_1(f, \pi) := \sum_{i=1}^{n-1} f(t_i)\,(t_{i+1} - t_i).$$

We can also define the sum S_2:

$$S_2(f, \pi) := \sum_{i=1}^{n-1} f(t_{i+1})\,(t_{i+1} - t_i).$$

In general, the Riemann sum is defined as

$$S(f, \pi, \{\tau_i\}) := \sum_{i=1}^{n-1} f(\tau_i)\,(t_{i+1} - t_i), \quad \text{where } t_i \leqslant \tau_i \leqslant t_{i+1}.$$

Let Π be the set of partitions of the interval $[a, b]$. Π is a set directed by the relation "to be a subpartition". The mappings

(1) $\Pi \ni \pi \to S_1(f, \pi) \in X$,

(2) $\Pi \ni \pi \to S_2(f, \pi) \in X$

define nets. On the other hand, the mapping

(3) $\Pi \ni \pi \to S(f, \pi, \{\tau_i\}) \in X$

defines a whole family of nets. Nets (1) and (2) are special representatives of that family.

DEFINITION. A function f is said to be *integrable in the Riemann sense* on $[a, b]$, if all the nets of the family (3) are identically convergent to a common limit. This limit is called the *integral of the function f* over the interval $[a, b]$:

$$\int_{[a,b]} f := \lim_{\pi \in \Pi} S(f, \pi, \{\tau_i\}).$$

Frequently we write

$$\int_{[a,b]} f = \int_a^b f(t)\,dt.$$

We shall now prove the counterpart of the Riemann Theorem III.4.3.

THEOREM IX.1.1. A function f which is continuous on the interval $[a, b]$ and assumes values in a Banach space X is integrable over $[a, b]$.

PROOF. By virtue of Theorem II.9.4 a continuous mapping on a compact set is uniformly continuous. Thus for every $\varepsilon > 0$ there exists a number $\delta > 0$ such that $\|f(t_1) - f(t_2)\| < \varepsilon$, provided that $|t_1 - t_2| < \delta$.

Let us take a partition $\pi_0 = (t_i)_{i = 1, \dots, n}$ of the interval $[a, b]$ with diameter less than δ. For $\pi \succ \pi_0$, $\pi = (p_j)_{j = 1, \dots, k}$, $k > n$, we have

$$\|S_1(f, \pi) - S_1(f, \pi_0)\|$$

$$= \left\| \sum_{j=1}^{k-1} f(p_j)(p_{j+1} - p_j) - \sum_{i=1}^{n-1} f(t_i)(t_{i+1} - t_i) \right\|$$

$$= \left\| \sum_{j=1}^{k-1} (f(p_j) - f(t_{ij}))(p_{j+1} - p_j) \right\|$$

$$\leqslant |b - a|\,\varepsilon, \quad \text{where } |t_{ij} - p_j| < \delta.$$

For $\pi_1, \pi_2 \succ \pi_0$ we therefore have

$$\|S_1(f, \pi_1) - S_1(f, \pi_2)\| \leqslant \|S_1(f, \pi_1) - S_1(f, \pi_0)\| +$$

$$+ \|S_1(f, \pi_0) - S_1(f, \pi_2)\| \leqslant 2\varepsilon|b - a|.$$

We have thus demonstrated that the net $\pi \to S_1(f, \pi)$ is a Cauchy net. The completeness of the space X implies that there exists a $\lim_{\pi \in \Pi}(S_1(f, \pi))$. We shall now show that

$$\lim_{\pi \in \Pi} S_1(f, \pi) = \lim_{\pi \in \Pi} S(f, \pi, \{\tau_i\})$$

for every $\{\tau_i\} = \{\tau_i\}(\pi)$. To this end it is sufficient to demonstrate that these nets are equivalent. Let us take $\pi \succ \pi_0$ and a fixed $\{\tau_i\}$. Then

$$\|S_1(f, \pi) - S(f, \pi, \{\tau_i\})\| = \left\| \sum_{i=1}^{k-1} (f(t_i) - f(\tau_i))(t_{i+1} - t_i) \right\|$$

$$\leqslant \varepsilon|b - a|. \qquad \square$$

1. INTEGRATION OF VECTOR-VALUED FUNCTIONS

The Properties of the Integral.

1. *Linearity:* f_1, f_2 are integrable over $[a, b]$; $\alpha, \beta \in R^1$; then $\alpha f_1 + \beta f_2$ is also integrable and

$$\int_a^b (\alpha f_1 + \beta f_2)(t) \, dt = \alpha \int_a^b f_1(t) \, dt + \beta \int_a^b f_2(t) \, dt.$$

2. If f is continuous on $[a, b]$, then

$$\left\| \int_a^b f(t) \, dt \right\| \leqslant \int_a^b \| f(t) \| \, dt.$$

3. If $b > a$, then

$$\int_b^a f(t) \, dt := - \int_a^b f(t) \, dt.$$

4. If $A \in L(X, Y)$ and $f \colon [a, b] \to X$ is integrable over $[a, b]$, the function $A \cdot f$ is also integrable and

$$\int_a^b A \cdot f(t) \, dt = A \left(\int_a^b f(t) \, dt \right).$$

5. If the function $t \to A(t) \in L(X, Y)$ is integrable over $[a, b]$, and $x \in X$, the function $t \to A(t) \cdot x \in Y$ is also integrable and

$$\left(\int_a^b A(t) \, dt \right) \cdot x = \int_a^b (A(t) \cdot x) \, dt.$$

6.

$$\int_a^c f(t) \, dt = \int_a^b f(t) \, dt + \int_b^c f(t) \, dt.$$

7. If f is a continuous function, then

$$\frac{d}{dt} \left(\int_a^t f(s) \, ds \right) = f(t), \quad t \in [a, b].$$

8. If f is differentiable over $]a, b[$ and if its derivative is continuous, then

$$\int_a^b f'(s) \, ds = f(b) - f(a).$$

Properties 1, 2, and 6 follow directly from the definition of the integral. Item 3 is a definition. Therefore, we prove 4, 5, 7, and 8.

PROOF OF PROPERTY 4. By the linearity and continuity of mapping A we have

$$A\left(\int\limits_a^b f(t)\,dt\right) = A\left(\lim_{\pi \in \Pi} S(f, \pi, \{\tau_i\})\right)$$

$$= \lim_{\pi \in \Pi} A \cdot (S(f, \pi, \{\tau_i\})) = \lim_{\pi \in \Pi} S(A \cdot f, \pi, \{\tau_i\}) = \int\limits_a^b A \cdot f(t)\,dt.$$

PROOF OF PROPERTY 5. We make use of the inequality

$$\left\|\left(\int\limits_a^b A(t)\,dt\right) \cdot x - S(A, \pi, \{\tau_i\}) \cdot x\right\|$$

$$\leqslant \left\|\int\limits_a^b A(t)\,dt - S(A, \pi, \{\tau_i\})\right\| \cdot \|x\|.$$

This inequality yields

$$\left(\lim_{\pi \in \Pi} S(A, \pi, \{\tau_i\})\right)x = \lim_{\pi \in \Pi}\left(S(A, \pi, \{\tau_i\})x\right).$$

Thus we arrive at

$$\left(\int\limits_a^b A(t)\,dt\right)x = \left(\lim_{\pi \in \Pi} S(A, \pi, \{\tau_i\})\right)x = \lim_{\pi \in \Pi}\left(S(A, \pi, \{\tau_i\})x\right)$$

$$= \lim_{\pi} S(A \cdot x, \pi, \{\tau_i\}) = \int\limits_a^b A(t) \cdot x\,dt.$$

PROOF OF PROPERTY 7. We have

$$\left\|\frac{1}{h}\left(\int\limits_a^{t+h} f(s)\,ds - \int\limits_a^t f(s)\,ds\right) - f(t)\right\| = \frac{1}{|h|}\left\|\int\limits_t^{t+h} (f(s) - f(t))\,ds\right\|$$

$$\leqslant \frac{1}{|h|}\int\limits_t^{t+h} \|f(s) - f(t)\|\,ds \leqslant \frac{1}{|h|}|h|\varepsilon = \varepsilon,$$

since $\|f(s) - f(t)\| < \varepsilon$ for $|h| < \delta$.

PROOF OF PROPERTY 8. By Property 7, we find that

$$\frac{d}{dt}\left(\int\limits_a^t f'(s)\,ds\right) = f'(t),$$

whence

$$\int_a^t f'(s)\,ds = f(t) + \text{const}.$$

Setting $t = a$, we obtain const $= -f(a)$. $\square$

LEMMA IX.1.2. Let T be a mapping of a Banach space X into a Banach space Y. Suppose that for an $h \in X$ there exists a directional derivative $\nabla_h T(x + t \cdot h)$, $t \in [a, b]$. If the mapping $[a, b] \ni t \to \nabla_h T(x + t \cdot h) \in Y$ is continuous, then

$$\int_a^b \nabla_h T(x + t \cdot h)\,dt = T(x + b \cdot h) - T(x + a \cdot h).$$

PROOF. In view of the equation

$$\nabla_h T(x + t \cdot h) = \frac{d}{dt} T(x + t \cdot h)$$

it is sufficient to invoke Property 8. $\square$

2. DIFFERENTIAL EQUATIONS. INITIAL-VALUE PROBLEMS

Let X be a Banach space and $I_n := \underset{i=1}{\overset{n}{\times}} [a_i, b_i] \subset R^n$. Let F be a mapping of the Cartesian product $I_n \times X^s$ into X. We define the mapping of the set I_n into X as follows:

We take a mapping $x(\cdot)$ defined on I_n with values in X having continuous partial derivative of order up to and including p and we define the mapping $F_x\colon I_n \to X$:

$$R^n \supset I_n \ni (t_1, \ldots, t_n) \to F_x(t_1, \ldots, t_n)$$

$$:= F\left(t_1, \ldots, t_n, x(t), \frac{\partial x(t)}{\partial t_1}, \ldots, \frac{\partial^p x(t)}{\partial t_n^p}\right) \in X.$$

We now seek an $x(\cdot)$ such that $F_x(t) \equiv 0$.

The equation (for the function $x(\cdot)$) $F_x = 0$ is called a (partial) differential equation of order p. If $I \subset R^1$, we speak of an ordinary differential equation. The order of the highest derivative appearing in the given equation is the order of the equation.

Henceforth we shall deal with ordinary differential equations of the first order. Equations of higher order, as well as systems of higher-order and first-order equations will be reduced to a single first-order equation.

Examples of Differential Equations. 1. The system of *equations of motion for one particle*

$$m \cdot \frac{d^2 x^i}{dt^2} = F_i\left(t, x^1, x^2, x^3, \frac{dx^1}{dt}, \frac{dx^2}{dt}, \frac{dx^3}{dt}\right), \quad i = 1, 2, 3.$$

2. The *equation for an electric circuit* containing a resistance R, inductance L, capacitance C, and an external electromotive force E:

$$L \frac{d^2 I}{dt^2} + R \frac{dI}{dt} + \frac{I}{C} = \frac{dE}{dt}.$$

3. The *system of Hamilton's equations*

$$\frac{dx_i}{dt} = \frac{\partial H(t, x_1, \dots, x_n, p_1, \dots, p_n)}{\partial p_i},$$

$$\frac{dp_j}{dt} = -\frac{\partial H(t, x_1, \dots, x_n, p_1, \dots, p_n)}{\partial x_j}, \qquad i, j = 1, 2, \dots, n.$$

This is a system of ordinary equations since the unknown functions $x_i(\cdot)$ and $p_j(\cdot)$ are functions of one variable. On the other hand, H is a known function.

4. The *wave equation*

$$\frac{\partial^2 u}{\partial x_1^2} + \frac{\partial^2 u}{\partial x_2^2} + \frac{\partial^2 u}{\partial x_3^2} - \frac{1}{c^2} \cdot \frac{\partial^2 u}{\partial t^2} = 0.$$

5. The *Laplace equation*

$$\frac{\partial^2 u}{\partial x_1^2} + \frac{\partial^2 u}{\partial x_2^2} + \frac{\partial^2 u}{\partial x_3^2} = 0.$$

6. The *heat equation*

$$\frac{\partial^2 u}{\partial x_1^2} + \frac{\partial^2 u}{\partial x_2^2} + \frac{\partial^2 u}{\partial x_3^2} = \frac{1}{a^2} \cdot \frac{\partial u}{\partial t}.$$

Equations 1, 2, and 3 are examples of ordinary equations, whereas 4, 5, and 6 are partial equations.

At this point we shall proceed to discuss ordinary differential equations of the first order. An equation of this kind is of the form

$$F\big(t, x(t), x'(t)\big) \underset{t}{\equiv} 0,$$

where F is a mapping of the Cartesian product $I \times X \times X$ into X. This form of the equation is, however, too general and thus most often we deal with equations written as

$$\frac{dx}{dt} = f(t, x),$$

where f is a mapping of the Cartesian product $I \times X$ into X. An equation of this kind is said to have been *reduced to the normal form*.

A question arises: When does the equation $\frac{dx}{dt} = f(t, x)$ have exactly one solution?

It turns out that in general such an equation has an infinite number of solutions. It is therefore purposeful to pose the following problem: "For the differential equation

$$\frac{dx}{dt} = f(t, x)$$

find a solution $x(\cdot)$ which satisfies the initial condition $x(t_0) = x_0$."

We shall show that with suitable assumptions as to the mapping f, this problem has exactly one solution.

First, we give a definition:

DEFINITION. A mapping f of a Cartesian product $R^1 \times X$ into X is a *Lipschitz mapping* with *Lipschitz constant* $L > 0$, if for all (t, x_1, x_2) $\in R^1 \times X \times X$

$$\|f(t, x_1) - f(t, x_2)\| \leqslant L\|x_1 - x_2\|.$$

Example. If a mapping f is differentiable with respect to x and if $\|f'_x(t, x)\| \leqslant L$ for $x \in X$ and $t \in R^1$, then it follows from the mean-value theorem that f satisfies the Lipschitz condition with the Lipschitz constant L.

The problem of finding for a differential equation a solution which would satisfy an initial condition is called the *initial-value problem* or, as it is often called, *Cauchy's problem*. We shall give a theorem on the existence and uniqueness of the solution of the initial-value problem.

THEOREM IX.2.1. Let V be an open set of a Banach space X and let $I = [a, b] \subset R^1$. Let f be a mapping of the Cartesian product $I \times V$ into X,

$$R^1 \times X \supset I \times V \ni (t, x) \to f(t, x) \in X,$$

satisfying the following conditions:

(a) the mapping f is continuous on $I \times V$;

(b) there exist balls $K_1(t_0, r_1) \subset I$, $K_2(x_0, r_2) \subset V$ and an $M > 0$, such that

$$\|f(t, x)\| \leqslant M \quad \text{for } (t, x) \in K_1 \times K_2$$

(that is, f is bounded on $K_1 \times K_2$);

(c) f satisfies the Lipschitz condition, i.e. there exists a constant $L > 0$ such that

$$\|f(t, x_1) - f(t, x_2)\| \leqslant L\|x_1 - x_2\|$$

for $t \in K_1$, $(x_1, x_2) \in K_2 \times K_2$.

Then there exists a $\tau > 0$ such that for $|t - t_0| < \tau$ the equation

(R) $$\frac{dx}{dt} = f(t, x)$$

has exactly one solution $t \to x(t)$ satisfying the initial condition

(P) $$x(t_0) = x_0.$$

PROOF. We begin with the equation

$$\frac{dx}{dt} = f(t, x).$$

Integration of both sides, when the initial condition $x(t_0) = x_0$ has been taken into account, yields

(1) $$x(t) = x_0 + \int_{t_0}^{t} f(s, x(s))\, ds.$$

We immediately verify that the problem (R), (P) is equivalent to the sought solution of equation (1).

We introduce the mapping P_{x_0}:

$$C(|t - t_0| < r_1, X) \ni g \to P_{x_0}(g) \in C(|t - t_0| < r_1, X),$$

where

(2) $$(P_{x_0}(g))(t) := x_0 + \int_{t_0}^{t} f(s, g(s))\, ds.$$

It follows from (1) and (2) that the solution of our initial-value problem is a fixed point of the mapping P_{x_0}. In order to prove the existence of a fixed point we shall invoke Banach's principle. The complete space

E will here be a closed ball in the space of functions continuous on $[t_0-\varrho, t_0+\varrho]$ (with values in X) with radius r_2 and centre at $g_0(t) \equiv x_0$:

$$E := \overline{K(g_0, r_2)} \subset C(|t-t_0| \leqslant \varrho, X).$$

We shall verify that

(i) $P_{x_0}: E \to E$,

(ii) the mapping P_{x_0} is a contraction.

(i) Let $g \in E$, then

$$\|P_{x_0}(g)-g_0\| = \sup_{|t-t_0|\leqslant\varrho} \left\| x_0 + \int_{t_0}^{t} f(s, g(s))\,ds - x_0 \right\|$$

$$\leqslant \sup_{|t-t_0|\leqslant\varrho} \int_{t_0}^{t} \|f(s, g(s))\|\,ds \leqslant \varrho \cdot M, \quad \text{when } \varrho < r_1.$$

Therefore, if we take $\varrho < \min(r_2/M, r_1)$, condition (i) will be satisfied.

(ii) We have

$$\|P_{x_0}(g_1)-P_{x_0}(g_2)\|$$

$$= \sup_{|t-t_0|\leqslant\varrho} \left\| x_0 + \int_{t_0}^{t} f(s, g_1(s))\,ds - x_0 - \int_{t_0}^{t} f(s, g_2(s))\,ds \right\|$$

$$\leqslant \sup_{|t-t_0|\leqslant\varrho} \int_{t_0}^{t} \|f(s, g_1(s)) - f(s, g_2(s))\|\,ds$$

$$\leqslant \sup_{|t-t_0|\leqslant\varrho} \sup_{s\in[t, t_0]} L\|g_1(s)-g_2(s)\| \cdot |t-t_0| \leqslant \varrho L\|g_1-g_2\|.$$

If $\varrho < 1/L$ the mapping P_{x_0} is a contraction.

Thus the hypotheses of the Banach principle are satisfied for

$$\varrho < \tau = \min(r_1, r_2/M, 1/L).$$

Thus, the initial-value problem has exactly one solution defined on the interval $]t_0-\tau, t_0+\tau[$. $\square$

It turns out that hypothesis (b) of Theorem IX.2.1 is not indispensable. For it follows from hypothesis (a).

LEMMA IX.2.2. Let $I = [a, b]$ and let X and Y be metric spaces. If the mapping $f: I \times X \to Y$ is continuous, then for every $x_0 \in X$ there exists a ball K with centre at x_0 and radius r such that the set $f(I \times K)$ is bounded in Y.

PROOF. The continuity of the mapping f implies that for every $t \in [a, b]$ there exists a ball K_t^1 (in $[a, b]$) centred on t and a ball $K_t^2 \subset X$

centred on x_0 such that the set $f(K_t^1 \times K_t^2)$ is bounded. The (open) balls K_t^1 cover the interval I. The compactness of I implies the existence of a finite number of balls $K_{t_1}^1, \ldots, K_{t_n}^1$ covering I. When out of the corresponding balls $K_{t_1}^2, \ldots, K_{t_n}^2$ we select the ball $K_{t_{i_0}}^2$ with the smallest radius, we find that the sets $f(K_{t_i}^1 \times K_{t_{i_0}}^2)$, $i = 1, 2, \ldots, n$, are bounded. Hence the set $f(I \times K_{t_{i_0}}^2)$ is bounded. $\square$

Now we can reformulate Theorem IX.2.1 with somewhat weaker hypotheses.

THEOREM IX.2.3. Let V be an open set of a Banach space X and let $I = [a, b] \subset R^1$. Let f be a mapping of the Cartesian product $I \times V$ into X which satisfies the assumptions that:

(a) f is continuous on $I \times V$;

(b) there exists a ball $K(x_0, r_1) \subset V$ such that on $I \times K$ the mapping satisfies the Lipschitz condition with Lipschitz constant $L > 0$.

Then there exists a $\tau > 0$ such that the initial-value problem

$$\frac{dx}{dt} = f(t, x), \quad x(t_0) = x_0,$$

has exactly one solution defined on $]t_0 - \tau, t_0 + \tau[$.

PROOF. By virtue of Lemma IX.2.2 there exists an $M > 0$ and an $r_2 > 0$ such that $\|f(t, x)\| \leqslant M$ for $(t, x) \in I \times K(x_0, r_2)$. The hypotheses of Theorem IX.2.1 are thus satisfied. Hence there exists exactly one solution defined on $]t_0 - \tau, t_0 + \tau[$, where

$$\tau = \min\left(\frac{r_1}{M}, \frac{r_2}{M}, \frac{1}{L}, |a - t_0|, |b - t_0|\right). \quad \square$$

The theorem above speaks of the existence of a solution on some interval $]t_0 - \tau, t_0 + \tau[$. In that interval take an arbitrary point t_1 lying a distance ε from its right end-point. The value of the function $x(\cdot)$ which is a solution of the initial-value problem at t_1 will be denoted by x_1; $x(t_1) = x_1$. In what follows we shall look for the solution of the equation satisfying the initial condition $x(t_1) = x_1$. If the hypotheses of Theorem IX.2.1 are satisfied for the point (t_1, x_1) there is a unique solution which satisfies the initial condition and is defined on the interval $]t_1 - \tau_1, t_1 + \tau_1[$, and which of course coincides with the original solution on the common part of the intervals $]t_0 - \tau, t_0 + \tau[$ and $]t_1 - \tau_1, t_1 + \tau_1[$. If $\tau_1 > \varepsilon$, we extend the solution to the interval $]t_0 - \tau, t_1 + \tau_1[$.

We can proceed further in analogous fashion and extend the solution (if this is possible) to the right of the point $t_1 + \tau_1$.

A similar procedure can be employed at the left-hand end-point of the interval $]t_0 - \tau, t_0 + \tau[$ and the integral curve passing through the point (t_0, x_0) can thus be extended to the maximal integral curve. In general, the solution does not admit extension to the entire interval $[a, b]$. For example, the equation $x' = x^2 + 1$ for $X = R^1$ has a unique solution $x(t) = \tan t$ satisfying the initial condition $x(0) = 0$. This solution is defined on the interval $]-\frac{1}{2}\pi, \frac{1}{2}\pi[$ and is not extendible beyond that interval since the tangent function has vertical asymptotes at the end-points of that interval.

We now introduce a convenient notation: the solution of the equation $x' = f(t, x)$ which assumes the value x_0 at the point $t = t_0$ will be denoted by the symbol $u(\cdot, t_0, x_0)$. We thus have $x(t) = u(t, t_0, x_0)$.

Thus it may be said that with t fixed the solution depends on an arbitrary vector x_0. If $X = R^n$, then n coordinates in a chosen (in general, canonical) basis are associated with the vector x_0. Accordingly, the solution of a first-order differential equation in an n-dimensional space depends on n constants. A solution dependent on n parameters is called the *general solution* as contrasted to a *particular solution* in which the constants have been replaced by particular numbers (substitution of a particular vector x_0). Solving an equation consists in finding the general solution and substituting constants such that the solution satisfies a given initial condition.

Theorem IX.2.1 not only states that a differential equation has a solution but also enables that solution to be found by the method of successive approximations. Namely, the solution $u(\cdot, t_0, x_0)$ is the limit of the sequence $P_{x_0}^n(x_0)$:

$$u(\cdot, t_0, x_0) = \lim_{n \to \infty} P_{x_0}^n(x_0).$$

We now give the general form of a first-order equation for $X = R^n$.

A system of n scalar functions $x_1(\cdot), \ldots, x_n(\cdot)$ corresponds to the function $t \to x(t) \in R^n$. A system of n scalar functions $f_1, \ldots, f_n$ also corresponds to the mapping f. The equation $x' = f(t, x)$ with an initial condition $x(t_0) = x_0$ thus reduces to a system of n scalar equations

$$\frac{dx_i}{dt} = f_i(t, x_1, \ldots, x_n), \quad i = 1, 2, \ldots, n.$$

The initial condition is of the form $x_i(t_0) = x_{0i}$, where x_{0i} are the components of the vector x_0 in the canonical basis $x_0 = (x_{01}, \ldots, x_{0n})$.

Examples. 1. $X = R^1$. We seek the solution of the equation

$$\frac{dx}{dt} = t + x$$

satisfying the initial condition $x(0) = 0$.

We employ the method of successive approximations and get

$$\overset{0}{x} = x_0 = 0;$$

$$\overset{1}{x}(t) = 0 + \int_0^t s\,ds = \frac{1}{2}t^2;$$

$$\overset{2}{x}(t) = \int_0^t \left(s + \frac{1}{2}s^2\right)ds = \frac{1}{2}t^2 + \frac{1}{2\cdot3}t^3.$$

It is easily seen that

$$\overset{n}{x}(t) = \frac{1}{2!}t^2 + \frac{1}{3!}t^3 + \ldots + \frac{1}{(n+1)!}t^{n+1}.$$

Hence

$$\overset{n}{x}(t) \xrightarrow[n\to\infty]{} e^t - t - 1.$$

Thus, finally

$$x(t) = e^t - t - 1.$$

2. $X = R^2$. Take a system of two scalar equations,

$$\frac{dx_1}{dt} = x_2, \quad \frac{dx_2}{dt} = -x_1;$$

$$x_1(0) = 0, \quad x_2(0) = 1.$$

The successive approximations are of the form

$$\overset{0}{x}_1 = 0, \qquad\qquad \overset{0}{x}_2 = 1,$$

$$\overset{1}{x}_1(t) = \int_0^t ds = t, \qquad \overset{1}{x}_2(t) = 1 + \int_0^t (-0)\,ds = 1,$$

$$\overset{2}{x}_1(t) = \int_0^t ds = t, \qquad \overset{2}{x}_2(t) = 1 - \int_0^t s\,ds = 1 - \frac{t^2}{2!},$$

$$\overset{3}{x}_1(t) = \int_0^t \left(1 - \frac{s^2}{2}\right) ds = t - \frac{t^3}{3!},$$

$$\overset{3}{x}_2(t) = 1 - \int_0^t s\, ds = 1 - \frac{t^2}{2!}.$$

It is readily seen that

$$x_1(t) = \sin t, \qquad x_2(t) = \cos t.$$

3. Let $X = R^1$. We look for the solution of the equation

$$\frac{dx}{dt} = 2tx,$$

with the initial condition $x(0) = 1$.

We evaluate the successive approximations:

$$\overset{0}{x}(t) = x_0 = 1,$$

$$\overset{1}{x}(t) = 1 + \int_0^t 2s\, ds = 1 + t^2,$$

$$\overset{2}{x}(t) = 1 + \int_0^t 2s(1 + s^2) ds = 1 + t^2 + \frac{1}{2!} t^4,$$

$$\overset{3}{x}(t) = 1 + \int_0^t 2s\left(1 + s^2 + \frac{1}{2!} s^4\right) ds = 1 + t^2 + \frac{1}{2!} t^4 + \frac{1}{3!} t^6.$$

It is easily seen that $\overset{n}{x}(t) \xrightarrow[n \to \infty]{} e^{t^2}$.

3. THE DEPENDENCE OF THE SOLUTION ON THE PARAMETER

LEMMA IX.3.1. Let X, Y, Z be Banach spaces and let $K \subset X$ be a compact set in X. If f is a continuous mapping of the product $K \times Y$ into Z, then f is uniformly continuous with respect to $x \in K$; that is, for every $\varepsilon > 0$ and every $y_0 \in Y$ there is a $\delta > 0$ such that for every $y \in K(y_0, \delta)$ and every $x \in K$

$$\|f(x, y) - f(x, y_0)\| < \varepsilon.$$

PROOF (a.a). Suppose that there exist an $\varepsilon > 0$, an $y_0 \in Y$, and a sequence (y_n) such that $\|y_0 - y_n\| < 1/n$, as well as a sequence $(x_n) \in K$ such that

(1) $\|f(x_n, y_n) - f(x_n, y_0)\| \geqslant \varepsilon$

for every n. By virtue of the compactness of the set K, a subsequence (x_{n_k}) convergent to $x_0 \in K$ can be extracted from the sequence (x_n). On selecting a subsequence (y_{n_k}) from the sequence (y_n) in the same way, we clearly have $y_{n_k} \to y_0$. The continuity of the mapping f implies that

$$\|f(x_{n_k}, y_{n_k}) - f(x_{n_k}, y_0)\|$$
$$\leqslant \|f(x_{n_k}, y_{n_k}) - f(x_0, y_0)\| + \|f(x_0, y_0) - f(x_{n_k}, y_0)\| < \varepsilon$$

for $k > N$. This, however, contradicts formula (1). $\square$

Given a family of differential equations

$$\frac{dx}{dt} = f(t, x, y),$$

where $y \in Y$, and Y is a Banach space. Solving these equations, we obtain a family of solutions dependent on the parameter y. A case of interest is that when every solution of this family satisfies the same initial conditions. It then turns out that under certain assumptions concerning the mapping f the solution is continuously dependent on the parameter.

THEOREM IX.3.2. Let $I = [a, b]$, let U be an open set in a Banach space X, and let V be an open set in a Banach space Y. Suppose that f is a continuous mapping of the product $I \times U \times V$ into X:

$$I \times U \times V \ni (t, x, y) \to f(t, x, y) \in X.$$

Suppose, furthermore, that f has a partial derivative with respect to X and let the mapping

$$I \times U \times V \ni (t, x, y) \to f'_x(t, x, y) \in L(X, X)$$

be continuous on $I \times U \times V$.

Then for any points $x_0 \in U$ and $y_0 \in V$ there are a $\tau > 0$ and an $r > 0$ such that for every $y \in K(y_0, r)$ there exists exactly one solution $t \to x(t, y)$ of the equation $dx/dt = f(t, x, y)$ satisfying the initial condition $x(t_0, y) = x_0$ and defined on the interval $J =]t_0 - \tau, t_0 + \tau[$. Moreover, the mapping

$$J \times K(y_0, r) \ni (t, y) \to x(t, y) \in X$$

is continuous and bounded on $J \times K(y_0, r)$.

PROOF. From the continuity of the mapping $(t, x, y) \to f(t, x, y)$ and $(t, x, y) \to f'_x(t, x, y)$ on $I \times U \times V$ and from Lemma IX.2.2 it follows that there are constants $M > 0$ and $L > 0$ as well as balls $K_1(x_0, r_1) \subset U$ and $K_2(y_0, r_2) \subset V$ such that

(1) $\qquad \|f(t, x, y)\| \leqslant M \qquad$ for $(t, x, y) \in I \times K_1 \times K_2$,

(2) $\qquad \|f'_x(t, x, y)\| \leqslant L \qquad$ for $(t, x, y) \in I \times K_1 \times K_2$.

Now, (2) implies that for every $y \in K_2(y_0, r_2)$ the mapping f satisfies the Lipschitz condition with respect to the variable x with Lipschitz constant equal to L. By Theorem IX.2.1, therefore, for every $y \in K_2(y_0, r_2)$ the equation $x' = f(t, x, y)$ has a single solution $t \to x(t, y)$ which is defined on the interval

$$J =]t_0 - \tau, t_0 + \tau[,$$

where $\tau = \min\left(\dfrac{r_1}{M}, \dfrac{1}{L}, |a - t_0|, |b - t_0|\right)$

and satisfies the initial condition $x(t_0, y) = x_0$.

To demonstrate that the mapping

$$J \times K_2(y_0, r_2) \ni (t, y) \to x(t, y) \in X$$

is continuous, it is sufficient to show it to be continuous on every set of the form $J_1 \times K_2$, where J_1 is an arbitrary closed interval contained in J. Thus, let J_1 be a closed interval contained in J and containing the points t and t_0. We fix $y \in K_2(y_0, r_2)$. For any $y_1 \in K_2(y_0, r_2)$ we have

$$\|x(t, y_1) - x(t, y)\|$$

$$\leqslant \left\| \int_{t_0}^{t} (f(s, x(s, y_1), y_1) - f(s, x(s, y), y)) ds \right\|$$

$$\leqslant \left\| \int_{t_0}^{t} (f(s, x(s, y_1), y_1) - f(s, x(s, y), y_1)) ds \right\| +$$

$$+ \left\| \int_{t_0}^{t} (f(s, x(s, y), y_1) - f(s, x(s, y), y)) ds \right\|$$

$$\leqslant |t - t_0| L \sup_{s \in [t_0, t]} \|x(s, y_1) - x(s, y)\| +$$

$$+ |t - t_0| \sup_{s \in [t_0, t]} \|f(s, x(s, y), y_1) - f(s, x(s, y), y)\|.$$

But Lemma IX.3.1 implies that

$$\sup_{s \in J_1} \|f(s, x(s, y), y_1) - f(s, x(s, y), y)\| < \varepsilon,$$

provided that $\|y - y_1\| < \delta$,

and consequently we have

$$\|x(t, y_1) - x(t, y)\| \leqslant |t - t_0| L \sup_{s \in J_1} \|x(s, y_1) - x(s, y)\| + \tau \varepsilon.$$

Taking the least upper bound of both sides with respect to $t \in J_1$, we obtain

$$\sup_{t \in J_1} \|x(t, y_1) - x(t, y)\| (1 - q \cdot L) \leqslant \tau \varepsilon,$$

where $q := \sup_{t \in J_1} |t - t_0|$. It is easily seen that $q < \tau$, and hence $q \cdot L < 1$, whereby we can write

$$\sup_{t \in J_1} \|x(t, y_1) - x(t, y)\| \leqslant \frac{\tau \varepsilon}{1 - q \cdot L}.$$

Finally, therefore,

$$(3) \qquad \|x(t, y_1) - x(t, y)\| \leqslant \frac{\tau \varepsilon}{1 - q \cdot L} \qquad \text{for } \|y_1 - y\| < \delta.$$

On the other hand,

$$(4) \qquad \|x(t_1, y_1) - x(t, y_1)\| \leqslant \left\| \int_t^{t_1} f(s, x(s, y_1), y_1) ds \right\| \leqslant |t_1 - t| M,$$

$$\|x(t_1, y_1) - x(t, y)\|$$
$$\leqslant \|x(t_1, y_1) - x(t, y_1)\| + \|x(t, y_1) - x(t, y)\|.$$

Use of (3) and (4) yields

$$\|x(t_1, y_1) - x(t, y)\| \leqslant \frac{\tau \varepsilon}{1 - q \cdot L} + |t - t_1| M$$

for $\|y_1 - y\| < \delta$ and any $t_1 \in J_1$.

The continuity has thus been proved. The boundedness of the mapping $(t, y) \rightarrow x(t, y)$ follows from the inequality

$$\|x(t, y)\| \leqslant \|x_0\| + M\tau, \qquad y \in K_2(y_0, r_2), \ t \in J. \qquad \square$$

LEMMA IX.3.3. Let $J = [t_0 - a, t_0 + a]$ and let $B(\cdot)$ be an operator-valued function defined on J and continuous on J: $J \ni t \rightarrow B(t) \in L(X, Y)$

(X is a Banach space). Let $C(J; X)$ be the Banach space of functions continuous on the interval J with values in X and norm $\|f\|_{C(J;X)} = \sup\limits_{t \in J} \|f(t)\|_X$.

Then the mapping T of the space $C(J; X)$ into itself,

$$C(J; X) \ni f \to T(f) \in C(J; X),$$

defined by the formula

$$(5) \qquad T(f)(t) := f(t) - \int_{t_0}^{t} B(s) \cdot f(s) ds,$$

is a topological isomorphism of the space $C(J; X)$.

PROOF. By virtue of formula (5), the mapping T may be written as

$$(6) \qquad T = I - A,$$

where I is the identity mapping in the space $C(J; X)$, and A is the mapping of $C(J; X)$ into itself:

$$(7) \qquad A(f)(t) := \int_{t_0}^{t} B(s) \cdot f(s) ds.$$

The mapping A is undoubtedly linear. It is also continuous. For we have

$$(8) \qquad \|A\|_{L(C(J;X), C(J;X))} = \sup\limits_{\|f\|_{C(J;X)} \leqslant 1} \|A(f)\|_{C(J;X)}$$

$$= \sup\limits_{\|f\|_{C(J;X)} \leqslant 1} \sup\limits_{t \in J} \left\| \int_{t_0}^{t} B(s) \cdot f(s) ds \right\|$$

$$\leqslant a \sup\limits_{\|f\|_{C(J;X)} \leqslant 1} \sup\limits_{t \in J} \|B(t)\|_{L(X,X)} \cdot \|f\|_{C(J;X)}$$

$$= a \cdot \sup\limits_{t \in J} \|B(t)\|_{L(X,X)} = aM,$$

where

$$M := \sup\limits_{t \in J} \|B(t)\|_{L(X, X)}.$$

By virtue of (6) the mapping T thus is also linear and continuous. The problem hence reduces to one of showing that the mapping $I - A$ is invertible. We know that this can always be done if $\|A\| < 1$. In our case, as is seen from formula (8), this is not so in general. We shall prove, however, that if into the space $C(J; X)$ we introduce a new norm equivalent to the previous one, we can get the condition $||A||_1 < 1$. Thus, for $f \in C(J; X)$ let us assume that

$$(9) \qquad ||f||_1 = \sup_{s \in J} e^{-\lambda|s-t_0|}||f(s)||, \qquad \lambda > 0.$$

It is not difficult to see that

$$(10) \qquad ||f||_1 \leqslant ||f||, \qquad ||f||_1 \geqslant e^{-\lambda a}||f||.$$

Formulae (10) show that the new norm is equivalent to the old one. The space $C(J, X)$ in this norm thus is also a Banach space. We denote it by the symbol $C_1(J; X)$. We have

$$||A||_{L(C_1(J;X),C_1(J;X))} = \sup_{||f||_1 \leqslant 1} ||A(f)||_1$$

$$\leqslant \sup_{||f||_1 \leqslant 1} \sup_{t \in J} \left|\left| e^{-\lambda|t-t_0|} \int_{t_0}^{t} B(s) \cdot f(s) ds \right|\right|_X$$

$$\leqslant \sup_{||f||_1 \leqslant 1} \sup_{t \in J} e^{-\lambda|t-t_0|} \left|\left| \int_{t_0}^{t} e^{\lambda|s-t_0|} B(s) \cdot f(s) e^{-\lambda|s-t_0|} ds \right|\right|$$

$$\leqslant \sup_{||f||_1 \leqslant 1} \sup_{t \in J} \left(M \left| \int_{t_0}^{t} e^{\lambda|s-t_0|} ds \right| \cdot ||f||_1 e^{-\lambda|t-t_0|} \right)$$

$$= M \sup_{t \in J} \left(e^{-\lambda|t-t_0|} \frac{1}{\lambda} |e^{\lambda|t-t_0|} - 1| \right)$$

$$= M \sup_{t \in J} \frac{1}{\lambda} (1 - e^{-\lambda|t-t_0|}) = \frac{M}{\lambda}(1 - e^{-\lambda a}).$$

Hence, on choosing suitably large λ, we have $||A||_1 < 1$. Accordingly, by the corollaries to Theorem VIII.1.3, there is a mapping $(I-A)^{-1}$ which is continuous in the norm $|| \cdot ||_1$. But, inasmuch as the two norms are equivalent, it is also continuous in the norm $|| \cdot ||$. $\square$

LEMMA IX.3.4. Let X and Y be Banach spaces, let $K_1(x_0, r_1)$ be an open ball in X, and let $K_2(y_0, r_2)$ be an open ball in Y. Now, suppose that f is a continuous mapping of the set $J \times K_1 \times K_2$ into X, where J is the open interval $]t_0 - b, t_0 + b[$:

$$J \times K_1 \times K_2 \ni (t, x, y) \rightarrow f(t, x, y) \in X.$$

Furthermore, suppose that the mapping f has the continuous partial derivatives $f'_x(t, x, y)$ and $f'_y(t, x, y)$ on $J \times K_1 \times K_2$. Let $C(I; X)$ be the Banach space of continuous functions on the interval $I = [t_0 - a, t_0 + a]$

3. DEPENDENCE OF SOLUTION ON PARAMETER

($a < b$) with values in the space X, with supremum norm. Let $K_3(g_0, r_1)$ be the ball in the space $C(I; X)$ with centre at the point g_0 ($g_0(t) \underset{t \in I}{\equiv} x_0$) and radius r_1. Then the mapping H

$$Y \times C(I, X) \supset K_2 \times K_3 \ni (y, g) \to H(y, g) \in C(I, X)$$

defined by the formula

$$H(y, g)(t) := g(t) - \int_{t_0}^{t} f(s, g(s), y) \, ds$$

is continuously differentiable, and

$$(H'_{C(I;X)}(y, g) \cdot h)(t) = h(t) - \int_{t_0}^{t} f'_x(s, g(s), y) \cdot h(s) \, ds,$$

$$h \in C(I; X).$$

Moreover, for any $(y, g) \in K_2 \times K_3$, the mapping $H'_{C(I;X)}(y, g)$ is a topological isomorphism of the space $C(I; X)$.

PROOF. In view of Theorem VII.6.2, it is sufficient to prove that the partial derivatives exist and are continuous. Thus, let $\tilde{y} \in K_2(y_0, r_2)$, and $\tilde{g} \in K(g_0, r_1)$. Let $z \in Y$ and $h \in C(I, X)$. We evaluate the increment

$$H(\tilde{y} + z, \tilde{g})(t) - H(\tilde{y}, \tilde{g})(t)$$

$$= - \int_{t_0}^{t} \left(f(s, \tilde{g}(s), \tilde{y} + z) - f(s, \tilde{g}(s), y) \right) ds$$

$$= - \int_{t_0}^{t} f'_Y(s, \tilde{g}(s), \tilde{y}) \cdot z \, ds - \int_{t_0}^{t} r_1(s, \tilde{g}(s), \tilde{y}; z) \, ds.$$

By reason of the existence of a continuous partial derivative with respect to Y, we can invoke Lemma VII.5.4:

(11) $\|r_1(s, \tilde{g}(s), \tilde{y}; z)\|$

$\leqslant \|z\| \sup_{0 \leqslant \theta \leqslant 1} \|f'_Y(s, \tilde{g}(s), \tilde{y} + \theta z) - f'_Y(s, \tilde{g}(s), \tilde{y})\|.$

Application of Lemma IX.3.1 to the mapping

$$(s, y) \to u(s, y) := f'_Y(s, \tilde{g}(s), y)$$

yields

(12) $\|f'_Y(s, \tilde{g}(s), y) - f'_Y(s, \tilde{g}(s), \tilde{y})\| < \varepsilon$

  for $\|y - \tilde{y}\| < \delta$ and every $s \in I$.

When we use (11) and (12), we obtain

(13) $\sup\limits_{s \in I} \|r_1(s, \tilde{g}(s), \tilde{y}; z)\| \leqslant \|z\| \varepsilon$ for $\|z\| < \delta$.

From (13) we have

(14) $\sup\limits_{t \in I} \left\| \int_{t_0}^{t} r_1(s, \tilde{g}(s), \tilde{y}; z)\, ds \right\| \leqslant a\|z\| \varepsilon$ for $\|z\| < \delta$.

We have thus evaluated the remainder and shown that

(15) $\left(H_Y'(\tilde{y}, \tilde{g}) \cdot z\right)(t) = - \int_{t_0}^{t} f_Y'(s, \tilde{g}(s), \tilde{y}) \cdot z\, ds.$

Now we go on to prove the existence of the other partial derivative:

$$H(\tilde{y}, \tilde{g}+h)(t) - H(\tilde{y}, \tilde{g})(t)$$

$$= h(t) - \int_{t_0}^{t} \left(f(s, \tilde{g}(s)+h(s), \tilde{y}) - f(s, \tilde{g}(s), \tilde{y}) \right) ds$$

$$= h(t) - \int_{t_0}^{t} f_X'(s, \tilde{g}(s), \tilde{y}) \cdot h(s)\, ds - \int_{t_0}^{t} r_2(s, \tilde{g}(s), \tilde{y}; h(s))\, ds.$$

In analogy to formula (11) we write

(16) $\|r_2(s, \tilde{g}(s), \tilde{y}; h(s))\|$

$\leqslant \|h(s)\| \sup\limits_{0 \leqslant \theta \leqslant 1} \|f_X'(s, \tilde{g}(s)+\theta h(s), \tilde{y}) - f_X'(s, g(s), \tilde{y})\|.$

When we apply Lemma IX.3.1 to the mapping

$$(s, x) \to v(s, x) := f_X'(s, \tilde{g}(s)+x, \tilde{y})$$

we have

(17) $\|f_X'(s, \tilde{g}(s)+h(s), \tilde{y}) - f_X'(s, \tilde{g}(s), \tilde{y})\| < \varepsilon$

for every $h \in C(I; X)$ such that $\|h\| < \delta$ and for every $s \in I$.

From (16) and (17) we obtain

(18) $\sup\limits_{s \in I} \|r_2(s, \tilde{g}(s), \tilde{y}; h(s))\| \leqslant \|h\| \varepsilon$ for $\|h\| < \delta$.

Formula (18) gives us

(19) $\sup\limits_{t \in I} \left\| \int_{t_0}^{t} r_2(s, \tilde{g}(s), \tilde{y}; h(s))\, ds \right\| \leqslant \|h\| \varepsilon a$ for $\|h\| < \delta$.

Thus, we have evaluated the remainder and we have the formula

$$(20) \qquad (H'_{C(I;X)}(\tilde{y}, \tilde{g}) \cdot h)(t) = h(t) - \int_{t_0}^{t} f'_x(s, \tilde{g}(s), \tilde{y}) \cdot h(s) ds.$$

Now we shall demonstrate the continuity of the second partial derivative; the continuity of the first partial derivative is proved in analogous manner.

When Lemma IX.3.1 is applied to the mapping $(s, x, y) \to w(s, x, y)$ $:= f'_x(s, \tilde{g}(s)+x, y)$ the result is

$$(21) \qquad \|w(s, x, y) - w(s, 0, \tilde{y})\| < \varepsilon$$

for $\|x\| < \delta_1$, $\|y - \tilde{y}\| < \delta_2$ and for every $s \in I$.

From formula (21) we have

$$(22) \qquad \|f'_x(s, \tilde{g}(s)+g_1(s), y) - f'_x(s, \tilde{g}(s), \tilde{y})\| < \varepsilon$$

for $\|g_1\| < \delta_1$, $\|y - \tilde{y}\| < \delta_2$, $s \in I$,

and thus

$$\|H'_{C(I;X)}(\tilde{y}+y_1, \tilde{g}+g_1) - H'_{C(I;X)}(\tilde{y}, \tilde{g})\|$$

$$= \sup_{\|h\| \leqslant 1} \sup_{t \in I} \left\| \int_{t_0}^{t} \left(f'_x(s, \tilde{g}(s)+g_1(s), \tilde{y}+y_1) \right. \right.$$

$$\left. \left. - f'_x(s, \tilde{g}(s), \tilde{y}) \right) h(s) ds \right\|.$$

On making use of formula (22), we have

$$\|H'_{C(I;X)}(\tilde{y}+y_1, \tilde{g}+g_1) - H'_{C(I;X)}(\tilde{y}, \tilde{g})\| \leqslant a\varepsilon$$

for $\|g_1\| < \delta_1$, $\|y_1\| < \delta_2$.

We have thus demonstrated the continuity of the second partial derivative. Once the first partial derivative has been shown to be continuous, we obtain the first part of the lemma from Theorem VII.6.2. The second part follows immediately from Lemma IX.3.3. $\square$

Now we shall prove a theorem on the differentiability of the solutions of a differential equation with respect to the parameter.

THEOREM IX.3.5. Let $I = [a, b]$, let U be an open set in a Banach space X, and let V be an open set in a Banach space Y. Let f be a continuous mapping of the product $I \times U \times V$ into the space X:

$$R^1 \times X \times Y \supset I \times U \times V \ni (t, x, y) \to f(t, x, y) \in X.$$

Furthermore, suppose that the mapping f has the continuous partial derivatives $f_X'(t, x, y)$ and $f_Y'(t, x, y)$ on $I \times U \times V$. In accordance with Theorem IX.3.2, for $x_0 \in U$, $y_0 \in V$, and $t_0 \in]a, b[$ let there exist a $\tau > 0$ and an $r > 0$ such that for an arbitrary $y \in K(y_0, r)$ there is a single solution $t \to x(t, y)$ of the equation

$$(23) \qquad \frac{dx}{dt} = f(t, x, y)$$

satisfying the initial condition $x(t_0, y) = x_0$ and defined on the interval $]t - \tau, t + \tau[$.

Then for every $\varrho < \tau$ there is an $r > r_1 > 0$ such that the mapping

$$I \times V \supset]t - \varrho, t + \varrho[\times K(y_0, r_1) \ni (t, y) \to x(t, y) \in X$$

is continuously differentiable on $]t - \varrho, t + \varrho[\times K(y_0, r_1)$. Moreover, for every fixed $y \in K(y_0, r_1)$ the mapping $t \to x_Y'(t, y) \in L(Y, X)$ satisfies the linear differential equation

$$\frac{d}{dt} x_Y'(t, y) = f_X'(t, x(t, y), y) \circ x_Y'(t, y) + f_Y'(t, x(t, y), y)$$

with the initial condition $x_Y'(t_0, y) = 0$.

PROOF. Let $J = [t - \varrho, t + \varrho]$. Let $K(g_0, r_2)$ be a ball in the space $C(J; X)$ with centre at the point g_0 ($g_0(t) \equiv x_0$) and radius r_2. Let H be the mapping introduced in Lemma IX.3.4:

$$Y \times C(J; X) \supset V \times K(g_0, r_2) \ni (y, g) \to H(y, g) \in C(J, X).$$

Then

$$(24) \qquad H(y, g)(t) := g(t) - \int_{t_0}^{t} f(s, g(s), y) \, ds.$$

Let F be a mapping of the set V into the space $C(J; X)$

$$(25) \qquad V \ni y \to F(y) \in C(J; X), \qquad F(y)(t) := x(t, y),$$

where $x(t, y)$ is a solution of equation (23), this solution being such that it satisfies the condition $x(t_0, y) = x_0$.

From (24) and (25) we see that

$$(26) \qquad H(y, F(y)) = x_0, \qquad y \in K(y_0, r).$$

It follows from formula (26) and Lemma IX.3.4 that the mapping H satisfies the hypotheses of the implicit-mapping theorems (Theorems

282

VIII.3.1, and VIII.3.2). Thus, there is an $r_1 > 0$ such that the mapping $K(y_0, r) \ni y \to F(y) \in C(J; X)$ is continuously differentiable on $K(y_0, r_1)$.

Now we shall show that the differentiability of the mapping $(t, y) \to x(t, y) \in X$ follows from this.

Equation (23) gives us

(27) $\qquad x_t'(t, y) = f\big(t, x(t, y), y\big).$

But Theorem IX.3.2 implies that the mapping $(t, y) \to x(t, y)$ is continuous on $J \times K(y_0, r_1)$ and hence, by formula (27) and the continuity of the mapping f, the partial derivative $x_t'(t, y)$ is continuous on $J \times K(y_0, r_1)$. Next we evaluate the partial derivative $x_Y'(t, y)$; we have

$$x(t, y+z) - x(t, y) = F(y+z)(t) - F(y)(t)$$
$$= \big(F'(y) \cdot z\big)(t) + r(y; z)(t),$$

but from the differentiability of the mapping F it follows that

$$\sup_{t \in J} \|r(y; z)(t)\| \leqslant \|z\|\varepsilon \quad \text{for } \|z\| < \delta.$$

We have thus demonstrated the existence of the partial derivative $x_Y'(t, y)$ and the formula

(28) $\qquad x_Y'(t, y) \cdot z = \big(F'(y) \cdot z\big)(t), \quad z \in Y.$

Let us prove the continuity of the partial derivative $x_Y'(t, y)$. We fix $t \in J$ and $y \in K(y_0, r_1)$; we have

(29) $\quad \|x_Y'(t_1, y) \cdot z - x_Y'(t, y) \cdot z\|_X = \|\big(F'(y) \cdot z\big)(t_1) - \big(F'(y) \cdot z\big)(t)\|_X$
$$\leqslant \|\big(F'(y) \cdot z\big)(t_1) - F(y+z)(t_1) + F(y)(t_1)\|_X +$$
$$+ \|F(y+z)(t_1) - F(y)(t_1) - F(y+z)(t) + F(y)(t)\|_X +$$
$$+ \|F(y+z)(t) - F(y)(t) - \big(F'(y) \cdot z\big)(t)\|_X.$$

From the differentiability of the mapping $y \to F(y)$ we have

(30) $\quad \sup_{t \in J} \|F(y+z)(t) - F(y)(t) - \big(F'(y) \cdot z\big)(t)\|_X \leqslant \varepsilon\|z\|$
$$\text{for } \|z\| < \delta_1.$$

We also have

(31) $\quad \|F(y+z)(t_1) - F(y)(t_1) - F(y+z)(t) + F(y)(t)\|_X$
$$\leqslant \|x(t_1, y+z) - x(t, y+z)\|_X + \|x(t, y) - x(t_1, y)\|_X.$$

The continuity of the mapping $(t, y) \rightarrow x(t, y)$, the continuity of the mapping f, and Lemma IX.2.2 imply that there exists an $M > 0$ such that

(32) $\qquad \|f(t, x(t, y+z), y+z)\|_x \leqslant M \quad$ for $t \in J, \|z\| < \delta_2.$

By the Mean-Value Theorem, and formulae (32) and (27) we have

(33) $\qquad \|x(t_1, y+z) - x(t, y+z)\|_x$

$\qquad \leqslant |t_1 - t| \sup_{0 \leqslant \theta \leqslant 1} \|x'_t(t + (t-t_1)\theta, y+z)\| \leqslant M|t_1 - t|$

$\qquad$ for $\|z\| < \delta_2;$

(34) $\qquad \|x(t_1, y) - x(t, y)\|_x \leqslant |t_1 - t| \sup_{0 \leqslant \theta \leqslant 1} \|x'_t(t + (t_1-t)\theta, y)\|_x$

$\qquad \leqslant M|t_1 - t|.$

Substituting (33) and (34) into (31), and then (30) and (31) into (29), we obtain

(35) $\qquad \|x'_Y(t_1, y) \cdot z - x'_Y(t, y) \cdot z\|_x \leqslant 2\varepsilon\|z\| + 2M|t_1 - t|$

$\qquad$ for $\|z\| < \delta_3 = \min(\delta_1, \delta_2).$

Therefore

$$\|x'_Y(t_1, y) - x'_Y(t, y)\|_{L(Y, X)} = \sup_{\|v\| \leqslant 1} \|x'_Y(t_1, y) \cdot v - x'_Y(t, y) \cdot v\|_x$$

$$= \frac{2}{\delta_3} \sup_{\|v\| \leqslant 1} \left\| x'_Y(t_1, y) \cdot v \frac{\delta_3}{2} - x'_Y(t, y) \cdot v \frac{\delta_3}{2} \right\|_x$$

and we can now invoke formula (35):

$$\leqslant \frac{2}{\delta_3} \left(2\varepsilon \cdot \frac{\delta_3}{2} + 2M|t_1 - t| \right) = 2\varepsilon + \frac{4M}{\delta_3}|t_1 - t|.$$

Putting $|t_1 - t| < \dfrac{\delta_3}{4M}\varepsilon = \delta_4$, we have

(36) $\qquad \|x'_Y(t_1, y) - x'_Y(t, y)\| \leqslant 3\varepsilon \quad$ for $|t_1 - t| < \delta_4.$

Now let $h \in Y$; then

(37) $\qquad \|x'_Y(t_1, y+h) - x'_Y(t_1, y)\|_{L(Y, X)} = \sup_{\|v\| \leqslant 1} \|x'_Y(t_1, y+h) \cdot v -$

$\qquad - x'_Y(t_1, y) \cdot v\|_x = \sup_{\|v\| \leqslant 1} \|(F'(y+h) \cdot v)(t_1) - (F'(y) \cdot v)(t_1)\|_x.$

As F is continuously differentiable, we have

(38) $\qquad \sup_{t \in J} \|(F'(y+h) \cdot v)(t) - (F'(y) \cdot v)(t)\|_x \leqslant \varepsilon\|v\|$

$\qquad$ for $\|h\| < \delta_5.$

284

3. DEPENDENCE OF SOLUTION ON PARAMETER

Inserting (38) into (37), we obtain

(39) $\|x'_Y(t_1, y+h) - x'_Y(t_1, y)\|_{L(Y,X)} \leqslant \varepsilon$

   for $\|h\| < \delta_5$ and every $t_1 \in J$.

Addition of both sides of inequalities (37) and (39) yields

(40) $\|x'_Y(t_1, y+h) - x'_Y(t, y)\|_{L(Y,X)} \leqslant 4\varepsilon$

   for $\|h\| < \delta_5$, $|t_1 - t| < \delta_4$.

Inequality (40) proves the continuity of the mapping $(t, y) \to x'_Y(t, y)$. By virtue of Theorem VII.6.2, the continuity of both partial derivatives implies the continuous differentiability of the mapping $J \times K(y_0, r_1)$ $\ni (t, y) \to x(t, y) \in X$.

We now prove the second part of the theorem. By formula (28) we have

$$x'_Y(t, y) \cdot z = (F'(y) \cdot z)(t).$$

From the implicit-mapping theorem it follows that

$$F'(y) = -\big(H'_{C(J;X)}(y, F(y))\big)^{-1} \circ H'_Y(y, F(y)).$$

This can be rewritten as

(41) $H'_{C(J;X)}(y, F(y)) \circ F'(y) = -H'_Y(y, F(y))$.

Formula (41) implies that for every $t \in J$ and any $z \in Y$,

(42) $\big(H'_{C(J;X)}(y, F(y)) \cdot F'(y) \cdot z\big)(t) = (-H'_Y(y, F(y)) \cdot z)(t)$.

From Lemma IX.3.4 we have

(43) $\big(H'_Y(y, F(y))z\big)(t) = -\int_{t_0}^{t} f'_Y(s, F(y)(s), y) \cdot z \, ds,$

(44) $\big(H'_{C(J;X)}(y, F(y))(F'(y) \cdot z)\big)(t)$

$$= (F'(y) \cdot z)(t) - \int_{t_0}^{t} f'_X(s, F(y)(s), y)(F'(y) \cdot z)(s) \, ds.$$

On inserting (43) and (44) into formula (42) and making use of (25) and (28), we obtain

(45) $x'_Y(t, y) \cdot z = \int_{t_0}^{t} f'_X(s, x(s, y), y) \circ x'_Y(s, y) \cdot z \, ds +$

$$+ \int_{t_0}^{t} f'_Y(s, x(s, y), y) \cdot z \, ds.$$

The right-hand side of equation (45) is differentiable with respect to t and hence the left-hand side is also differentiable with respect to t. Differentiation of both sides yields

$$(46) \quad \frac{d}{dt}\big(x'_Y(t, y) \cdot z\big)$$

$$= f'_X\big(t, x(t, y), y\big) \circ x'_Y(t, y) \cdot z + f'_Y\big(t, x(t, y), y\big) \cdot z.$$

Since equation (46) holds for every $z \in Y$ and in view of Lemma IX.3.6 (which we shall give below), we can write:

$$(47) \quad \frac{d}{dt} x'_Y(t, y) = f'_X\big(t, x(t, y), y\big) \circ x'_Y(t, y) + f'_Y\big(t, x(t, y), y\big).$$

When in formula (45) we set $t = t_0$ we get

$$x'_Y(t_0, y) \cdot z = 0 \quad \text{for every } z \in Y,$$

whence $x'_Y(t_0, y) = 0$. $\square$

LEMMA IX.3.6. Let X and Y be Banach spaces and let A be a mapping of the interval $[a, b]$ into the space $L(X, Y)$; $[a, b] \ni t \to A(t) \in L(X, Y)$. If for every $x \in X$ the mapping $t \to A(t) \cdot x \in Y$ is differentiable on $[a, b]$ and if $\frac{d}{dt}(A(t) \cdot x) = B(t) \cdot x$, where $B(t) \in L(X, Y)$, and the mapping $t \to B(t) \in L(X, Y)$ is continuous on $[a, b]$, then A is differentiable and

$$\frac{dA}{dt}(t) = B(t), \quad t \in [a, b].$$

PROOF. Let $x \in X$. By the Mean-Value Theorem

$$\left\| A(t+h) \cdot x - A(t) \cdot x - h\frac{d}{dt}\big(A(t) \cdot x\big) \right\|$$

$$\leqslant |h| \cdot \sup_{0 \leqslant \theta \leqslant 1} \left\| \frac{d}{dt}\big(A(t+\theta h) \cdot x\big) - \frac{d}{dt}\big(A(t) \cdot x\big) \right\|.$$

Therefore

$$\|A(t+h) \cdot x - A(t) \cdot x - hB(t) \cdot x\|$$
$$\leqslant |h| \sup_{0 \leqslant \theta \leqslant 1} \|B(t+\theta h) \cdot x - B(t) \cdot x\|.$$

Taking the least upper bound with respect to $||x|| \leqslant 1$ and making use of the continuity of the mapping $t \to B(t)$, we have

$$\sup_{||x|| \leqslant 1} ||A(t+h) \cdot x - A(t) \cdot x - hB(t) \cdot x|| \leqslant |h| \cdot \varepsilon$$

for $|h| < \delta$. $\square$

We now give a generalized version of Theorem IX.3.5.

THEOREM IX.3.7. Suppose that the hypotheses of Theorem IX.3.5 are satisfied and, moreover, let the mapping f have all partial derivatives of order up to p with respect to t, X, and Y (including mixed derivatives), these derivatives being continuous on $I \times U \times V$.

There then exist a $\tau > 0$, and an $r > 0$ such that the mapping $(t, y) \to x(t, y)$ is p times continuously differentiable on $K(t_0, \tau) \times K(y_0, r)$.

PROOF. By the methods used in proving Lemma IX.3.4 it may be shown that the mapping H is p times continuously differentiable. It follows from the formula for the derivative of an implicit function that the mapping F is also p times continuously differentiable and hence the mapping $(t, y) \to x(t, y)$ may also be shown to be p times differentiable. $\square$

4. THE DEPENDENCE OF THE SOLUTION ON THE INITIAL CONDITIONS

It will now be demonstrated that the solution of a differential equation depends continuously on the initial conditions.

THEOREM IX.4.1. Suppose that the hypotheses of Theorem IX.2.1 are satisfied for the point $(\tilde{t}, \tilde{x}) \in [a, b] \times X$, that is

(a) f is a continuous mapping on $[a, b] \times X$,

(b) there exist balls $K_1(\tilde{t}, r_1) \subset [a, b]$ and $K_2(\tilde{x}, r_2) \subset X$, such that $||f(t, x)|| \leqslant M$, $(t, x) \in K_1 \times K_2$,

(c) f satisfies the Lipschitz condition on $K_1 \times K_2$ with Lipschitz constant equal to L.

Then:

(i) There is a ball $K_3(\tilde{t}, r_3) \subset K_1(\tilde{t}, r_1)$ and a ball $K_4(\tilde{x}, r_4) \subset K_2(\tilde{x}, r_2) \subset X$ such that for any point $(t_0, x_0) \in K_3(\tilde{t}, r_3) \times K_4(x, r_4)$ the equation $dx/dt = f(t, x)$ has exactly one solution satisfying the initial condition $x(t_0) = x_0$ and defined on the entire ball $K_3(\tilde{t}, r_3)$.

(ii) Let $t \to u(t, t_0, x_0)$ be a solution satisfying the above initial condition. Then the mapping $K_3(\tilde{t}, r_3) \times K_3(\tilde{t}, r_3) \times K_4(\tilde{x}, r_4) \ni (t, t_0, x_0)$ $\to u(t, t_0, x_0) \in X$ is uniformly continuous on $K_3 \times K_3 \times K_4$.

(iii) There exists a ball $K_5(\tilde{x}, r_5) \subset K_4(\tilde{x}, r_4) \subset X$ such that for every point $(t, t_0, x_0) \in K_3 \times K_3 \times K_5$ the equation $x_0 = u(t_0, t, x)$ has exactly one solution $x \in K_4(\tilde{x}, r_4)$ equal to $u(t, t_0, x_0)$. The solution is then said to be soluble for the constants of integration.

PROOF. From Theorem IX.2.1 we know that there exists a unique solution of the equation $dx/dt = f(t, x)$ defined on the interval $]\tilde{t} - \tau, \tilde{t} + \tau[$, where $\tau = \min(r_2/M, 1/L, r_1)$, and satisfying the initial condition $x(\tilde{t})$ $= \tilde{x}$. We also know that

$$||x(t) - \tilde{x}|| < r_2 \quad \text{for } t \in K(\tilde{t}, \tau).$$

We shall now show that for any point $(t_0, x_0) \in K(\tilde{t}, \frac{1}{4}\tau) \times K(\tilde{x}, \frac{1}{2}r_2)$ there exists exactly one solution satisfying the initial condition $x(t_0)$ $= x_0$ and defined on the interval $]\tilde{t} - \frac{1}{4}\tau, \tilde{t} + \frac{1}{4}\tau[$. In order to show this, note that

$$||f(t, x)|| \leqslant M \quad \text{for } (t, x) \in K(t_0, \frac{1}{2}\tau) \times K(x_0, \frac{1}{2}r_2).$$

It is also seen that f satisfies on $K(t_0, \frac{1}{2}\tau) \times K(x_0, \frac{1}{2}r_2)$ the Lipschitz condition with constant equal to L. Thus, there exists exactly one solution assuming the value x_0 at t_0. This solution is defined on the interval $]t_0 - \varrho, t + \varrho[$ where

$$\varrho = \min\left(\frac{r_2}{2M}, \frac{1}{L}, \frac{1}{2}\tau\right) = \frac{1}{2}\tau.$$

It is thus seen that for every point $(t_0, x_0) \in K(t_0, \frac{1}{4}\tau) \times K(x_0, \frac{1}{2}r_2)$ the solution is defined on the interval $]t_0 - \frac{1}{2}\tau, t_0 + \frac{1}{2}\tau[$ which contains the interval $]\tilde{t} - \frac{1}{4}\tau, \tilde{t} + \frac{1}{4}\tau[$. Item (i) of the theorem has thus been proved.

(ii) Let $t_1, t_2, t_{01}, t_{02} \in K(\tilde{t}, \frac{1}{4}\tau)$ and $x_{01}, x_{02} \in K(\tilde{x}, \frac{1}{2}r_2)$. Let the functions

$$u_1(t_1, t_{01}, x_{01}) = u_1(t_1),$$
$$u_2(t_2, t_{02}, x_{02}) = u_2(t_2)$$

be solutions of the equation $x' = f(t, x)$ such that they satisfy the initial conditions $u_1(t_{01}) = x_{01}, u_2(t_{02}) = x_{02}$.

4. DEPENDENCE OF SOLUTION ON INITIAL CONDITIONS

In accordance with item (i) of this theorem, these functions are certainly defined on the interval $]\tilde{t}-\frac{1}{4}\tau, \tilde{t}+\frac{1}{4}\tau[$.

We may also write

$$u_1(t_1) = u_1(t_1, t_{01}, x_{01}) = x_{01} + \int_{t_{01}}^{t_1} f(s, u_1(s))\,ds,$$

$$u_2(t_2) = u_2(t_2, t_{02}, x_{02}) = x_{02} + \int_{t_{02}}^{t_2} f(s, u_2(s))\,ds.$$

When these formulae are used it is easily shown that

(1) $$\sup_{t \in K(\tilde{t},\tau/4)} \|u_2(t)-u_1(t)\| \leqslant \|x_{02}-x_{01}\| + 2M\tfrac{1}{2}\tau.$$

We make the evaluation

$$\|u_2(t)-u_1(t)\|$$

$$\leqslant \|x_{02}-x_{01}\| + \left\|\int_{t_{02}}^{t_{01}} f(s, u_2(s))\,ds\right\| +$$

$$+ \left\|\int_{t_{01}}^{t} \big(f(s, u_2(s)) - f(s, u_1(s))\big)\,ds\right\|$$

$$\leqslant \|x_{02}-x_{01}\| + |t_{01}-t_{02}| \cdot M +$$

$$+ |t-t_{01}| \cdot L \cdot \sup_{s \in K(\tilde{t},\tau/4)} \|u_2(s)-u_1(s)\|.$$

But it follows from formula (1) that $\sup_{s \in K(\tilde{t},\tau/4)} \|u_2(s)-u_1(s)\|$ is finite.

Thus we take the least upper bound of both sides with respect to $t \in K(\tilde{t}, \frac{1}{4}\tau)$:

(2) $$\sup_{t \in K(\tilde{t},\tau/4)} \|u_2(t)-u_1(t)\|$$

$$\leqslant \|x_{02}-x_{01}\| + |t_{01}-t_{02}| \cdot M + \tfrac{1}{2}\tau L \sup_{s \in K(\tilde{t},\tau/4)} \|u_2(s)-u_1(s)\|.$$

From (2) we have

(3) $$(1-\tfrac{1}{2}\tau L) \sup_{t \in K(\tilde{t},\tau/4)} \|u_2(t)-u_1(t)\|$$

$$\leqslant \|x_{02}-x_{01}\| + |t_{01}-t_{02}| \cdot M.$$

However, bear in mind that $\tau L \leqslant 1$, whereby $(1-\tfrac{1}{2}\tau L) \geqslant \tfrac{1}{2}$. Thus (3) yields

(4) $$\sup_{t \in K(\tilde{t},\tau/4)} \|u_2(t)-u_1(t)\| \leqslant 2\|x_{02}-x_{01}\| + 2M \cdot |t_{02}-t_{01}|.$$

289

Hence

(5) $\|u_2(t) - u_1(t)\| \leqslant 2\|x_{02} - x_{01}\| + 2M \cdot |t_{02} - t_{01}|$

for $t \in K(\bar{t}, \frac{1}{4}\tau)$.

On the other hand, we have

(6) $\|u_1(t_2) - u_1(t_1)\| \leqslant \left\|\int_{t_1}^{t_2} f(s, u_1(s)) ds\right\| \leqslant M \cdot |t_2 - t_1|$

for $t_1, t_2 \in K(\bar{t}, \frac{1}{4}\tau)$.

Adding both sides of (5) and (6), we obtain

(7) $\|u_2(t_2, t_{02}, x_{02}) - u_1(t_1, t_{01}, x_{01})\|$

$\leqslant \|u_2(t_2, t_{02}, x_{02}) - u_1(t_2, t_{01}, x_{01})\| +$

$+ \|u_1(t_2, t_{01}, x_{01}) - u_1(t_1, t_{01}, x_0)\|$

$\leqslant 2\|x_{02} - x_{01}\| + 2M|t_{02} - t_{01}| + M|t_2 - t_1|.$

Inequality (7) shows the uniform continuity of the mapping

$$K(\bar{t}, \tfrac{1}{4}\tau) \times K(\bar{t}, \tfrac{1}{4}\tau) \times K(\tilde{x}, \tfrac{1}{2}r_2) \ni (t, t_0, x_0)$$

$$\to u(t, t_0, x_0) \in X.$$

(iii) Let us set $r_5 = \frac{1}{4}r_2$ and $r_3 = \frac{1}{8}\tau$. By (i) for any point (t_0, x_0) $\in K(\bar{t}, r_3) \times K(\tilde{x}, r_5)$ there is a unique solution $t \to u(t, t_0, x_0)$ defined on $K(\bar{t}, r_3)$ and satisfying the initial condition $u(t_0) = x_0$. It turns out that the set of values of the solution on $K(\bar{t}, r_3)$ is contained in a ball with centre at the point $\tilde{x}$ and radius $\frac{1}{2}r_2$. For we have

$$\|u(t, t_0, x_0) - \tilde{x}\| \leqslant \|u(t, t_0, x_0) - x_0\| + \|x_0 - \tilde{x}\|$$

$$\leqslant \left\|\int_{t_0}^{t} f(s, u(s)) ds\right\| + \|x_0 - \tilde{x}\| \leqslant M2r_3 + r_5 \leqslant \tfrac{1}{4}r_2 + \tfrac{1}{4}r_2 = \tfrac{1}{2}r_2.$$

Accordingly, let $t, t_0 \in K(\bar{t}, r_3)$, $x_0 \in K(\tilde{x}, r_5)$. Let $x := u(t, t_0, x_0)$. From what has been said above, it follows that $x \in K(\tilde{x}, r_4)$. Item (i) implies that the mapping $s \to u(s, t, x)$ is the only solution of the differential equation which satisfies the condition $u(t) = x$. Further-more, this solution is defined on $K(\bar{t}, r_3)$. The function $s \to u(s, t_0, x_0)$ is also a solution of the equation defined on the whole of $K(\bar{t}, r_3)$ and satisfying the condition $u(t) = u(t, t_0, x_0) = x$. From the uniqueness of the solution, $u(s, t, x) \underset{s}{\equiv} u(s, t_0, x_0)$. In particular, on setting $s = t_0$, we obtain

$$u(t_0, t, x) = u(t_0, t_0, x_0) = x_0.$$

Thus we have proved the solvability of the equation $x_0 = u(t_0, t, x)$. For the solution is a vector $x = u(t, t_0, x_0)$.

We now prove the uniqueness of the solution. Suppose that there is a vector $y \in K(\tilde{x}, r_4)$ such that $x_0 = u(t_0, t, y)$. The function $s \to u(s, t, y)$ thus is a solution of the differential equation defined on $K(t, r_3)$ and assuming the value x_0 at the point t_0. From the uniqueness of the solution of the initial-value problem we have:

$$u(s, t, y) \underset{s}{\equiv} u(s, t_0, x_0).$$

When in particular we set $s = t$, the result is

$$y = u(t, t, y) = u(t, t_0, x_0) = x. \quad \square$$

The theorem above may be generalized to the case of a differentiable dependence on initial conditions.

THEOREM IX.4.2. Suppose that the hypotheses of Theorem IX.4.1 are satisfied and, furthermore, let the mapping f be p times continuously differentiable on $K_1(\tilde{t}, r_1) \times K_2(\tilde{x}, r_2) \subset [a, b] \times X$. Then there exist a $\tau_1 > 0$ and an $r > 0$ such that the mapping $(t, t_0, x_0) \to u(t, t_0, x_0) \in X$ is p times continuously differentiable on $K(\tilde{t}, \tau_1) \times K(\tilde{t}, \tau_1) \times K(\tilde{x}, r)$.

PROOF. The proof consists in reducing the problem to Theorem IX.3.7. Let

$$v(t, t_0, x_0) := u(t + t_0, t_0, x_0) - x_0.$$

Then

$$\frac{dv(t)}{dt} = \frac{du(t + t_0, t_0, x_0)}{d(t + t_0)} = f(t + t_0, u(t + t_0, t_0, x_0))$$
$$= f(t + t_0, v(t) + x_0).$$

The function $t \to v(t)$ thus satisfies the differential equation $\dfrac{dv}{dt}$ $= f(t + t_0, v + x_0)$ with the initial condition $v(0) = 0$. Treating t_0 and x_0 as parameters, from Theorem IX.3.7 we find that the mapping $(t, t_0, x_0) \to v(t, t_0, x_0)$ is p times (continuously) differentiable and hence also the mapping $(t, t_0, x_0) \to u(t, t_0, x_0)$ is p times (continuously) differentiable. $\square$

5. SYSTEMS OF DIFFERENTIAL EQUATIONS

Let X_i, $i = 1, 2, ..., n$, be Banach spaces and let f_i be mappings of the product $R^1 \times X_1 \times ... \times X_n$ into X_i. Suppose that the mappings f_i are bounded on the set $V = K(t_0, r_0) \times K(x_{01}, r_1) \times ... \times K(x_{0n}, r_n)$, that is, that there exist $M_1, ..., M_n$, such that

$$\|f_i(t, x_1, ..., x_n)\| \leqslant M_i \quad \text{for } (t, x_1, ..., x_n) \in V.$$

Suppose, moreover, that the mappings f_i are continuous on V and that they are Lipschitz mappings, i.e. that there are constants $L_1, ..., L_n$ such that

$$\|f_i(t, x_1, ..., x_n) - f_i(t, x_1', ..., x_n')\| \leqslant L_i \sum_{k=1}^{n} \|x_k - x_k'\|$$

for $(t, x_1, ..., x_n) \in V$, $(t, x_1', ..., x_n') \in V$.

We consider the initial-value problem:

$$\frac{dx_1}{dt} = f_1(t, x_1, ..., x_n),$$

$$\frac{dx_2}{dt} = f_2(t, x_1, ..., x_n),$$

(1) $\quad \cdot \cdot \cdot \cdot \cdot \cdot \cdot \cdot \cdot \cdot \cdot \cdot \cdot$

$$\frac{dx_n}{dt} = f_n(t, x_1, ..., x_n),$$

$$x_1(t_0) = x_{01}, \quad x_2(t_0) = x_{02}, \quad ..., \quad x_n(t_0) = x_{0n}.$$

Next, we introduce a Banach space $X = X_1 \times X_2 \times ... \times X_n$, $\|x\| = \sum_{k=1}^{n} \|x_k\|$; $x = (x_1, ..., x_n)$,

$$f(t, x) := \big(f_1(t, x), ..., f_n(t, x)\big).$$

Clearly, f is a continuous, bounded, and Lipschitz mapping. For we have

$$\|f(t, x)\| = \sum_{k=1}^{n} \|f_k(t, x)\| \leqslant \sum_k M_k = M,$$

$$\|f(t, x) - f(t, x')\| \leqslant \sum_{i=1}^{n} L_i \sum_{k=1}^{n} \|x_k - x_k'\| = L \|x - x'\|.$$

The system of equations (1) reduces to a single equation

(2) $\qquad \dfrac{dx}{dt} = f(t, x), \quad x(t_0) = x_0 = (x_{01}, \ldots, x_{0n}).$

By virtue of Theorem IX.2.1 the initial-value problem (2) has exactly one solution.

In this way we have reduced a system of n first-order equations to a single first-order equation.

6. EQUATIONS OF HIGHER ORDER

Let f be a mapping of the Cartesian product $R^1 \times X \times \ldots \times X$ into X. Suppose that f is continuous on $V = K(t_0, r) \times K(x_0, r_0) \times \ldots \times K(x_{n-1}, r_{n-1})$, bounded on V (that is, $\|f(t, x)\| \leqslant M, (t, x) \in V$) and a Lipschitz mapping (that is, $\|f(t, x) - f(t, x')\| \leqslant L\|x - x'\|$).

Let us consider a differential equation of order n,

$$\frac{d^n x}{dt^n} = f\left(t, x, \frac{dx}{dt}, \frac{d^2 x}{dt^2}, \ldots, \frac{d^{n-1} x}{dt^{n-1}}\right),$$

with the initial condition

$$\frac{d^k x(t_0)}{dt^k} = x_{(k)}, \quad x_{(k)} \in X, \quad k = 0, 1, \ldots, n-1.$$

We reduce this equation to a system of n first-order equations. We make the substitutions

$$x_1(t) := x(t),$$

$$x_2(t) := \frac{dx_1(t)}{dt} = \frac{dx(t)}{dt},$$

$$x_3(t) := \frac{dx_2(t)}{dt} = \frac{d^2 x(t)}{dt^2},$$

$$\cdot\;\cdot\;\cdot\;\cdot\;\cdot\;\cdot\;\cdot\;\cdot\;\cdot\;\cdot\;\cdot\;\cdot\;\cdot$$

$$x_n(t) := \frac{dx_{n-1}(t)}{dt} = \frac{d^{n-1} x(t)}{dt^{n-1}}.$$

Of course, $\dfrac{dx_n}{dt} = \dfrac{d^n x}{dt^n}$. We obtain the system of equations

$$\frac{dx_1}{dt} = x_2 =: f_1,$$

$$\frac{dx_2}{dt} = x_3 =: f_2,$$

$$\cdot \quad \cdot \quad \cdot \quad \cdot \quad \cdot \quad \cdot \quad \cdot \quad \cdot \quad \cdot \quad \cdot \quad \cdot$$

$$\frac{dx_{n-1}}{dt} = x_n =: f_{n-1},$$

$$\frac{dx_n}{dt} = f(t, x_1, ..., x_n) =: f_n,$$

with the initial condition

$$x_k(t_0) = x_{(k-1)}, \quad k = 1, ..., n.$$

The mappings f are, as is seen, continuous, bounded, and Lipschitz mappings. From Section 5 it follows that this system has exactly one solution satisfying the initial condition.

We have thus reduced an equation of order n to a system of n first-order equations which we can, in turn, reduce to a single first-order equation.

7. EQUATIONS WITH RIGHT-HAND SIDE ANALYTIC

Let us consider an initial-value problem with right-hand side analytic:

$$\frac{dx}{dt} = \sum_{j,k=0}^{\infty} a_{jk}(t-t_0)^j (x-x_0)^k =: f(t, x), \quad x(t_0) = x_0.$$

We assume that the series on the right-hand side of the equation is absolutely convergent for $|t-t_0| < r_1$, $|x-x_0| < r_2$. Clearly, in this case $X = C^1$. It turns out that the solution of the equation with right-hand side analytic is also an analytic function.

THEOREM IX.7.1 (Cauchy). The initial-value problem with right-hand side analytic (in complex variables)

$$\frac{dx}{dt} = \sum_{j,k=0}^{\infty} a_{jk}(t-t_0)^j (x-x_0)^k, \quad x(t_0) = x_0$$

has a unique solution which is an analytic function.

PROOF. Plainly the right-hand side of the equation is bounded, i.e. there is an $M > 0$ such that

$$\|f(t, x)\| \leqslant M \quad \text{for } (t, x) \in K(t_0, \varrho_1) \times K(x_0, \varrho_2),$$

where $\varrho_1 < r_1$, $\varrho_2 < r_2$. Moreover, $\partial f / \partial x$ is bounded on $K(t_0, \varrho_1) \times K(x_0, \varrho_2)$, and hence the function f satisfies the Lipschitz condition. Thus, by Theorem IX.2.1 we can employ the method of successive approximations:

$$\overset{n+1}{x}(t) = x_0 + \int_{t_0}^{t} \sum_{j,k=0}^{\infty} a_{jk}(s-t_0)^j \left(\overset{n}{x}(s) - x_0\right)^k ds.$$

It is thus seen that if the n-th approximation is analytic, then so is the $(n+1)$-th. Since x_0 is analytic as a constant function, we see that all approximations are analytic functions. Making use of the Weierstrass theorem stating that the uniform limit of analytic functions is an analytic function (Corollary XIV.6, Part II), we obtain the analyticity of the solution. $\square$

We know now that the solution of an equation with right-hand side analytic is an analytic function. It has exactly one representation in the form of a power series

$$(1) \qquad x(t) = \sum_{j=0}^{\infty} b_j(t-t_0)^j.$$

The solution of the equation may be sought by substituting series (1) into the equation and, by comparing the coefficients of the corresponding powers of $t-t_0$, evaluating the numbers b_j.

Example. Let

$$\frac{dx}{dt} = 2tx, \qquad x(0) = 1.$$

We are looking for a solution in the form $x(t) = \sum_{j=0}^{\infty} b_j t^j$; we have

$$\frac{dx}{dt} = \sum_{j=1}^{\infty} jb_j t^{j-1}.$$

Substitution into the equation yields

$$\sum_{j=1}^{\infty} jb_j t^{j-1} = \sum_{j=0}^{\infty} 2b_j t^{j+1}.$$

This equation can be rewritten as

$$\sum_{j=-1}^{\infty} (j+2)b_{j+2} t^{j+1} = \sum_{j=0}^{\infty} 2b_j t^{j+1}.$$

On comparing the coefficients of the same powers of t, we obtain:

$$b_1 = 0,$$
$$2b_2 = 2b_0,$$
$$3b_3 = 2b_1,$$
$$4b_4 = 2b_2,$$
$$. \quad . \quad . \quad . \quad . \quad . \quad .$$

This system of equations gives us

$$b_1 = b_3 = b_5 = \ldots = 0, \quad b_2 = b_0, \quad b_4 = \frac{b_0}{2},$$

$$b_6 = \frac{b_0}{2 \cdot 3}, \ldots$$

Generally

$$b_{2j+1} = 0, \quad b_{2j} = \frac{b_0}{j!}.$$

Thus we have

$$x(t) = b_0 \sum_{j=0}^{\infty} \frac{t^{2j}}{j!} = b_0 e^{t^2},$$

and from the initial condition we find $b_0 = 1$.

8. THE PEANO THEOREM

THEOREM IX.8.1 (The Schauder Principle). Let $S = \overline{K(x_0, r)} \subset X$ be a closed ball in a Banach space X, and let P be a continuous mapping of the set S into S. If the set $\overline{P(S)}$ is compact, the mapping P has at least one fixed point (i.e. there is an $x \in S$ such that $P(x) = x$).

The proof of the Schauder principle (Theorem XIII.14, Part II) consists in reducing it to Brouwer's fixed-point theorem.

Note, moreover, that the uniqueness of a fixed point does not follow from the Schauder principle and that in general there may be many fixed points.

THEOREM IX.8.2 (Ascoli). Let $(g_\alpha)_{\alpha \in A}$ be a family of functions on $[a, b]$ with values in R^n: $[a, b] \ni t \to g_\alpha(t) \in R^n$. Assume that

296

8. THE PEANO THEOREM

(i) the functions $(g_\alpha)_{\alpha \in A}$ are uniformly bounded, i.e. there exists an $M > 0$ such that $\|g_\alpha(t)\| \leqslant M$ for $\alpha \in A$ and $t \in [a, b]$;

(ii) $(g_\alpha)_{\alpha \in A}$ is a family of equicontinuous functions, i.e. for every $\varepsilon > 0$ and every $t_0 \in [a, b]$ there is a $\delta > 0$ such that for every $\alpha \in A$ it follows from the inequality $|t - t_0| < \delta$ that $\|g_\alpha(t) - g_\alpha(t_0)\| < \varepsilon$.

Then that family is a relatively compact set in the space $C([a, b], R^n)$, that is, its closure is compact in $C([a, b], R^n)$ (in a topology of uniform convergence).

This fact is a special case of the general Ascoli theorem XI.1.4.

The two theorems above will be used to prove the Peano theorem (obviously Peano could not have employed the Schauder principle!).

THEOREM IX.8.3 (Peano). Let $I = [a, b]$ and let f be a mapping of the Cartesian product $I \times R^n$ into R^n. If there is a ball $K(x_0, r)$ such that f is continuous on $I \times K(x_0, r)$, there exists a $\tau > 0$ such that the initial-value problem

$$\frac{dx}{dt} = f(t, x), \quad x(t_0) = x_0, \ t_0 \in \,]a, b[,$$

has at least one solution defined on the interval $]t_0 - \tau, t_0 + \tau[$.

PROOF. Let $r_1 < r$. Then $\overline{K(x_0, r_1)} \subset K(x_0, r)$. Being continuous on the compact set $I \times \overline{K(x_0, r_1)}$, the mapping f is bounded. Thus, there is an $M > 0$ such that $\|f(t, x)\| \leqslant M$ for $(t, x) \in I \times \overline{K(x_0, r_1)}$. Let $g_0 \in C(|t - t_0| \leqslant \varrho, R^n)$, $g_0(t) = x_0$, $t \in I$. We define the mapping

$$P_{x_0} : C(|t - t_0| \leqslant \varrho, R^n) \to C(|t - t_0| \leqslant \varrho, R^n),$$

$$(P_{x_0} g)(t) := x_0 + \int_{t_0}^{t} f(s, g(s)) \, ds.$$

It is easily shown (cf. the proof of Theorem IX.2.1) that

$$\text{if } \varrho < \min\left(\frac{r_1}{M}, |a - t_0|, |b - t_0|\right) \text{ then}$$

$$P_{x_0} \overline{K(g_0, r_1)} \subset \overline{K(g_0, r_1)}.$$

We now demonstrate that the mapping P_{x_0} is continuous. Let us take g_1, g_2 such that

(1) $$\|g_1 - g_2\| = \sup_{|t - t_0| \leqslant \varrho} \|g_1(t) - g_2(t)\| < \delta.$$

Since a mapping continuous on the compact set $I \times \overline{K(x_0, r_1)}$ is uniformly continuous, we have

(2) $\|f(t, x_1) - f(t, x_2)\| < \varepsilon$ for $t \in I$, $\|x_1 - x_2\| < \delta$.

From (1) and (2) we have

(3) $\sup\limits_{|s-t_0| \leqslant \varrho} \|f(s, g_1(s)) - f(s, g_2(s))\| < \varepsilon.$

Therefore

$$\|P_{x_0}(g_1) - P_{x_0}(g_2)\| = \sup\limits_{|t-t_0| \leqslant \varrho} \|(P_{x_0}g_1)(t) - (P_{x_0}g_2)(t)\|$$

$$= \sup\limits_{|t-t_0| \leqslant \varrho} \left\| \int\limits_{t_0}^{t} \left(f(s, g_1(s)) - f(s, g_2(s)) \right) ds \right\|$$

$$\leqslant \varrho \cdot \sup\limits_{|t-t_0| \leqslant \varrho} \sup\limits_{s \in [t_0, t]} \|f(s, g_1(s)) - f(s, g_2(s))\| \leqslant \varrho\varepsilon.$$

We have thus demonstrated the continuity of the mapping P_{x_0}. Using the Ascoli theorem, we shall show that the set $\overline{P_{x_0}(K(g_0, r_1))}$ is compact in the space $C(|t-t_0| \leqslant \varrho, R^n)$.

We verify that the hypotheses of the Ascoli theorem are satisfied:
(i) Uniform boundedness

$$\|P_{x_0}(g)\| = \sup\limits_{|t-t_0| \leqslant \varrho} \|(P_{x_0}g)(t)\| = \sup\limits_{|t-t_0| \leqslant \varrho} \left\| x_0 + \int\limits_{t_0}^{t} f(s, g(s)) ds \right\|$$

$$\leqslant \|x_0\| + \varrho M.$$

(ii) Equicontinuity

$$\|(P_{x_0} \cdot g)(t_1) - (P_{x_0}g)(t_2)\| = \left\| \int\limits_{t_0}^{t} f(s, g(s)) ds \right\|$$

$$\leqslant |t_1 - t_2| \cdot M.$$

The hypotheses of the Ascoli theorem thus are satisfied and hence the set $\overline{P_{x_0}(K(g_0, r_1))}$ is compact in the space $C(|t-t_0| \leqslant \varrho, R^n)$. Therefore, we can apply the Schauder principle. Accordingly, the mapping P_{x_0} has at least one fixed point which, as is known, is a solution of our initial-value problem. The solution is defined on the interval $]t - \tau, t + \tau[$, where

$$\varrho < \tau = \min\left(\frac{r_1}{M}, |a-t_0|, |b-t_0| \right). \quad \square$$

It is seen, therefore, that for $X = R^n$ the hypotheses of the Peano theorem are weaker than those of Theorem IX.2.1. They lack any assumption about the right-hand side of the equation satisfying the Lipschitz condition. This lack results in the solution of the initial-value problem being nonunique.

Example.

$$\frac{dx}{dt} = 2\sqrt{|x|}, \quad x(0) = 0.$$

The right-hand side of the equation is seen not to satisfy the Lipschitz condition. The functions

$$x_{ab}(t) := \begin{cases} -(t+b)^2 & \text{for } t \leqslant -b, \\ 0 & \text{for } -b \leqslant t \leqslant a,\ a \geqslant 0,\ b \geqslant 0, \\ (t-a)^2 & \text{for } t \geqslant a \end{cases}$$

are solutions.

The Peano theorem is not valid if X is an infinite-dimensional Banach space.

In both Theorem IX.2.1 and the Peano theorem we assumed that the initial point t_0 lay inside the interval $[a, b]$ and we obtained a solution on the interval $]t-\tau, t+\tau[\subset]a, b[$. If $t_0 = a$, we can also find a solution defined on $[a, a+\tau[$; if $t = b$, we have a solution on $]b-\tau, b]$. It is thus seen that the initial point t_0 may be any point of the interval $[a, b]$.

9. LINEAR DIFFERENTIAL EQUATIONS

Let $b(\cdot)$ be a mapping of the interval $[a, b]$ into a Banach space X: $[a, b] \ni t \to b(t) \in X$, and let A be a mapping of the interval $[a, b]$ into the space of continuous linear mappings of the space X into itself:

$$[a, b] \ni t \to A(t) \in L(X, X).$$

We consider an initial-value problem of the form

$$(1) \qquad \frac{dx}{dt} = A(t) \cdot x + b(t), \quad x(t_0) = x_0.$$

The differential equation (1) is also called a *nonhomogeneous linear differential equation*. If $b(\cdot) = 0$, we speak of a *homogeneous linear equation*.

THEOREM IX.9.1. If the mappings $[a, b] \ni t \to A(t) \in L(X, X)$ and $[a, b] \ni t \to b(t) \in X$ are continuous on $[a, b]$, the initial-value problem

$$\frac{dx}{dt} = A(t) \cdot x + b(t), \quad x(t_0) = x_0$$

has, for any $(t_0, x_0) \in [a, b] \times X$, exactly one solution defined on the entire interval $[a, b]$.

PROOF. We confine ourselves to the case when t_0 is an interior point of the interval $[a, b]$. The changes for other cases are obvious. We verify the hypotheses of Theorem IX.2.1.

(i) The right-hand side of the equation is a mapping continuous on $[a, b] \times X$.

(ii) The mappings $t \to A(t)$ and $t \to b(t)$ are continuous on I, and hence there exist constants $L > 0$ and $R > 0$ such that

$$L = \sup_{t \in I} \|A(t)\|_{L(X, X)}, \quad R = \sup_{t \in I} \|b(t)\|_X.$$

Thus we have

$$\|A(t) \cdot x + b(t)\|_X \leqslant LR + 2L\|x_0\| + R = M$$
$$\text{for } (t, x) \in I \times K(x_0, r),$$

where $r = R + \|x_0\|$, and

$$\|A(t) \cdot x_1 + b(t) - A(t) \cdot x_2 - b(t)\|_X \leqslant L\|x_1 - x_2\|_X.$$

By virtue of Theorem IX.2.1, the initial-value problem has a single solution defined on the parts common to the intervals $[a, b]$ and $]t_0 - \tau, t_0 + \tau[$, where

$$\tau = \min\left(\frac{r}{M}, \frac{1}{L}\right) \doteq \min\left(\frac{R + \|x_0\|}{LR + 2L\|x_0\| + R}, \frac{1}{L}\right)$$

$$\geqslant \min\left(\frac{R + \|x_0\|}{2L(R + \|x_0\|) + R + \|x_0\|}, \frac{1}{L}\right) = \frac{1}{2L + 1}.$$

Since τ does not depend on the choice of the point (t_0, x_0), the solution may be extended and after a finite number of steps we have a solution defined over the entire interval $[a, b]$. $\square$

Let us now consider the homogeneous linear equation:

$$\frac{dx}{dt} = A(t) \cdot x, \quad x(t_0) = x_0.$$

The mapping $t \rightarrow A(t)$ will, of course, be assumed to be continuous on $[a, b]$. Then, by Theorem IX.9.1 the initial-value problem has a unique solution defined on $[a, b]$.

We recall that the symbol $u(\cdot, t_0, x_0)$ denotes the solution satisfying the initial condition $x(t_0) = x_0$. We thus have $u(\cdot, t_0, x_0) \in C([a, b], X)$; $x(t) = u(t, t_0, x_0)$; $u(t_0, t_0, x_0) = x_0$.

Now we give the fundamental theorem of the theory of homogeneous linear equations.

THEOREM IX.9.2. Let A be a mapping of the interval $I = [a, b]$ into the space of continuous linear mappings of a Banach space X into itself. Consider the differential equation

$$(2) \qquad \frac{dx}{dt} = A(t) \cdot x.$$

Let $I \ni t \rightarrow x(t) = u(t, t_0, x_0) \in X$ be a solution satisfying the initial condition $x(t_0) = x_0$.

Then:

(i) The set of solutions of (2) is a subspace of the linear space $C([a, b], X)$. Denoting this subspace by J, we have $J \subset C([a, b], X)$.

(ii) For all $t, t_0 \in I$, the mapping $X \ni x_0 \rightarrow u(t, t_0, x_0) \in X$ is a continuous linear mapping. In other words, we can write

$$u(t, t_0, x_0) = R(t, t_0)x_0, \qquad \text{where } R(t, t_0) \in L(X, X).$$

The mapping $R(t, t_0)$ is called the *resolvent of the homogeneous linear differential equation.*

(iii) The operator-valued function $[a, b] \ni t \rightarrow R(t, t_0) \in L(X, X)$ satisfies the differential equation

$$\frac{dR(t, t_0)}{dt} = A(t) \circ R(t, t_0)$$

with the initial condition $R(t_0, t_0) = I$.

(iv) For all $t, t_0, s \in [a, b]$

$$R(t, s)R(s, t_0) = R(t, t_0).$$

Hence $R(t_0, t) = (R(t, t_0))^{-1}$, and thus the mapping $R(t, t_0)$ is a topological isomorphism of the space X.

(v) The mapping $X \ni x \rightarrow u(\cdot, t_0, x) \in J \subset C([a, b], X)$ is a topological isomorphism of the space X and the space J of the solutions of equation (2).

PROOF. (i) Let the functions $t \rightarrow x_1(t)$, $t \rightarrow x_2(t)$ be solutions of equation (2) and let $\lambda_1, \lambda_2 \in R^1$. The function $t \rightarrow \lambda_1 \cdot x_1(t) + \lambda_2 \cdot x_2(t)$ is also a solution of equation (2) since

$$\frac{d(\lambda_1 \cdot x_1 + \lambda_2 \cdot x_2)}{dt} = \lambda_1 \frac{dx_1}{dt} + \lambda_2 \frac{dx_2}{dt}$$

$$= \lambda_1 A(t) \cdot x_1 + \lambda_2 A(t) \cdot x_2 = A(t) \cdot (\lambda_1 x_1 + \lambda_2 x_2).$$

(ii) It follows from item (i) that the function $t \rightarrow \lambda_1 \cdot u(t, t_0, x_{01}) + \lambda_2 \cdot u(t, t_0, x_{02})$ is a solution of equation (2). This solution assumes the value $\lambda_1 \cdot x_{01} + \lambda_2 \cdot x_{02}$ at the point $t = t_0$. Since the solution of the initial-value problem is unique, we infer that

$$u(t, t_0, \lambda_1 \cdot x_{01} + \lambda_2 \cdot x_{02})$$
$$\underset{t \in I}{\equiv} \lambda_1 \cdot u(t, t_0, x_{01}) + \lambda_2 \cdot u(t, t_0, x_{02}).$$

Thus, we may write $u(t, t_0, x_0) = R(t, t_0) \cdot x_0$. The mapping $R(t, t_0)$ is linear; its continuity will be demonstrated in item (iii).

(iii) Let $I \ni t \rightarrow V(t) \in L(X, X)$ be a function with values in the space of continuous linear mappings of the Banach space X. We consider the differential equation

$$\frac{dV}{dt} = A(t) \circ V \quad \text{with initial condition } V(t_0) = I.$$

In accordance with Theorem IX.9.1, this problem has only one solution defined on the interval $[a, b]$. Let the function $t \rightarrow V(t)$ be a solution of the problem. Then for any $x_0 \in X$

$$(3) \qquad \frac{dV(t)}{dt} \cdot x_0 = A(t) \circ V(t) \cdot x_0.$$

Since

$$\frac{dV(t)}{dt} \cdot x_0 = \frac{d(V(t) \cdot x_0)}{dt},$$

we have

$$\frac{d(V(t) \cdot x_0)}{dt} = A(t) \left(V(t) \cdot x_0 \right).$$

Thus the function $[a, b] \ni t \rightarrow V(t) \cdot x_0 \in X$ is a solution of equation (2). This solution satisfies the initial condition $x(t_0) = x_0$. On account of the uniqueness of the solution of the initial-value problem, we have

$$u(t, t_0, x_0) \underset{t \in I}{\equiv} V(t) \cdot x_0,$$

and consequently

$$R(t, t_0) x_0 \underset{t \in I}{\equiv} V(t) \cdot x_0.$$

However, x_0 was chosen arbitrarily, and therefore

$$R(t, t_0) = V(t) \in L(X, X).$$

Item (iii) has been proved. At the same time we have shown that the mapping $R(t, t_0)$ is continuous.

(iv) The function $t \to R(t, t_0) \cdot x_0$ is a solution satisfying the initial condition $x(t_0) = x_0$. At the point $s \in I$ this solution assumes the value $R(s, t_0) \cdot x_0$. On the other hand, the function $t \to R(t, s) \cdot R(s, t_0) \cdot x_0$ is, for equation (2), a solution satisfying the initial condition $x(s) = R(s, t_0) \cdot x_0$. The uniqueness of the solution of the initial-value problem implies that

$$R(t, t_0) \cdot x_0 \underset{t \in I}{\equiv} R(t, s) \cdot R(s, t_0) \cdot x_0.$$

From the fact that x_0 is arbitrary it follows that for any $t, t_0, s \in [a, b]$

$$R(t, t_0) = R(t, s) \cdot R(s, t_0).$$

Hence, immediately

$$R(t_0, t) = \left(R(t, t_0) \right)^{-1}.$$

(v) In item (ii) we proved that the mapping

$$(4) \qquad X \ni x_0 \to u(\cdot, t_0, x_0) \in J \subset C([a, b], X)$$

is linear. We shall show that it is a bounded mapping:

$$\|u(\cdot, t_0, x_0)\|_J = \sup_{t \in I} \|u(t, t_0, x_0)\|_X$$

$$= \sup_{t \in I} \|R(t, t_0) \cdot x_0\|_X \leqslant \sup_{t \in I} \|R(t, t_0)\|_{L(X, X)} \|x_0\|_X$$

$$= M \cdot \|x_0\|_X.$$

Use has been made here of the fact that the function $t \to R(t, t_0)$ is continuous (since it is differentiable) on $[a, b]$. Plainly, the mapping (4) is a mapping of X onto J, since every solution satisfies some initial condition. If $u(\cdot, t_0, x_0) = 0$, then $u(t_0, t_0, x_0) = 0 = x_0$. The mapping (4) thus is a one-to-one mapping and hence is invertible. Furthermore,

$$\|x_0\|_X \leqslant \sup_{t \in I} \|u(t, t_0, x_0)\|_X = \|u(\cdot, t_0, x_0)\|_J.$$

Hence it follows that the inverse mapping is continuous. $\square$

We now proceed to solve the nonhomogeneous linear equation. We seek the solution of the equation

$$\frac{dx}{dt} = A(t) \cdot x + b(t)$$

with an initial condition $x(t_0) = x_0$. We assume that the given mappings $[a, b] \ni t \to A(t) \in L(X, X)$ and $[a, b] \ni t \to b(t) \in X$ are continuous. By virtue of Theorem IX.9.1 there exists exactly one solution of the initial-value problem.

We look for a solution in the form

$$(5) \qquad x(t) = R(t, t_0) \cdot z(t),$$

where $R(t, t_0)$ is the resolvent of the homogeneous equation and $z(\cdot)$ is a differentiable function satisfying the condition $z(t_0) = x_0$. Thus we seek the function $z(\cdot)$. We substitute (5) in the equation:

$$\frac{dx(t)}{dt} = \frac{dR(t, t_0)}{dt} z(t) + R(t, t_0) \frac{dz(t)}{dt}$$
$$= A(t) \circ R(t, t_0) z(t) + b(t).$$

On account of item (iii) of Theorem IX.9.2 two terms cancel, and therefore

$$R(t, t_0) \frac{dz(t)}{dt} = b(t),$$

whence

$$\frac{dz(t)}{dt} = \left(R(t, t_0) \right)^{-1} b(t).$$

On making use of item (iv) of Theorem IX.9.2, we obtain

$$\frac{dz(t)}{dt} = R(t_0, t) b(t).$$

Integration of both sides of this equation with account for the condition $z(t_0) = x_0$ yields

$$(6) \qquad z(t) = x_0 + \int_{t_0}^{t} R(t_0, s) b(s) ds.$$

Inserting (6) into (5), we arrive at

$$x(t) = R(t, t_0) \cdot x_0 + R(t, t_0) \int_{t_0}^{t} R(t_0, s) b(s) ds.$$

But $R(t, t_0)$ can be put under the sign of integral. Next, invoking item (iv) of Theorem IX.9.2, we have

$$x(t) = R(t, t_0) \cdot x_0 + \int_{t_0}^{t} R(t, s)b(s)\,ds.$$

This is the general form of a solution of a nonhomogeneous linear differential equation, which satisfies the initial condition $x(t_0) = x_0$. This solution consists of two members. The first is a general solution of the homogeneous equation whereas the second is a particular solution (assuming the value 0 for $t = t_0$) of the nonhomogeneous equation. To convince ourselves of this let us differentiate:

$$\frac{d}{dt}\left(\int_{t_0}^{t} R(t, s)b(s)\,ds\right) = \frac{d}{dt}\left(R(t, t_0)\int_{t_0}^{t} R(t_0, s)b(s)\,ds\right)$$

$$= \frac{dR(t, t_0)}{dt}\int_{t_0}^{t} R(t_0, s)b(s)\,ds + R(t, t_0) \circ R(t_0, t)b(t)$$

$$= A(t)\int_{t_0}^{t} R(t, s)b(s)\,ds + b(t).$$

Suppose we have the general solution of the homogeneous equation $x(t) = R(t, t_0) \cdot x_0$ and a particular solution of the nonhomogeneous equation $y(t)$ such that $y(t_0) = y_0$. For the nonhomogeneous equation the solution satisfying the initial condition $x(t_0) = x_0$ is of the form

$$x(t) = R(t, t_0)(x_0 - y_0) + y(t).$$

Thus, once we know the general solution of the homogeneous solution and any solution of the nonhomogeneous equation, we can find the general solution of the nonhomogeneous equation. This method will be pressed into service on many occasions in the sequel.

The solution of scalar linear equation. If $X = R^1$, then $L(X, X) \equiv R^1$, and hence the equation is of the form

$$\frac{dx}{dt} = a(t) \cdot x + b(t), \qquad x(t_0) = x_0,$$

$$t \to a(t) \in R^1, \qquad t \to b(t) \in R^1.$$

305

We seek the resolvent of the homogeneous equation. Therefore we must solve the equation

$$(7) \qquad \frac{dR(t, t_0)}{dt} = a(t)R(t, t_0)$$

with the initial condition $R(t_0, t_0) = 1$.

Rearrangement of equation (7) gives us

$$\frac{dR(t, t_0)}{R(t, t_0)} = a(t)\,dt.$$

On integrating both sides, we have

$$\ln R(t, t_0) = \int_{t_0}^{t} a(s)\,ds + \ln C.$$

Setting $t = t_0$, we obtain $C = 1$. Hence

$$R(t, t_0) = \exp \int_{t_0}^{t} a(s)\,ds.$$

Finally, the solution of the initial-value problem is of the form

$$x(t) = x_0 \cdot \exp \int_{t_0}^{t} a(s)\,ds + \int_{t_0}^{t} \left(\exp \int_{s}^{t} a(k)\,dk \right) b(s)\,ds.$$

Remark. The method of finding the solution of a nonhomogeneous equation, given earlier, is known as the *method of variation of parameters*. The name comes from the fact that the constant vector x_0 is replaced by a new unknown function $z(\cdot)$.

10. THE MAPPING $A \to \exp A$

A familiar concept is that of the exponential function $R^1 \ni t \to e^t \in R^1$. It is an analytic function, i.e. it is series-expandible. This expansion is of the form

$$e^t = \sum_{n=0}^{\infty} \frac{t^n}{n!}$$

and converges on the whole straight line. At the same time, the function exp is the unique solution of the initial-value problem

$$\frac{du}{dt} = u, \quad u(0) = 1.$$

We shall now generalize the concept of exponential function. Let A be a continuous linear operator in a Banach space X, $A \in L(X, X)$. We define

$$(1) \qquad \exp A = e^A := \sum_{n=0}^{\infty} \frac{A^n}{n!}, \qquad A^0 := I.$$

The series on the right-hand side should be inspected for convergence. We examine whether the sequence of partial sums is a Cauchy sequence:

$$\|S_{k+p} - S_k\| = \left\| \sum_{n=0}^{k+p} \frac{A^n}{n!} - \sum_{n=0}^{k} \frac{A^n}{n!} \right\| = \left\| \sum_{n=k+1}^{k+p} \frac{A^n}{n!} \right\|$$

$$\leqslant \sum_{n=k+1}^{k+p} \frac{\|A\|^n}{n!}.$$

The expression on the right-hand side is the difference of the partial sums of the series

$$\sum_{n=0}^{\infty} \frac{\|A\|^n}{n!} = e^{\|A\|}.$$

The sequence of partial sums thus is a Cauchy sequence. The completeness of the space $L(X, X)$ thus implies that a limit of the series (1) exists. The operator $\exp A$ has thus been correctly defined.

LEMMA IX.10.1. The operator-valued function $R^1 \ni t \to \exp(t \cdot A) \in L(X, X)$ is the unique solution of the initial-value problem

$$\frac{dx}{dt} = A \cdot x, \quad x(0) = I.$$

PROOF. Show that $d\exp(t \cdot A)/dt = A \cdot \exp(t \cdot A)$. Clearly the initial condition is satisfied. Show that the series $\sum_{n=0}^{\infty} \frac{(t \cdot A)^n}{n!}$ may be differentiated term by term. Let $S_k(t) = \sum_{n=0}^{k} A \frac{(t \cdot A)^n}{n!}$ be a partial sum of the expansion of the function $A \exp(t \cdot A)$. Then

$$\left\| \int_0^t (S_k(s) - A \exp(s \cdot A)) \, ds \right\|$$

$$\leqslant \|A\| \int_0^t \sum_{n=k+1}^{\infty} \frac{(|s| \cdot \|A\|)^n}{n!} \, ds < \varepsilon, \quad k > N.$$

Thus we have

$$\lim_{k \to \infty} \int_0^t S_k(s) \, ds = \int_0^t A \exp(s \cdot A) \, ds.$$

When we integrate the left-hand side of this equation we have

$$-I + \sum_{n=0}^{\infty} \frac{(t \cdot A)^n}{n!} = \int_0^t A \exp(s \cdot A) \, ds$$

whereby, on differentiating both sides, we get

$$\frac{d}{dt} \sum_{n=0}^{\infty} \frac{(t \cdot A)^n}{n!} = \frac{d}{dt} \exp(t \cdot A) = A \exp(t \cdot A). \quad \square$$

Lemma IX.10.1 leads immediately to the following

COROLLARY IX.10.2.

$$\frac{d^n \exp(t \cdot A)}{dt^n} = A^n \exp(t \cdot A) = \exp(t \cdot A) \cdot A^n.$$

LEMMA IX.10.3. If the operators A and B commute, that is, $A \cdot B = B \cdot A$, then

$$\exp(A + B) = \exp A \cdot \exp B = \exp B \cdot \exp A.$$

PROOF. This lemma may be proved directly, by starting from the definition (cf. the analogous result for functions, Chapter V, Section 6). However, we shall give a more elegant proof. Note that if A and B commute, then A commutes with $\exp B$ and B commutes with $\exp A$. The function $t \to \exp(t \cdot (A + B))$ satisfies the differential equation

$$\frac{d \exp(t \cdot (A + B))}{dt} = (A + B) \exp(t \cdot (A + B))$$

with the initial condition $x(0) = I$. On the other hand

$$\frac{d\big(\exp(t \cdot A) \cdot \exp(t \cdot B)\big)}{dt}$$

$$= A\exp(t \cdot A) \cdot \exp(t \cdot B) + \exp(t \cdot A) \cdot \exp(t \cdot B) \cdot B$$

$$= (A+B)\exp(t \cdot A) \cdot \exp(t \cdot B).$$

In the foregoing rearrangements we have made use of the commutativity of the operators A and B.

We thus see that the function $t \to \exp(t \cdot A) \cdot \exp(t \cdot B)$ satisfies the same equation as the function $t \to \exp\big(t \cdot (A+B)\big)$ and the same initial condition. Therefore $\exp\big(t \cdot (A+B)\big) = \exp(t \cdot A) \cdot \exp(t \cdot B)$, $t \in R^1$. On setting $t = 1$, we immediately obtain the thesis. $\square$

COROLLARY IX.10.4. The following equations hold:

$$\exp\big((t+s) \cdot A\big) = \exp(t \cdot A) \cdot \exp(s \cdot A), \qquad t, s \in R^1,$$

$$\exp(-A) = (\exp A)^{-1}.$$

11. THE GENERAL FORM OF THE RESOLVENT OF A HOMOGENEOUS EQUATION

We have demonstrated above that in the case $X = R^1$, the resolvent of a linear equation is of the form

$$R(t, t_0) = e^{\int_{t_0}^{t} a(s)\,ds}.$$

Now let us generalize this formula to the case of an arbitrary Banach space. Henceforth in our discussion in this section we shall invoke the theory of multiple integrals. This theory will be presented in Chapter XIII.

We start with the equation

$$(1) \qquad \frac{dR(t, t_0)}{dt} = A(t) \circ R(t, t_0)$$

with the initial condition $R(t_0, t_0) = I$.

We look for the solution by the method of successive approximations, and we get

$$R_0(t, t_0) = I,$$

$$R_1(t, t_0) = I + \int_{t_0}^{t} A(s)\,ds,$$

$$R_2(t, t_0) = I + \int_{t_0}^{t} A(s_1)\,ds_1 + \int_{t_0}^{t} ds_2 A(s_2) \int_{t_0}^{s_2} A(s_1)\,ds_1,$$

$$R_n(t, t_0) = I + \int_{t_0}^{t} A(s) \circ R_{n-1}(s, t_0)\,ds.$$

The solution satisfying the initial condition $R(t_0, t_0) = I$ thus is in the form of a series

$$(2) \qquad R(t, t_0) = I + \int_{t_0}^{t} A(s_1)\,ds_1 + \int_{t_0}^{t} ds_2 A(s_2) \int_{t_0}^{s_2} A(s_1)\,ds_1 +$$

$$+ \int_{t_0}^{t} ds_3 A(s_3) \int_{t_0}^{s_3} ds_2 A(s_2) \int_{t_0}^{s_2} A(s_1)\,ds_1 +$$

$$+ \ldots + \int_{t_0}^{t} ds_n A(s_n) \int_{t_0}^{s_n} ds_{n-1} A(s_{n-1}) \ldots \int_{t_0}^{s_2} A(s_1)\,ds_1 + \ldots$$

We now rearrange formula (2). Let us consider the third term in the expansion. It is readily seen that

$$(3) \qquad \int_{t_0}^{t} ds_2 A(s_2) \int_{t_0}^{s_2} A(s_1)\,ds_1 = \iint_{\Delta} A(s_2) \cdot A(s_1)\,ds_2\,ds_1,$$

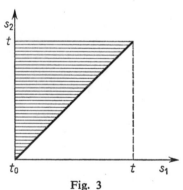

Fig. 3

where Δ is the triangle indicated in the figure. Fubini's theorem enables us to change the order of integration, and we then have

$$\iint_{\Delta} A(s_2) \cdot A(s_1)\,ds_2\,ds_1 = \int_{t_0}^{t} \left(\int_{s_1}^{t} A(s_2)\,ds_2 \right) \cdot A(s_1)\,ds_1.$$

When the variables in the integral on the right-hand side are relabelled, we obtain

$$(4) \qquad \iint_{\Delta} A(s_2) \cdot A(s_1) ds_2 ds_1 = \int_{t_0}^{t} \left(\int_{s_2}^{t} A(s_1) ds_1 \right) \cdot A(s_2) ds_2.$$

From (3) and (4) we have

$$(5) \qquad \frac{1}{2} \left(\int_{t_0}^{t} ds_2 A(s_2) \int_{t_0}^{s_2} A(s_1) ds_1 + \int_{t_0}^{t} \left(\int_{s_2}^{t} A(s_1) s \, ds_1 \right) \cdot A(s_2) ds_2 \right)$$

$$= \iint_{\Delta} A(s_2) A(s_1) ds_2 ds_1.$$

Now let us introduce the chronological ordering operation:

$$TA(t) := A(t);$$

$$TA(t_1) \cdot A(t_2) := \begin{cases} A(t_1) \cdot A(t_2), & \text{when } t_1 \geqslant t_2, \\ A(t_2) \cdot A(t_1), & \text{when } t_1 \leqslant t_2. \end{cases}$$

In general

$$TA(t_1) \cdot A(t_2) \cdot \ldots \cdot A(t_n) := A(t_{\alpha_1}) \cdot A(t_{\alpha_2}) \cdot \ldots \cdot A(t_{\alpha_n}),$$

where $t_{\alpha_1} \geqslant t_{\alpha_2} \geqslant t_{\alpha_3} \geqslant \ldots \geqslant t_{\alpha_n}$.

On using the definition of the chronological product and formulae (3) and (5), we obtain[1]

$$\int_{t_0}^{t} ds_2 A(s_2) \int_{t_0}^{s_2} A(s_1) ds_1 = \frac{1}{2!} \int_{t_0}^{t} \int_{t_0}^{t} TA(s_1) \cdot A(s_2) ds_1 ds_2.$$

Similarly, it may be shown that

$$\int_{t_0}^{t} ds_3 A(s_3) \int_{t_0}^{s_3} ds_2 A(s_2) \int_{t_0}^{s_2} A(s_1) ds_1$$

$$= \frac{1}{3!} \int_{t_0}^{t} \int_{t_0}^{t} \int_{t_0}^{t} TA(s_1) A(s_2) \cdot A(s_3) ds_1 ds_2 ds_3.$$

Generally

$$\int_{t_0}^{t} ds_n A(s_n) \int_{t_0}^{s_n} ds_{n-1} A(s_{n-1}) \ldots \int_{t_0}^{s_2} A(s_1) ds_1$$

---

[1] For $t < t_0$ there should be used the opposite ordering.

$$= \frac{1}{n!} \int_{t_0}^{t} \int_{t_0}^{t} \cdots \int_{t_0}^{t} TA(s_1) \cdot A(s_2) \cdots A(s_n)\, ds_1\, ds_2 \cdots ds_n.$$

Substitution of the formulae derived above into (2) yields

$$(6) \qquad R(t, t_0) = I + \frac{1}{1!} \int_{t_0}^{t} TA(s)\, ds + \frac{1}{2!} \int_{t_0}^{t} \int_{t_0}^{t} TA(s_1) \cdot A(s_2)\, ds_1\, ds_2 +$$

$$+ \frac{1}{3!} \int_{t_0}^{t} \int_{t_0}^{t} \int_{t_0}^{t} TA(s_1) \cdot A(s_2) \cdot A(s_3)\, ds_1\, ds_2\, ds_3 + \cdots +$$

$$+ \frac{1}{n!} \int_{t_0}^{t} \int_{t_0}^{t} \cdots \int_{t_0}^{t} TA(s_1) \cdot A(s_2) \cdots A(s_n)\, ds_1\, ds_2 \cdots ds_n + \cdots$$

$$=: T \exp \left(\int_{t_0}^{t} A(s)\, ds \right).$$

The last definition was introduced for notation convenience. Thus we obtain

$$(7) \qquad R(t, t_0) = T \exp \left(\int_{t_0}^{t} A(s)\, ds \right).$$

In the particular case when the operators $A(s)$ commute, that is, $A(s_1) \cdot A(s_2) = A(s_2) \cdot A(s_1)$, $s_1, s_2 \in [a, b]$, formula (7) takes the form

$$R(t, t_0) = \exp \left(\int_{t_0}^{t} A(s)\, ds \right).$$

When $A(s) \underset{s}{\equiv} A$ (the case of an equation with constant coefficients), the formula is even simpler:

$$R(t, t_0) = \exp \left((t - t_0) \cdot A \right).$$

The method of successive approximations was employed throughout the discussion above. It is known, however, that formulae obtained by this method are valid only for some neighbourhood of the point t_0. We shall, however, show these formulae to be valid over the entire interval $[a, b]$. Thus, series (2) and (6) must be shown to converge for $t \in [a, b]$. It is easily seen that the partial sums of series (6) are dominated by the partial sums of the exponential numerical function. For we have

$$\left\| \sum_{n=0}^{k} \frac{1}{n!} \int_{t_0}^{t} \dots \int_{t_0}^{t} TA(s_1) \dots A(s_n) ds_1 \dots ds_n \right\|$$

$$\leqslant \sum_{n=0}^{k} \frac{M^n |t - t_0|^n}{n!} ,$$

where $M = \sup\limits_{s \in [a,b]} \|A(s)\|$.

The convergence of series (6) and the equivalent series (2) has thus been demonstrated. In order to show that series (2) and (6) satisfy equation (1), it is sufficient to differentiate expansion (2) term by term. To justify this passage, reasoning similar to that in the proof of Lemma IX.10.1 must be carried out (this reasoning is in actual fact a transfer of Theorem V.4.10 on the differentiation of function sequences to the vector case).

We end this section with a discussion of a certain property of the resolvent in Hilbert space.

LEMMA IX.11.1. Let X be a Banach space with scalar product $(\cdot | \cdot)$, that is, X is a Hilbert space. Let $t \to A(t)$ be a continuous mapping of the interval $[a, b]$ in the space $L(X, X)$. If for every $t \in [a, b]$ the operator $A(t)$ is antisymmetric (anti-Hermitian), i.e.

$$(A(t) \cdot x | y) = -(x | A(t) \cdot y), \quad x, y \in X,$$

the operator $R(t, t_0)$ is orthogonal (unitary) (i.e. $(R(t, t_0) \cdot x | R(t, t_0) \cdot y) = (x|y)$ and $R(t, t_0)$ is a bijection).

PROOF. We have

$$\frac{d}{dt} \left(R(t, t_0) \cdot x | R(t, t_0) \cdot y \right)$$

$$= \left(\frac{dR(t, t_0)}{dt} \cdot x | R(t, t_0) \cdot y \right) + \left(R(t, t_0) \cdot x \left| \frac{dR(t, t_0)}{dt} \cdot y \right. \right)$$

$$= \left(A(t) \cdot R(t, t_0) \cdot x | R(t, t_0) \cdot y \right) +$$
$$+ \left(R(t, t_0) \cdot x | A(t) \cdot R(t, t_0) \cdot y \right)$$
$$= - \left(R(t, t_0) \cdot x | A(t) \cdot R(t, t_0) \cdot y \right) +$$
$$+ \left(R(t, t_0) \cdot x | A(t) \cdot R(t, t_0) \cdot y \right) = 0.$$

313

Therefore $(R(t, t_0) \cdot x | R(t, t_0) \cdot y) = $ const. On setting $t = t_0$, we have const $= (x|y)$. From Theorem IX.9.2 (iv) we know that $R(t, t_0)$ is a bijection. $\square$

12. LINEAR EQUATIONS IN FINITE-DIMENSIONAL SPACE

Let $\dim X = n$. Consider the homogeneous linear equation

(1) $\qquad \dfrac{dx}{dt} = A(t) \cdot x \quad$ with initial condition $x(t_0) = x_0$.

Let the set of vectors $e_1, \ldots, e_n$ be a basis in a space X. In this basis, equation (1) reduces to a system of n scalar equations:

(2) $\qquad \dfrac{dx_i}{dt} = \sum_{j=1}^{n} a_{ij}(t) \cdot x_j, \quad i = 1, 2, \ldots, n,$

where

(3) $\qquad x(t) = \sum_{i=1}^{n} x_i(t) \cdot e_i.$

The initial condition assumes the form

(4) $\qquad x_i(t_0) = \overset{0}{x}_i, \quad i = 1, 2, \ldots, n, \quad$ where $x_0 = \sum_{i=1}^{n} \overset{0}{x}_i \cdot e_i.$

We define

(5) $\qquad u_j(t) := R(t, t_0) \cdot e_j \quad (j = 1, 2, \ldots, n).$

Thus

(6) $\qquad x(t) = R(t, t_0) \cdot x_0 = R(t, t_0) \sum_{j=1}^{n} \overset{0}{x}_j \cdot e_j = \sum_{j=1}^{n} u_j(t) \cdot \overset{0}{x}_j.$

It is thus seen that the vector-valued functions u_j, $j = 1, 2, \ldots, n$, form a basis of the space of solutions. This is in accordance with item (v) of Theorem IX.9.2 which states that the space X is isomorphic to the space of solutions of the homogeneous equation.

Decomposing the vectors $u_j(t)$ into basis vectors, we obtain a system of n^2 scalar functions $u_{kj}(\cdot)$:

(7) $\qquad u_j(t) = \sum_{k=1}^{n} u_{kj}(t) \cdot e_k, \quad j = 1, 2, \ldots, n.$

It is readily seen that these functions satisfy the system of equations

$$(8) \qquad \frac{du_{kj}}{dt} = \sum_{i=1}^{n} a_{ki}(t) \cdot u_{ij}, \qquad k, j = 1, 2, ..., n,$$

with the initial condition $u_{kj}(t_0) = \delta_{kj}$. On making use of (5) and (7), we arrive at

$$R(t, t_0) \cdot e_j = \sum_{k=1}^{n} u_{kj}(t) \cdot e_k.$$

This equation shows that the matrix $u_{kj}(t)$ may be identified with the matrix of the operator $R(t, t_0)$. The determinant of this matrix is called the *Wronskian determinant* or simply *Wronskian*:

$$W(t) = \det\big(u_{kj}(t)\big) = \det\big(R(t, t_0)\big).$$

Differentiation yields

$$\frac{dW}{dt} = \sum_{k,j=1}^{n} \frac{\partial W}{\partial u_{kj}} \frac{du_{kj}}{dt}.$$

But $\partial W/\partial u_{kj} = Du_{kj}$, where the symbol Du_{kj} denotes the cofactor of the element u_{kj}. When we use (8), we obtain

$$\frac{dW}{dt} = \sum_{i,k=1}^{n} \sum_{j=1}^{n} (Du_{kj}u_{ij}) a_{ki}(t).$$

However

$$\sum_{j=1}^{n} Du_{kj}u_{ij} = \delta_{ki}\det(u_{kj}) = \delta_{ki}W.$$

Hence

$$\frac{dW}{dt} = W \sum_{i,k=1}^{n} \delta_{ki}a_{ki}(t) = W \sum_{k=1}^{n} a_{kk}(t).$$

Integration of this equation gives

$$(9) \qquad W(t) = W(t_0)\exp\left(\int_{t_0}^{t} \sum_{k=1}^{n} a_{kk}(s)\,ds\right).$$

Making use of the initial conditions (8), we have $W(t_0) = 1$ and

$$(10) \qquad W(t) = \exp\left(\int_{t_0}^{t} \sum_{k=1}^{n} a_{kk}(s)\,ds\right).$$

Formulae (9) and (10) are known as the *Liouville formulae*. They show that the Wronskian formed from the system of functions g_{kj} which satisfy the system of equations

$$\frac{dg_{kj}}{dt} = \sum_{i=1}^{n} a_{ki}(t)g_{ij}, \quad i,j = 1, 2, ..., n,$$

does not vanish at any point, provided it does not vanish at $t = t_0$. This means that the vectors formed by the columns of the matrix $[g_{kj}(t)]$ are linearly independent at every point, if they are linearly independent at the point $t = t_0$.

Thus we can formulate the following corollary:

COROLLARY IX.12.1. A system of n vector-valued functions $t \to u_j(t) \in X$, $j = 1, 2, ..., n$, which are solutions of the equation

$$\frac{dx}{dt} = A(t) \cdot x$$

form a basis of the solution space J of that equation if and only if the vectors $u_1(t_0), ..., u_n(t_0)$ are linearly independent. Moreover, for every $t \in [a, b]$ the vectors $u_1(t), ..., u_n(t)$ are then linearly independent.

The basis of the solution space formed by the system of functions $u_1, ..., u_n$ is called the *fundamental system of solutions*.

Let us now go on to discuss the nonhomogeneous equation. We have

$$\frac{dx}{dt} = A(t) \cdot x + b(t), \quad x(t_0) = x_0.$$

In the basis $e_1, ..., e_n$ this equation reduces to a system of scalar equations

$$\frac{dx_i}{dt} = \sum_{j=1}^{n} a_{ij}(t) \cdot x_j + b_i(t),$$

where

$$x(t) = \sum_{i=1}^{n} x_i(t) \cdot e_i, \quad b(t) = \sum_{i=1}^{n} b_i(t) \cdot e_i;$$

$$(11) \qquad A(t) \cdot e_j = \sum_{k=1}^{n} a_{kj}(t) \cdot e_k.$$

316

The initial condition is of the form

$$(12) \qquad x(t_0) = x_0, \qquad \text{where } x_0 = \sum_{i=1}^{n} \overset{0}{x_i} e_i.$$

We know that the solution of a nonhomogeneous equation which satisfies the above initial condition is of the form

$$x(t) = R(t, t_0) \cdot x_0 + R(t, t_0) \int_{t_0}^{t} (R(s, t_0))^{-1} b(s) \, ds.$$

On making use of relations (11) and (12) and bearing in mind that

$$R(t, t_0) = (u_{kj}(t)), \qquad u_j(t) = \sum_{k=1}^{n} u_{kj}(t) \cdot e_k,$$

where $(u_j(t))$ is the fundamental system of solutions of the homogeneous equation satisfying the initial conditions

$$u_j(t_0) = e_j, \qquad j = 1, 2, \ldots, n,$$

we obtain

$$x_i(t) = \sum_{j=1}^{n} u_{ij}(t) \cdot \overset{0}{x_j} + \sum_{j=1}^{n} u_{ij}(t) \int_{t_0}^{t} \sum_{k=1}^{n} \frac{D u_{kj}(s)}{W(s)} b_k(s) \, ds.$$

It is thus seen that, knowing the fundamental system of solutions of the homogeneous equation, we can solve the initial-value problem for the inhomogeneous equation.

13. THE SCALAR EQUATION OF THE n-TH ORDER. THE WRONSKIAN

Let $X = R^1$. Consider the equation

$$\frac{d^n x}{dt^n} = \sum_{k=0}^{n-1} a_k(t) \frac{d^k x}{dt^k} + b(t)$$

with the initial condition

$$x(t_0) = \overset{0}{x_0}, \qquad \frac{dx(t_0)}{dt} = \overset{0}{x_1}, \qquad \ldots, \qquad \frac{d^{n-1} x(t_0)}{dt^{n-1}} = \overset{0}{x_{n-1}}.$$

As we know (cf. Sections 5 and 6), an equation of the n-th order is reducible to a first-order vector equation. On making the substitution

$$x_1(t) := x(t), \qquad \frac{d^k x(t)}{dt^k} := x_{k+1}(t), \qquad k = 1, 2, \ldots, n-1,$$

we obtain a system of first-order equations,

$$\frac{dx_1}{dt} = x_2,$$

$$\frac{dx_2}{dt} = x_3,$$

$$\cdot \quad \cdot \quad \cdot \quad \cdot \quad \cdot \quad \cdot \quad \cdot$$

$$\frac{dx_n}{dt} = \sum_{k=0}^{n-1} a_k(t) x_{k+1} + b(t),$$

with the initial condition $x_k(t_0) = \overset{0}{x}_{k-1}, k = 1, 2, \ldots, n$. It is also known that this system is equivalent to one first-order equation in the space R^n:

$$\frac{dy}{dt} = A(t) \cdot y + B(t), \qquad y(t_0) = (\overset{0}{x}_0, \ldots, \overset{0}{x}_{n-1}),$$

where

$$y(t) = \big(x_1(t), \ldots, x_n(t)\big),$$
$$B(t) = \big(0, 0, \ldots, 0, b(t)\big).$$

On the other hand, the matrix

$$\begin{bmatrix} 0 & 1 & 0 & \ldots & & 0 \\ 0 & 0 & 1 & \ldots & & 0 \\ \cdot & \cdot & \cdot & \cdot & \cdot & \cdot \\ 0 & 0 & 0 & \ldots & & 1 \\ a_0(t) & a_1(t) & a_2(t) & \ldots & a_{n-1}(t) \end{bmatrix}$$

corresponds in the canonical basis $e_1, \ldots, e_n$ to the operator $A(t)$ $\in L(R^n, R^n)$. Let the system of functions $t \to w_j(t) \in R^n$, $j = 1, 2, \ldots, n$, form a fundamental system of solutions of the homogeneous equation. The vectors $w_j(t_0), j = 1, 2, \ldots, n$, form a basis (in general, not canonical) of the space R^n. Forming suitable linear combinations of the functions w, we can obtain the fundamental system of solutions of the homogeneous equation $u_j(\,\cdot\,)$ satisfying the initial conditions $u_j(t_0) = e_j$, $j = 1, 2, ..., n$. Decomposing the functions $u_j(\,\cdot\,)$ into basis vectors, we obtain a system of n^2 scalar functions which satisfy the system of equations:

$$u_j(t) = \sum_{k=1}^{n} u_{kj}(t) \cdot e_k, \quad j = 1, 2, ..., n,$$

$$\frac{du_{kj}}{dt} = u_{k+1, j}, \quad k = 1, 2, ..., n-1,$$

$$\frac{du_{nj}}{dt} = \sum_{l=1}^{n} a_{l-1}(t) \cdot u_{lj}, \quad j = 1, 2, ..., n,$$

$$u_{kj}(t_0) = \delta_{kj}, \quad k, j = 1, 2, ..., n.$$

It is readily seen that knowing the system of functions u_{kj} is equivalent to knowing for the n-th order equation

(1) $$\frac{d^n x}{dt^n} = \sum_{k=0}^{n-1} a_k(t) x^{(k)},$$

n solutions $v_j, j = 1, 2, ..., n$, which satisfy the initial condition $v_j^{(k-1)}(t_0) = \delta_{jk}$.

To convince oneself of this, it is sufficient to take $u_{kj}(t) = v_j^{(k-1)}(t)$. We can then say that the fundamental system of solutions of equation (1) consists of n functions $v_j(\,\cdot\,)$ which satisfy that equation and the initial condition $v_j^{(k-1)}(t_0) = \delta_{kj}, k, j = 1, 2, ..., n$. This system constitutes the basis of the space of solutions of equation (1). Let us form the Wronskian

$$W(t) = \begin{vmatrix} u_{11}(t) & u_{12}(t) & \cdots & u_{1n}(t) \\ u_{21}(t) & u_{22}(t) & \cdots & u_{2n}(t) \\ \cdot & \cdot & \cdot & \cdot \\ u_{n-1,1}(t) & u_{n-1,2}(t) & \cdots & u_{n-1,n}(t) \\ u_{n1}(t) & u_{n2}(t) & \cdots & u_{nn}(t) \end{vmatrix}$$

$$= \begin{vmatrix} v_1^{(0)}(t) & v_2^{(0)}(t) & \cdots & v_n^{(0)}(t) \\ v_1^{(1)}(t) & v_2^{(1)}(t) & \cdots & v_n^{(1)}(t) \\ \cdot & \cdot & \cdot & \cdot \\ v_1^{(n-2)}(t) & v_2^{(n-2)}(t) & \cdots & v_n^{(n-2)}(t) \\ v_1^{(n-1)}(t) & v_2^{(n-1)}(t) & \cdots & v_n^{(n-1)}(t) \end{vmatrix}.$$

By virtue of the Liouville formula, $W(t) = W(t_0)\exp\int_{t_0}^{t} a_{n-1}(s)ds$.

COROLLARY IX.13.1. In order that a system of functions $v_j(t)$, $j = 1, 2, ..., n$ satisfying equation (1) be a fundamental system of solutions it is necessary and sufficient that the Wronskian of this system be nonzero at any point t_0.

Once we have a fundamental system of solutions of equation (1) which furthermore satisfies the condition $v_j^{(k-1)}(t_0) = \delta_{kj}$; $k, j = 1, 2, ..., n$, we can solve the initial-value problem for the nonhomogeneous equation. Proceeding as in the previous section, we find that the solution of the complete equation satisfying the initial condition $x^{(k)}(t_0) = \overset{0}{x_k}$, $k = 0, 1, ..., n-1$ is of the form

$$x(t) = \sum_{k=1}^{n} v_k(t) \cdot \overset{0}{x_{k-1}} + \sum_{k=1}^{n} v_k(t) \int_{t_0}^{t} \frac{Dv_k^{(n-1)}(s)}{W(s)} b(s) ds.$$

14. LINEAR EQUATIONS WITH CONSTANT COEFFICIENTS

Let A be a square matrix of order n with complex coefficients. We seek vectors $x \in C^n$ ($x \neq 0$) and (complex) numbers λ such that

(1) $\qquad Ax = \lambda \cdot x \qquad$ or $\qquad (A - \lambda I) \cdot x = 0.$

This is the *eigenvalue problem* for the matrix X. Vectors satisfying equations (1) are called the *eigenvectors* of the matrix A, and the numbers λ corresponding to them are referred to as *eigenvalues*. Note that the necessary condition for a solution of equations (1) to exist for $x \neq 0$ is that the determinant of the matrix $A - \lambda I$ vanish. The equation $\det(A - \lambda I) = 0$ is actually an algebraic equation of the n-th degree in the unknown λ. As is known from the fundamental theorem of algebra, an equation of this kind has n roots. They are called the *characteristic roots* of the matrix A. Let the numbers $\lambda_1, \lambda_2, ..., \lambda_k$ be distinct characteristic roots, and let the integers $n_1, n_2, ..., n_k$ denote the mult plicity of the individual roots. Of course, $\sum_{i=1}^{k} n_i = n$.

LEMMA IX.14.1. For every matrix A (square matrix of degree n) k linear subspaces $X_i \subset C^n$, $i = 1, 2, ..., k$ (k is the number of distinct characteristic roots), exist in the space C^n. These subspaces have the following properties:

   (i) $AX_i \subset X_i$ (*invariance property*),

   (ii) $X_i \cap X_j = \{0\}$, $i \neq j$,

   (iii) $C^n = \overset{k}{\underset{i=1}{\oplus}} X_i$,

   (iv) $X_i = \{x \in C^n; (A - \lambda_i I)^{n_i} \cdot x = 0\}$,

   (v) $\dim X_i = n_i$.

This lemma is given here without the proof, which could be found in most textbooks on linear algebra.

Items (ii) and (iii) of the lemma state that every vector $x \in C^n$ is decomposable into a sum of vectors from the subspaces X_i,

$$x = \sum_{i=1}^{k} x_i,$$

this decomposition being unique.

After that algebraic introduction, let us go on to solve a homogeneous linear equation with constant coefficients in the space R^n. We have the equation

$$\frac{dx}{dt} = A \cdot x,$$

where A does not depend on t and is a constant operator (matrix). We assume the initial condition $x(t_0) = x_0$.

In accordance with general theory, the solution of this initial-value problem is of the form

$$x(t) = R(t, t_0) \cdot x_0, \quad \text{where } R(t, t_0) = T e^{\int_{t_0}^{t} A\, ds} = e^{A \cdot (t - t_0)},$$

and hence $x(t) = e^{A(t - t_0)} \cdot x_0$. The solution thus obtained is written as a series and thus is not "convenient". We shall demonstrate that it can be rewritten as a finite sum.

To this end we decompose the vector $x_0 \in R^n \subset C^n$ into invariant subspaces of the matrix $A \cdot x_0 = \sum_{i=1}^{k} \overset{0}{x_i}$. Using the identity

$$e^{A(t-t_0)} = e^{(t-t_0)\lambda_i I} e^{(t-t_0)(A - \lambda_i I)},$$

we can write

$$x(t) = e^{(t-t_0)\cdot A} \sum_{i=1}^{k} \overset{0}{x_i} = \sum_{i=1}^{k} e^{(t-t_0)\lambda_i I} e^{(t-t_0)(A-\lambda_i I)} \overset{0}{x_i}.$$

Expansion of the exponential function in a series yields

$$x(t) = \sum_{i=1}^{k} e^{(t-t_0)\lambda_i I} \left(\sum_{j=0}^{\infty} \frac{(t-t_0)^j}{j!} (A - \lambda_i I)^j \cdot \overset{0}{x_i} \right).$$

Item (iv) of Lemma IX.14.1 implies that

$$(A - \lambda_i I)^j \cdot \overset{0}{x_i} = 0 \quad \text{for } j = n_i.$$

Thus, for $j > n_i$ we also have

$$(A - \lambda_i I)^j \cdot \overset{0}{x_i} = 0,$$

The series breaks off and we get

$$x(t) = \sum_{i=1}^{k} e^{(t-t_0)\lambda_i} \left(\sum_{j=0}^{n_i-1} \frac{(t-t_0)^j}{j!} (A - \lambda_i I)^j \cdot \overset{0}{x_i} \right).$$

Example. Let $n = 3$. We have a system of equations

$$\frac{dx_1}{dt} = x_1 + x_2 + 2x_3,$$

$$\frac{dx_2}{dt} = x_2 + x_3,$$

$$\frac{dx_3}{dt} = 2x_3;$$

$$x_1(0) = 1, \quad x_2(0) = 2, \quad x_3(0) = 1.$$

The matrix A of this system is of the form

$$A = \begin{bmatrix} 1 & 1 & 2 \\ 0 & 1 & 1 \\ 0 & 0 & 2 \end{bmatrix}.$$

We seek the characteristic roots of A:

$$\det \begin{bmatrix} 1-\lambda & 1 & 2 \\ 0 & 1-\lambda & 1 \\ 0 & 0 & 2-\lambda \end{bmatrix} = 0.$$

Thus we get the equation $(1-\lambda)^2(2-\lambda) = 0$. The elements of this equation are the numbers

$$\lambda_1 = 1 \quad \text{multiplicity } n_1 = 2,$$
$$\lambda_2 = 2 \quad \text{multiplicity } n_2 = 1.$$

We look for the invariant subspaces

$$(A - \lambda_1 I)^{n_1} \cdot y = \begin{bmatrix} 0 & 1 & 2 \\ 0 & 0 & 1 \\ 0 & 0 & 1 \end{bmatrix}^2 \begin{bmatrix} y_1 \\ y_2 \\ y_3 \end{bmatrix} = 0,$$

whence

$$\begin{bmatrix} 0 & 0 & 3 \\ 0 & 0 & 1 \\ 0 & 0 & 1 \end{bmatrix} \cdot \begin{bmatrix} y_1 \\ y_2 \\ y_3 \end{bmatrix} = 0.$$

We obtain a system of equations

$$3y_3 = 0,$$
$$y_3 = 0,$$
$$y_3 = 0,$$

whereby $y_3 = 0$, and y_1 and y_2 are arbitrary. Therefore

$$X_1 = \{x \in C^3 : x = (x_1, x_2, 0) : x_1, x_2 \in C^1\}.$$

We seek X_2:

$$(A - \lambda_2 I)^{n_2} \cdot z = \begin{bmatrix} -1 & 1 & 2 \\ 0 & -1 & 1 \\ 0 & 0 & 0 \end{bmatrix} \cdot \begin{bmatrix} z_1 \\ z_2 \\ z_3 \end{bmatrix} = 0.$$

We obtain the system of equations

$$-z_1 + z_2 + 2z_3 = 0,$$
$$-z_2 + z_3 = 0.$$

Hence $z_2 = z_3, z_1 = 3z_3$ and

$$X_2 = \{x \in C^3 : x = (3z, z, z) : z \in C^1\}.$$

We decompose the initial condition into vectors from the invariant subspaces:

$$\begin{bmatrix} 1 \\ 2 \\ 1 \end{bmatrix} = \begin{bmatrix} y_1 \\ y_2 \\ 0 \end{bmatrix} + \begin{bmatrix} 3z \\ z \\ z \end{bmatrix}.$$

We obtain a system of equations:

$$y_1 \quad +3z = 1,$$
$$y_2 + \ z = 2,$$
$$z = 1.$$

We thus have $y_1 = -2, y_2 = 1, z = 1$. Consequently we obtain the decomposition

$$\begin{bmatrix} 1 \\ 2 \\ 1 \end{bmatrix} = \begin{bmatrix} -2 \\ 1 \\ 0 \end{bmatrix} + \begin{bmatrix} 3 \\ 1 \\ 1 \end{bmatrix}.$$

In accordance with the general theory, the solution is of the form

$$x(t) = e^t \cdot (I + t \cdot (A - \lambda I)) \cdot \overset{0}{x}_1 + e^{2t} \cdot \overset{0}{x}_2.$$

When the previously calculated values are inserted into this formula, the result is

$$x(t) = e^t \cdot \left[\begin{bmatrix} -2 \\ 1 \\ 0 \end{bmatrix} + t \cdot \begin{bmatrix} 0 & 1 & 2 \\ 0 & 0 & 1 \\ 0 & 0 & 1 \end{bmatrix} \cdot \begin{bmatrix} -2 \\ 1 \\ 0 \end{bmatrix} \right] + e^{2t} \cdot \begin{bmatrix} 3 \\ 1 \\ 1 \end{bmatrix}.$$

Finally, we have

$$x(t) = e^t \cdot \left(\begin{bmatrix} -2 \\ 1 \\ 0 \end{bmatrix} + t \begin{bmatrix} 1 \\ 0 \\ 0 \end{bmatrix} \right) + e^{2t} \cdot \begin{bmatrix} 3 \\ 1 \\ 1 \end{bmatrix}.$$

On writing out the individual components, we have

$$x_1(t) = (t-2) \cdot e^t + 3e^{2t},$$
$$x_2(t) = e^t + e^{2t},$$
$$x_3(t) = e^{2t}.$$

In deriving the formula for the solution of the equation $dx/dt = A \cdot x$, we tacitly disregarded the fact that the characteristic roots of the matrix A may be complex numbers. Thus, a linear combination of complex expressions appears in the solution. The question is: is the result obtained real? Of course it is; this follows immediately from two theorems already proved: (i) on the *existence* of a solution in the real domain, and (ii) on the *uniqueness* of the solution in the complex domain.

We shall demonstrate somewhat more:

Note that from the fact that the elements of matrix A are real it follows that if λ_i is a characteristic root of A, then so is $\bar{\lambda}_i$ and it has the same multiplicity as λ_i. If the vector $x = (x_1, \ldots, x_n) \in C^n$ satisfies the equation

$$(A - \lambda_i I)^{n_i} \cdot x = 0,$$

hen the vector $\bar{x} = (\bar{x}_1, \ldots, \bar{x}_n)$ satisfies the equation

$$(A - \bar{\lambda}_i I)^{n_i} \cdot \bar{x} = 0.$$

Thus, if X_i is an invariant space corresponding to the element λ_i, then $\bar{X}_i := \{x \in C^n : \bar{x} \in X_i\}$ is an invariant space corresponding to the element $\bar{\lambda}_i$. Suppose that the matrix A has $2s$ distinct complex roots and $k - 2s$ real roots. Hence we have k distinct roots.

Let us label these roots so that

$$\bar{\lambda}_{2l-1} = \lambda_{2l}, \quad 1 \leqslant l \leqslant s,$$

$$\lambda_i = \bar{\lambda}_i, \quad 2s+1 \leqslant i \leqslant k.$$

On denoting the multiplicity of the root λ_i by n_i, we have

$$n_{2l-1} = n_{2l}, \quad 1 \leqslant l \leqslant s,$$

Then

$$\bar{X}_{2l-1} = X_{2l}, \quad 1 \leqslant l \leqslant s.$$

$$\bar{X}_i = X_i, \quad 2s+1 \leqslant i \leqslant k.$$

Let us decompose the real initial vector x_0 into invariant subspaces:

$$x_0 = \overset{0}{x}_1 + \overset{0}{x}_2 + \ldots + \overset{0}{x}_{2s-1} + \overset{0}{x}_{2s} + \overset{0}{x}_{2s+1} + \ldots \overset{0}{x}_k.$$

Conjugation of both sides of this equation gives us

$$x_0 = \overset{\bar{0}}{x}_1 + \overset{\bar{0}}{x}_2 + \ldots + \overset{\bar{0}}{x}_{2s-1} + \overset{\bar{0}}{x}_{2s} + \overset{\bar{0}}{x}_{2s+1} + \ldots + \overset{\bar{0}}{x}_k.$$

However, if $\overset{0}{x}_{2l-1} \in \bar{X}_{2l-1}$, then $\overset{\bar{0}}{x}_{2l-1} \in X_{2l}, 1 \leqslant l \leqslant s$, and if $\overset{0}{x}_i \in \bar{X}_i$, then $\overset{\bar{0}}{x}_i \in X_i$, $2s+1 \leqslant i \leqslant k$. From the foregoing relations and from the uniqueness of the decomposition of the vector x_0 into the subspaces X_i it follows that

$$\overset{\bar{0}}{x}_{2l-1} = \overset{0}{x}_{2l}, \quad 1 \leqslant l \leqslant s,$$

$$\overset{\bar{0}}{x}_i = \overset{0}{x}_i, \quad 2s+1 \leqslant i \leqslant k.$$

Thus, we see that the formula for the solution contains components of the form

(a) $\quad e^{(t-t_0)\cdot\lambda_i}\sum_{j=0}^{n_i-1}\dfrac{(t-t_0)^j}{j!}(A-\lambda_i I)^j\cdot\overset{0}{x_i},\quad 2s+1\leqslant i\leqslant k,$

where λ_i is real; then this expression is real, and of the form

(b) $\quad e^{(t-t_0)\lambda_{2l-1}}\sum_{j=0}^{n_{2l-1}-1}\dfrac{(t-t_0)^j}{j!}(A-\lambda_{2l-1}I)^j\cdot\overset{0}{x}_{2l-1}$

and

$$e^{(t-t_0)\lambda_{2l}}\sum_{j=0}^{n_{2l}-1}\dfrac{(t-t_0)^j}{j!}(A-\lambda_{2l}I)^j\cdot\overset{0}{x}_{2l}$$

for $1\leqslant l\leqslant s$. These expressions are conjugate to each other. Accordingly, the solution contains a sum of expressions of the form

$$c_{ij}e^{\lambda_i(t-t_0)}(t-t_0)^j,\quad 1\leqslant i\leqslant k,\ 0\leqslant j\leqslant n_i-1,$$

where

$$c_{ij}=\bar{c}_{ij}\quad\text{for }2s+1\leqslant i\leqslant k,$$
$$c_{2lj}=\bar{c}_{2l-1,j}\quad\text{for }1\leqslant l\leqslant s.$$

For $1\leqslant l\leqslant s$, $\lambda_{2l}=\bar\lambda_{2l-1}$ and we may write

$$\lambda_{2l-1}=a_l+b_l i,\quad b\neq 0,$$
$$\lambda_{2l}=a-b_l i.$$

On adding together the terms which correspond to the roots λ_{2l} and λ_{2l-1}, we obtain

$$c_{2l-1,j}e^{(a_l+b_l i)(t-t_0)}(t-t_0)^j+\bar{c}_{2l-1,j}e^{(a_l-b_l i)(t-t_0)}(t-t_0)^j$$
$$=2\,\mathrm{Re}\,c_{2l-1,j}e^{a_l(t-t_0)}\cos b_l(t-t_0)\cdot(t-t_0)^j-$$
$$-2\,\mathrm{Im}\,c_{2l-1,j}e^{a_l(t-t_0)}\sin b_l(t-t_0)\cdot(t-t_0)^j.$$

Finally, it can be said that the basis of the solution space of a system of n linear equations with constant coefficients is formed by the functions

$$e^{a_l(t-t_0)}\cos b_l(t-t_0)\cdot(t-t_0)^j,$$
$$e^{a_l(t-t_0)}\sin b_l(t-t_0)\cdot(t-t_0)^j,$$

where

$$\lambda_{2l-1} = a_l + b_l i, \quad 1 \leqslant l \leqslant s,$$
$$\lambda_{2l} = a_l - b_l i, \quad 0 \leqslant j \leqslant n_{2l} - 1,$$

and the functions

$$e^{a_i(t-t_0)}(t-t_0)^j,$$

where

$$\lambda_i = a_i, \quad 2s+1 \leqslant i \leqslant k, \ 0 \leqslant j \leqslant n_i - 1.$$

The integers n_i are the multiplicities of the respective roots.

The solution $x(t)$ is a linear combination of the basis functions with the corresponding real coefficients.

Having found the solution of the homogeneous equation, we proceed to the full equation

$$\frac{dx}{dt} = A \cdot x + b(t), \quad x(t_0) = x_0.$$

In accordance with the general theory, the solution of this initial-value problem is of the form

$$x(t) = R(t, t_0) \cdot x_0 + \int_{t_0}^{t} R(t, s) \cdot b(s)\, ds,$$

where $R(t, t_0) = e^{A(t-t_0)}$. We decompose the vectors x_0 and $b(s)$ into invariant subspaces

$$x_0 = \sum_{i=1}^{k} \overset{0}{x}_i, \quad b(s) = \sum_{i=1}^{k} b_i(s).$$

Proceeding as when solving the homogeneous equation, we obtain

$$R(t, t_0) \cdot x_0 = \sum_{i=1}^{k} e^{\lambda_i(t-t_0)} \left(\sum_{j=0}^{n_i-1} \frac{(t-t_0)^j}{j!} (A - \lambda_i I)^j \cdot \overset{0}{x}_i \right),$$

$$R(t, s) \cdot b(s) = \sum_{i=1}^{k} e^{\lambda_i(t-t_0)} \left(\sum_{j=0}^{n_i-1} \frac{(t-s)^j}{j!} (A - \lambda_i I)^j \cdot b_i(s) \right).$$

Substitution into the formula yields

$$x(t) = \sum_{i=1}^{k} e^{\lambda_i(t-t_0)} \left(\sum_{j=0}^{n_i-1} \frac{(t-t_0)^j}{j!} (A - \lambda_i I)^j \cdot \overset{0}{x}_i \right) +$$

$$+ \int_{t_0}^{t} \sum_{i=1}^{k} e^{\lambda_i(t-s)} \left(\sum_{j=0}^{n_i-1} \frac{(t-s)^j}{j!} (A - \lambda_i I)^j \cdot b_i(s) \right) ds.$$

This is the general formula for the solution of an initial-value problem for a nonhomogeneous equation. A linear combination of complex expressions appears in this formula. However, with a real initial condition and a real vector $b(s)$ the result is real.

15. SCALAR EQUATIONS OF THE n-TH ORDER WITH CONSTANT COEFFICIENTS

Let $X = R^1$. We consider the equation

$$\frac{d^n x}{dt^n} = \sum_{k=0}^{n-1} a_k \frac{d^k x}{dt^k} + b(t)$$

with the initial condition

$$x(t_0) = \overset{0}{x}_0, \qquad \frac{dx(t_0)}{dt} = \overset{0}{x}_1, \qquad \frac{d^2 x(t_0)}{dt^2} = \overset{0}{x}_2, \dots,$$

$$\frac{d^{n-1} x(t_0)}{dt^{n-1}} = \overset{0}{x}_{n-1}.$$

By substituting

$$x_1(t) := x(t),$$
$$x_2(t) := x'(t),$$
$$\cdots \cdots \cdots \cdots$$
$$x_n(t) := x^{(n-1)}(t),$$

we reduce the equation to a system of n first-order equations

$$\frac{dx_1}{dt} = x_2,$$

$$\frac{dx_2}{dt} = x_3,$$

$$\cdots \cdots \cdots \cdots$$

$$\frac{dx_{n-1}}{dt} = x_n,$$

$$\frac{dx_n}{dt} = \sum_{k=0}^{n-1} a_k x_{k+1} + b(t).$$

The initial condition takes the form

$$x_1(t_0) = \overset{0}{x_0}, \quad x_2(t_0) = \overset{0}{x_1}, \quad \ldots, \quad x_n(t_0) = \overset{0}{x_{n-1}}.$$

The matrix A of this system is of the form:

$$A = \begin{bmatrix} 0 & 1 & 0 & 0 & \ldots & 0 \\ 0 & 0 & 1 & 0 & \ldots & 0 \\ 0 & 0 & 0 & 1 & \ldots & 0 \\ \multicolumn{6}{c}{\cdots \cdots \cdots \cdots} \\ 0 & 0 & 0 & 0 & \ldots & 1 \\ a_0 & a_1 & a_2 & a_3 & \ldots & a_{n-1} \end{bmatrix}.$$

The characteristic polynomial is

$$F(\lambda) = \begin{bmatrix} -\lambda & 1 & 0 & 0 & \ldots & 0 \\ 0 & -\lambda & 1 & 0 & \ldots & 0 \\ 0 & 0 & -\lambda & 1 & \ldots & 0 \\ \multicolumn{6}{c}{\cdots \cdots \cdots \cdots} \\ 0 & 0 & 0 & 0 & \ldots & 1 \\ a_0 & a_1 & a_2 & a_3 & \ldots & a_{n-1} - \lambda \end{bmatrix} = 0.$$

Expansion of the determinant yields

(1) $$F(\lambda) = \lambda^n - \sum_{k=0}^{n-1} a_k \lambda^k = 0.$$

Thus we see that the characteristic polynomial is obtained from the homogeneous equation when the derivatives are replaced by the appropriate powers of λ.

Once the roots of equation (1) have been determined and the invariant subspaces found, in accordance with the preceding section we can write the solution of the homogeneous equation as

$$\begin{bmatrix} x_1(t) \\ x_2(t) \\ \vdots \\ x_n(t) \end{bmatrix} = x(t) = \sum_{i=1}^{k} e^{\lambda_i(t-t_0)} \left(\sum_{j=0}^{n_i-1} \frac{(t-t_0)^j}{j!} (A - \lambda_i I)^j x^i \right),$$

329

$$\text{where } \overset{0}{x} = \sum_{i=1}^{k} x^i.$$

Only $x_1(t)$ is of interest to us. The results of Section 14 can be used to write that the general solution (i.e. the solution dependent on n arbitrary constants) of the equation

$$(2) \qquad \frac{d^n x}{dt^n} = \sum_{k=0}^{n-1} a^k \frac{d^k x}{dt^k}$$

is of the form

$$(3) \qquad x(t) = \sum_{i=1}^{p} e^{\lambda_i(t-t_0)} \Big(\sum_{j=0}^{n_i-1} c_{ji}(t-t_0)^j \Big) +$$

$$+ \sum_{k=1}^{s} \Big\{ e^{\alpha_k(t-t_0)} \cdot \cos \beta_k(t-t_0) \Big(\sum_{j=0}^{n_k-1} d_{jk}(t-t_0)^j \Big) +$$

$$+ e^{\alpha_k(t-t_0)} \cdot \sin \beta_k(t-t_0) \Big(\sum_{j=0}^{n_k-1} f_{jk}(t-t_0)^j \Big) \Big\}$$

where p is the number of distinct real roots of the characteristic equation (1) and $2s$ is the number of distinct complex roots of the form $\alpha_k \pm \beta_k i$. The numbers c_{ij}, d_{jk}, f_{jk} are real constants which are uniquely determined by the initial conditions.

We have thus obtained a method of solving a homogeneous equation of the n-th order. To this end, we find the characteristic roots from equation (1) and write out the general solution dependent on n constants (formula (3)). In order to determine the constants, we differentiate the result a suitable number of times and find the expressions for the first, second, ..., $(n-1)$-st derivative of the function $x(\cdot)$. On setting $t = t_0$, we get a system of n equations

$$x(t_0) = \overset{0}{x}_0,$$

$$\frac{dx(t_0)}{dt} = \overset{0}{x}_1,$$

$$\cdot \ \cdot \ \cdot \ \cdot \ \cdot \ \cdot \ \cdot \ \cdot$$

$$\frac{d^{n-1}x(t_0)}{dt^{n-1}} = \overset{0}{x}_{n-1}$$

for the unknown constants. In view of the uniqueness of the solution of the initial-value problem, this system has exactly one solution.

Examples. 1. Let us consider the equation

$$\frac{d^2x}{dt^2} + a^2 \cdot x = 0, \qquad x(0) = 0, \; x'(0) = 1.$$

This is the *harmonic oscillator equation*. The characteristic equation $\lambda^2 + a^2 = 0$ has the solutions $\lambda_1 = a \cdot i$, and $\lambda_2 = -a \cdot i$.

The general solution:

$$x(t) = C_1 \cos at + C_2 \sin at;$$

differentiating it, we obtain

$$x'(t) = -C_1 a \sin at + C_2 a \cos at,$$

and hence $x(0) = C_1$, $x'(0) = C_2 a$.

When we take the initial conditions into account, we have $C_1 = 0$, $C_2 = 1/a$. Ultimately, the solution of the initial-value problem is of the form

$$x(t) = \frac{1}{a} \sin at.$$

2. Consider the equation

$$\frac{d^2x}{dt^2} - a^2x = 0; \qquad x(0) = 1, \; x'(0) = 0.$$

The characteristic equation: $\lambda^2 - a^2 = 0$, where $\lambda_1 = a$, $\lambda_2 = -a$.

The general solution:

$$x(t) = C_1 e^{at} + C_2 e^{-at};$$

differentiation of this solution leads to

$$x'(t) = C_1 a e^{at} - C_2 a e^{-at}.$$

Therefore

$$x(0) = C_1 + C_2, \qquad x'(0) = a(C_1 - C_2).$$

On taking the initial conditions into account, we obtain $C_1 = C_2 = \frac{1}{2}$. Finally, therefore,

$$x(t) = \frac{1}{2}(e^{at} + e^{-at}).$$

3. The *equation of the damped harmonic oscillator*:

$$\frac{d^2x}{dt^2} + 2b\frac{dx}{dt} + a^2x = 0, \quad b > 0;$$

$$x(0) = 0, \quad x'(0) = 1.$$

The characteristic equation:

$$\lambda^2 + 2b\lambda + a^2 = 0.$$

We consider three cases:

1) $b < a$; then

$$\lambda_1 = -b + i\sqrt{a^2 - b^2}, \quad \lambda_2 = -b - i\sqrt{a^2 - b^2}.$$

The general solution:

$$x(t) = e^{-bt}(C_1 \cos \sqrt{a^2 - b^2} \cdot t + C_2 \sin \sqrt{a^2 - b^2} \cdot t).$$

From the initial conditions we have

$$C_1 = 0, \quad C_2 = \frac{1}{(a^2 - b^2)^{1/2}}.$$

2) $b = a$; we then have one double root, $\lambda = -b$.
The general solution:

$$x(t) = e^{-bt}(C_1 + C_2 t).$$

It follows from the initial conditions that $C_1 = 0, C_2 = 1$.

3) $b > a$. Then

$$\lambda_1 = -b + \sqrt{b^2 - a^2}, \quad \lambda_2 = -b - \sqrt{b^2 - a^2}.$$

The general solution:

$$x(t) = e^{-bt}(C_1 e^{\sqrt{b^2 - a^2} \cdot t} + C_2 e^{-\sqrt{b^2 - a^2} \cdot t}).$$

From the initial conditions we have $C_1 = -C_2 = 1/2(b^2 - a^2)^{1/2}$, and therefore

$$x(t) = e^{-bt} \frac{1}{(b^2 - a^2)^{1/2}} \sinh \sqrt{b^2 - a^2} \cdot t.$$

Let us now take up the nonhomogeneous equation. We know a method for finding the solution of the nonhomogeneous equation, as discussed in the preceding section. In the case of an equation of the n-th order, this method requires relatively long calculations. We also know a different method which consists in adding an arbitrary particular solution

332

of the nonhomogeneous equation to the general solution of the homogeneous equation. To be able to employ this method, one must know how to find particular solutions of nonhomogeneous equations.

We now give a method of finding a particular solution of a scalar nonhomogeneous equation of the *n*-th order with an inhomogeneity of the form

$$b(t) = e^{zt} P_l(t),$$

where $P_l(t)$ is a polynomial of degree l with complex coefficients and z is an arbitrary complex number.

Let $F(\lambda) = \lambda^n + \sum_{k=0}^{n-1} a_k \lambda^k$ be the characteristic polynomial of the equation

$$\text{(4)} \qquad \frac{d^n x}{dt^n} + \sum_{k=0}^{n-1} a_k \frac{d^k x}{dt^k} = b(t).$$

We consider two cases:

Case 1. The number z is not a root of the characteristic equation, that is, $F(z) \neq 0$. Suppose that the inhomogeneity is of the form

$$b(t) = e^{zt} P_l(t), \quad \text{where } P_l(t) = \sum_{m=0}^{l} p_m t^m.$$

We seek a particular solution of equation (4) in the form

$$x(t) = e^{zt} Q_l(t),$$

where $Q_l(t)$ is a polynomial of degree l with complex coefficients, $Q_l(t) = \sum_{m=0}^{l} q_m t^m$. We substitute them into equation (4):

$$\frac{d^n}{dt^n}\left(e^{zt}\sum_{m=0}^{l} q_m t^m\right) + \sum_{k=0}^{n-1} a_k \frac{d^k}{dt^k}\left(e^{zt}\sum_{m=0}^{l} q_m t^m\right) = e^{zt}\sum_{m=0}^{l} p_m t^m.$$

The left-hand side is equal to

$$\sum_{m=0}^{l} q_m \frac{d^n}{dt^n}(e^{zt} t^m) + \sum_{m=0}^{l} q_m \sum_{k=0}^{n-1} a_k \frac{d^k}{dt^k}(e^{zt} t^m).$$

333

Making use of the formulae

(5)
$$\frac{d^j}{dt^j}(e^{zt}) = z^j e^{zt},$$

(6)
$$\frac{d^j}{dt^j}(t^m) = \begin{cases} \dfrac{m!}{(m-j)!}t^{m-j} & \text{for } m \geqslant j, \\ 0 & \text{for } m < j, \end{cases}$$

we obtain

(7)
$$\frac{d^k}{dt^k}(e^{zt}t^m) = \sum_{j=0}^{k}\binom{k}{j}\frac{d^{k-j}}{dt^{k-j}}(e^{zt})\frac{d^j}{dt^j}(t^m).$$

When (5) and (6) are inserted into (7), we obtain

(8)
$$\frac{d^k}{dt^k}(e^{zt}t^m) = e^{zt}\sum_{j=0}^{h}\binom{m}{j}\frac{k!}{(k-j)!}t^{m-j}z^{k-j},$$

where $h = \min(m, k)$. Moreover,

(9)
$$F^{(j)}(z) = \frac{n!}{(n-j)!}z^{n-j} + \sum_{k=j}^{n-1}a_k\frac{k!}{(k-j)!}z^{k-j}, \quad j = 0, 1, \dots, n.$$

Setting $a_n = 1$ and using (8), we have

$$\sum_{m=0}^{l}q_m\sum_{k=0}^{n}a_k\sum_{j=0}^{h}\binom{m}{j}\frac{k!}{(k-j)!}t^{m-j}z^{k-j} = \sum_{m=0}^{l}p_m t^m,$$

wherefrom, by changing the order of summation, we get

$$\sum_{m=0}^{l}q_m\sum_{j=0}^{s}\binom{m}{j}t^{m-j}\cdot\sum_{k=j}^{n}\frac{k!}{(k-j)!}a_k z^{k-j} = \sum_{m=0}^{l}p_m t^m,$$

where $s = \min(m, n)$. Use of (9) leads to

(10)
$$\sum_{m=0}^{l}q_m\sum_{j=0}^{s}\binom{m}{j}t^{m-j}F^{(j)}(z) = \sum_{m=0}^{l}p_m t^m.$$

Comparison of the coefficients of identical powers of t in formula (10) gives us a system of $l+1$ equations for $l+1$ unknowns $q_0, q_1, \dots, q_l$:

(11)
$$\sum_{k=0}^{m}\binom{l-k}{m-k}F^{(m-k)}(z)\cdot q_{l-k} = p_{l-m}, \quad m = 0, 1, \dots, l.$$

334

On being written out, this system becomes:

$$F(z) \cdot q_l = p_l,$$

$$F(z) \cdot q_{l-1} + \binom{l}{1} F'(z) \cdot q_l = p_{l-1},$$

$$F(z) \cdot q_{l-2} + \binom{l-1}{1} F'(z) \cdot q_{l-1} + \binom{l}{2} F'(z) \cdot q_l = p_{l-2},$$

$$\cdot \quad \cdot \quad \cdot \quad \cdot \quad \cdot \quad \cdot \quad \cdot \quad \cdot \quad \cdot \quad \cdot \quad \cdot \quad \cdot \quad \cdot \quad \cdot \quad \cdot \quad \cdot \quad \cdot \quad \cdot \quad \cdot$$

$$F(z) \cdot q_0 + F'(z) \cdot q_1 + F'(z) \cdot q_2 + \ldots + F^{(l)}(z) \cdot q_l = p_0.$$

As is readily seen, this system has exactly one solution. Its determinant is equal to $(F(z))^{l+1} \neq 0$ (since z is not a root of the characteristic polynomial).

Case 2. If z is a root of the characteristic equation and has a multiplicity of r, then

$$\text{(12)} \qquad \begin{aligned} F^{(j)}(z) &= 0, \quad j = 0, 1, \ldots, r-1, \\ F^{(r)}(z) &\neq 0. \end{aligned}$$

To demonstrate this, we write the characteristic polynomial as $F(\lambda) = (\lambda - z)^r G(\lambda)$, where $G(\cdot)$ is a polynomial of order $n-r$. Differentiation of this equation j times with respect to λ and the substitution $\lambda = z$ yield formulae (12).

We seek a particular solution of equation (4) in the form

$$x(t) = t^r e^{zt} Q_l(t),$$

where $Q_l(t)$ is a polynomial of degree l. Inserting it into the equation, we have

$$\sum_{m=0}^{l} q_m \sum_{k=0}^{n} a_k \frac{d^k}{dt^k} (t^{r+m} e^{zt}) = e^{zt} \sum_{m=0}^{l} p_m t^m.$$

Next, we make use of formulae (5), (6), and (7) and cancel by e^{zt}, thus obtaining $(h = \min(m+r, k))$

$$\sum_{m=0}^{l} q_m \sum_{k=0}^{n} a_k \sum_{j=0}^{h} \binom{r+m}{j} \frac{k!}{(k-j)!} t^{r+m-j} z^{k-j} = \sum_{m=0}^{l} p_m t^m.$$

When we change the order of summation, we have

$$\sum_{m=0}^{l} q_m \sum_{j=0}^{s} t^{r+m-j}\binom{r+m}{j}\sum_{k=j}^{n} a_k \frac{k!}{(k-j)!}z^{k-j} = \sum_{m=0}^{l} p_m t^m,$$

$$s = \min(m+r, n).$$

Use of formulae (9) and (12) results in

$$\sum_{m=0}^{l} q_m \sum_{j=0}^{s} t^{r+m-j}\binom{r+m}{j}F^{(j)}(z) = \sum_{m=0}^{l} p_m t^m.$$

On comparing the coefficients of corresponding powers of t, we obtain a system of $l+1$ equations for $l+1$ unknowns $q_0, q_1, \ldots, q_l$:

(13) $$\sum_{k=0}^{m} q_{l-k}\binom{l+r-k}{m+r-k}F^{(m+r-k)}(z) = p_{l-m},$$

where $m = 0, 1, \ldots, l$. Writing out this system, we have:

$$\binom{l+r}{r}F^{(r)}(z)\cdot q_l = p_l,$$

$$\binom{l+r-1}{r}F^{(r)}(z)\cdot q_{l-1} + \binom{l+r}{r+1}F^{(r+1)}(z)\cdot q_l = p_{l-1},$$

$$\cdots\cdots\cdots\cdots\cdots\cdots\cdots\cdots$$

$$F^{(r)}(z)\cdot q_0 + F^{(r+1)}(z)\cdot q_1 + \cdots + F^{(r+l)}(z)\cdot q_l = p_0.$$

The determinant of this system is thus seen to be equal to

$$\binom{l+r}{r}\binom{l+r-1}{r}\cdots\binom{r}{r}(F^{(r)}(z))^{l+1}.$$

From formulae (12) it follows that this determinant is nonzero and hence the system (13) has exactly one solution.

It may thus be said that the particular solution of equation (4) with inhomogeneity of the form

$$b(t) = e^{zt}\cdot P_l(t)$$

is of the form

(i) $y(t) = e^{zt}\cdot Q_l(t)$, if z is not a root of the characteristic polynomial;

(ii) $y(t) = t^r e^{zt}\cdot Q_l(t)$ if z is a root of the characteristic polynomial and has a multiplicity of r. (The polynomials $P_l(t)$ and $Q_l(t)$ are of order l and have complex coefficients.)

The coefficients of the polynomial Q_l can be determined from the system of equations (11) or (13). In practice, however, Q is written with letters for coefficients and solutions of the form of (i) or (ii) are substituted into equation (4). By comparing the coefficients of the corresponding powers of t, we find their numerical values. This method is known as the *method of undetermined coefficients*.

Note that if the functions $y_1, y_2, \ldots, y_n$ are solutions of equation (4) with inhomogeneities $b_1, b_2, \ldots, b_n$, respectively, the function $y(t)$

$$= \sum_{i=1}^{n} y_i(t) \text{ is a solution of equation (4) with inhomogeneity } b(t)$$

$$= \sum_{i=1}^{n} b_i(t).$$

The coefficients a_k, $k = 0, 1, \ldots, n-1$, of equation (4) are real, and thus if the particular solution

$$y(t) = t^r e^{zt} Q_l(t)$$

corresponds to the inhomogeneity $b(t) = e^{zt} P_l(t)$, the solution

$$y_1(t) = \overline{y(t)} = t^r e^{\bar{z}t} \overline{Q_l(t)}$$

corresponds to the inhomogeneity $b_1(t) = \overline{b(t)} = e^{\bar{z}t} \overline{P_l(t)}$.

Now we seek the particular solution of equation (4) with inhomogeneity of the form

$$b(t) = e^{ct} \big(P_l(t) \cos d \cdot t + R_s(t) \sin d \cdot t \big),$$

where the numbers c and d are real, and $P_l(t)$ and $R_s(t)$ are polynomials of degree l and s, respectively, with real coefficients.

On introducing complex powers, we have

$$b(t) = \frac{1}{2} \big(\exp(c+di)t + \exp(c-di)t \big) P_l(t) +$$

$$+ \frac{1}{2i} \big(\exp(c+di)t - \exp(c-di)t \big) R_s(t).$$

We are looking for the particular solutions corresponding to the inhomogeneities

$$b_1(t) = \frac{1}{2} e^{(c+di)t} P_l(t),$$

$$b_2(t) = \frac{1}{2} e^{(c-di)t} P_l(t),$$

$$b_3(t) = \frac{1}{2i} e^{(c+di)t} R_s(t),$$

$$b_4(t) = -\frac{1}{2i} e^{(c-di)t} R_s(t).$$

In accordance with the remarks above and inasmuch as the polynomials P_l and R_s corresponding to these inhomogeneities have real coefficients, the solutions are of the form

$$y_1(t) = t^r e^{(c+di)t} Q_l(t),$$
$$y_2(t) = t^r e^{(c-di)t} \overline{Q_l(t)},$$
$$y_3(t) = t^r e^{(c+di)t} S_s(t),$$
$$y_4(t) = t^r e^{(c-di)t} \overline{S_s(t)}.$$

Adding them together, we get

$$y(t) = \overline{y(t)} = t^r 2 e^{ct} (\operatorname{Re} Q_l(t) + \operatorname{Re} S_s(t)) \cos d \cdot t -$$
$$- t^r 2 e^{ct} (\operatorname{Im} Q_l(t) + \operatorname{Im} S_s(t)) \sin d \cdot t.$$

In the formulae above $Q_l(t)$ and $S_s(t)$ are polynomials of degree l and s with complex coefficients, and $r \geqslant 0$ is equal to the multiplicity of the root $c+di$ (or $c-di$) of the characteristic equation (if $c+di$ is not a root of the characteristic equation, then $r = 0$).

In summary, it may be said that for an equation of the n-th order with constant, real coefficients and inhomogeneity

$$b(t) = e^{ct} \{ P_l(t) \cos d \cdot t + R_s(t) \sin d \cdot t \},$$

where c and d are real numbers and $P_l(t)$ and $R_s(t)$ are polynomials of degree l and s, respectively, the particular solution is of the form

$$y(t) = t^r \cdot e^{ct} \cdot \left(\underset{1}{W_m}(t) \cos d \cdot t + \underset{2}{W_m}(t) \sin d \cdot t \right),$$

where $\underset{1}{W_m}(t)$ and $\underset{2}{W_m}(t)$ are polynomials of degree $m = \max(l, s)$ with real coefficients, and $r \geqslant 0$ is the multiplicity of the root $c+di$ (or the root $c-di$) of the characteristic equation.

In practice the coefficients of the polynomials $\underset{1}{W}$ and $\underset{2}{W}$ are determined by substituting the expression for $y(t)$ into the equation and comparing the coefficients of the same powers of t.

338

Examples. 1. The *equation of a harmonic oscillator with perturbing force*:

$$\frac{d^2x}{dt^2} = -a^2x + A\sin b \cdot t.$$

Characteristic equation: $\lambda^2 + a^2 = 0$.

Roots: $\lambda_1 = ia$, $\lambda_2 = -ia$.

The general solution of the homogeneous equation is of the form

$$x(t) = C_1\cos at + C_2\sin at.$$

To find a particular solution of the nonhomogeneous equation, we consider two cases:

(i) $a \neq b$. We seek a solution of the nonhomogeneous equation in the form

$$y(t) = D_1\cos bt + D_2\sin bt.$$

This is inserted into the equation

$$-D_1 b^2\cos bt - D_2 b^2\sin bt + D_1 a^2\cos bt + D_2 a^2\sin bt$$

$$= A\sin bt.$$

Hence,

$$D_1 a^2 - D_1 b^2 = 0, \qquad D_2(a^2 - b^2) = A,$$

and consequently $D_1 = 0$, $D_2 = A/(a^2 - b^2)$.

The general solution of the nonhomogeneous equation thus is of the form

$$x(t) = C_1\cos at + C_2\sin at + \frac{A}{a^2 - b^2}\sin bt.$$

(ii) $a = b$. We seek a particular solution of the nonhomogeneous equation in the form

$$y(t) = t(B_1\cos at + B_2\sin at).$$

When we substitute it into the equation we get

$$-2B_1 a\sin at + 2B_2 a\cos at = A\sin at,$$

whereby $B_1 = -A/2a$, $B_2 = 0$. The solution thus is of the form

$$x(t) = C_1\cos at + C_2\sin at - \frac{A}{2a}t\cos at.$$

339

2. The *equation of a damped oscillator with periodic perturbing force*:

$$\frac{d^2x}{dt^2} + 2d\frac{dx}{dt} + a^2 \cdot x = A \sin bt.$$

We consider only the case $d < a$. The roots of the characteristic equation are $\lambda_{1,2} = -d \pm i(a^2 - d^2)^{1/2}$. The general solution of the homogeneous equation is of the form

$$x(t) = e^{-dt}(C_1 \cos \sqrt{a^2 - d^2}\, t + C_2 \sin \sqrt{a^2 - d^2}\, t).$$

We seek a particular solution of the nonhomogeneous equation in the form

$$y(t) = D_1 \cos bt + D_2 \sin bt.$$

We insert it into the equation and obtain

$$-D_1 b^2 \cos bt - D_2 b^2 \sin bt - 2dD_1 b \sin bt +$$
$$+2dD_2 b \cos bt + a^2 D_1 \cos bt + a^2 D_2 \sin bt = A \sin bt.$$

Thus we get a system of equations

$$-2bdD_1 + (a^2 - b^2)D_2 = A,$$
$$(a^2 - b^2)D_1 + 2dbD_2 = 0.$$

On solving this system, we have

$$D_1 = \frac{-2dbA}{(a^2 - b^2)^2 + 4d^2b^2}, \qquad D_2 = \frac{(a^2 - b^2)A}{(a^2 - b^2)^2 + 4d^2b^2}.$$

Thus finally, we have

$$x(t) = e^{-dt}(C_1 \cos \sqrt{a^2 - d^2}\, t + C_2 \sin \sqrt{a^2 - d^2}\, t) + D_1 \cos bt +$$
$$+ D_2 \sin bt.$$

16. FIRST INTEGRALS

DEFINITION. Given a system of n differential equations of the first order

(1) $$\frac{dx_i}{dt} = f(t, x_1, \ldots, x_n), \qquad i = 1, 2, \ldots, n.$$

Suppose that these are scalar equations, that is, $x_i(t) \in R^1$. A function $R^1 \times R^n \ni (t, x_1, \ldots, x_n) \to g(t, x_1, \ldots, x_n) \in R^1$, different from a constant function, is said to be a *first integral* of system (1) if it is

340

constant on the integral curves of system (1); i.e. if the functions $t \to x_1(t), ..., t \to x_n(t)$ are solutions of system (1), then the function $t \to g(t, x_1(t), ..., x_n(t))$ is constant.

In other words, the integral curves of system (1) are contained in the level surfaces of the function g.

The definition given here specifies the concept of first integral only for systems of first-order equation. As is readily seen, this definition is immediately transferable to systems of equations of any order. This is not necessary, however, inasmuch as every system of a higher order is reducible to a system of first-order equations.

We shall now demonstrate that the existence of n first integrals follows from knowledge of the solution of system (1). Suppose that system (1) satisfies the hypotheses of the theorems on the existence and uniqueness of a solution. Let the vector-valued function $t \to u(t, t_0, x_0)$ $\in R^n$ be a solution satisfying some initial condition for system (1);

$$u(t, t_0, x_0) = \sum_{i=1}^{n} x_i(t) e_i \text{ (where } e_1, ..., e_n \text{ is a canonical basis in } R^n).$$

We define the mapping

$$R^1 \times R^n \ni (t, x) \to g(t, x) := u(t_0, t, x) \in R^n.$$

By virtue of item (iii) of Theorem IX.4.1, the mapping g is constant on the integral curve of system (1). For we have

$$g(t, u(t, t_0, x_0)) \underset{t}{\equiv} x_0.$$

The components $g_j(t, x_1, ..., x_n)$, $j = 1, 2, ..., n$, of the mapping g in the canonical basis $(g(t, x) = \sum_{j=1}^{n} g_j(t, x) e_j)$ thus are first integrals of system (1).

It has therefore been shown that, knowing the solution of system (1), we have n first integrals. The question is whether knowledge of n first integrals of system (1) enables the solution of the system to be found? The answer is provided by the following theorem:

THEOREM IX.16.1. Given a system of n scalar first-order equations satisfying the hypotheses of the theorem on the unique solvability of the initial-value problem (Theorem IX.2.1)

$$(2) \qquad \frac{dx_i}{dt} = f_i(t, x_1, ..., x_n), \quad i = 1, 2, ..., n, \quad x_i(t_0) = \overset{0}{x_i}.$$

341

Given also k functions of $n+1$ variables $t, x_1, \ldots, x_n$ continuously differentiable in a neighbourhood of the point $(t_0, \overset{0}{x}_1, \ldots, \overset{0}{x}_n)$,

$$g_j(t, x_1, \ldots, x_n), \quad j = 1, 2, \ldots, k, \; k \leqslant n,$$

which are first integrals of system (2) and let

$$\frac{\partial(g_1, \ldots, g_k)}{\partial(x_1, \ldots, x_k)} \neq 0 \text{ at the point } (t_0, \overset{0}{x}_1, \ldots, \overset{0}{x}_n),$$

Then in a neighbourhood of the point $(t_0, \overset{0}{x}_1, \ldots, \overset{0}{x}_n)$ system (2) reduces to a system of $n-k$ scalar equations of the first order.

PROOF. Since the corresponding Jacobian is nonzero at the point $(t_0, \overset{0}{x}_1, \ldots, \overset{0}{x}_n)$, by the implicit-function theorem it follows that from the system of equations

$$g_j(t, x_1, \ldots, x_n) = c_j, \quad j = 1, 2, \ldots, k,$$

we can, in a certain neighbourhood of the point $(t_0, \overset{0}{x}_1, \ldots, \overset{0}{x}_n)$, determine $x_j(\,\cdot\,), j = 1, 2, \ldots, k$, as functions of the variables $t, x_{k+1}, \ldots, x_n$ and parameters $c_1, \ldots, c_k$:

$$(3) \qquad x_j = h_j(t, x_{k+1}, \ldots, x_n, c_1, \ldots, c_k), \quad j = 1, 2, \ldots, k.$$

The functions h_j are continuously differentiable. System (2) thus reduces to a system of $n-k$ equations

$$(4) \qquad \frac{dx_{k+j}}{dt} = f_{k+j}\big(t, h_1(t, x_{k+1}, \ldots, x_n), \ldots$$

$$\ldots, h_k(t, x_{k+1}, \ldots, x_n), x_{k+1}, \ldots, x_n\big), \quad j = 1, 2, \ldots, n-k,$$

with the initial condition $x_{k+j}(t) = \overset{0}{x}_{k+j}$.

Since system (2) satisfied the hypotheses of the uniqueness theorem and since the functions h are continuously differentiable, system (4) also satisfies the hypotheses of Theorem IX.2.1. On solving this system, we find the functions $x_{k+j}(\,\cdot\,), j = 1, 2, \ldots, n-k$, and by means of them we determine the functions $x_j(\,\cdot\,), j = 1, 2, \ldots, k$, from formulae (3). In this way we get the solution of the initial-value problem (2). $\square$

The procedure carried out in the proof of Theorem IX.16.1 is known as the *reduction of a problem by means of first integrals*. It is seen that, once we know k first integrals of a given system of n differential equations, we can reduce that system to one of $n-k$ equations, which is in general much easier to solve. Precisely therein consists the profit from first integrals in the practical solution of systems of equations. We see also that, once we know n independent first integrals, we in fact have the solution of the system. The solution of a system of differential equations can therefore be found either in explicit form, or in implicit form in terms of first integrals.

Now we shall show that a first integral of a system of equations satisfies a certain differential equation in partial derivatives. Let $g(t, x_1, \ldots, x_n)$ be a first integral of a system of n differential equations

$$(5) \qquad \frac{dx_i}{dt} = f_i(t, x_1, \ldots, x_n), \qquad i = 1, 2, \ldots, n.$$

If the functions $x_1(\cdot), \ldots, x_n(\cdot)$ are solutions of system (5), then the function

$$t \to g(t, x_1(t), \ldots, x_n(t))$$

is constant and hence its derivative vanishes identically

$$0 = \frac{dg}{dt} = \sum_{i=1}^{n} \frac{\partial g}{\partial x_i} \frac{dx_i}{dt} + \frac{\partial g}{\partial t}.$$

Making use of system (5), we have

$$0 = \frac{\partial g}{\partial t}(t, x_1, \ldots, x_n) + \sum_{i=1}^{n} \frac{\partial g(t, x_1, \ldots, x_n)}{\partial x_i} f_i(t, x_1, \ldots, x_n).$$

It is readily seen that the formula above is a necessary and sufficient condition for the function g to be a first integral of system (5).

In conclusion, let us emphasize that there is no universal method for finding first integrals. Usually, we find them by performing certain algebraic operations on the system of equations. In physics, the laws of conservation (of energy, momentum, angular momentum) are first integrals of the equations of motion.

EXAMPLE. *Euler's equation for asymmetrical rotator.*

Let $I_1 > I_2 > I_3$ be the principal moments of inertia of the top and let $\omega = (\omega_1, \omega_2, \omega_3)$ be the angular velocity vector. Then the equations of motion are of the form

$$I_1 \frac{d\omega_1}{dt} = (I_2 - I_3)\omega_2\omega_3,$$

$$(6) \qquad I_2 \frac{d\omega_2}{dt} = (I_3 - I_1)\omega_1\omega_3,$$

$$I_3 \frac{d\omega_3}{dt} = (I_1 - I_2)\omega_1\omega_2.$$

Multiplying the first equation through by ω_1, the second by ω_2, and the third by ω_3, and then adding the results, we have

$$\sum_{k=1}^{3} I_k \omega_k \frac{d\omega_k}{dt} = 0.$$

This may be rewritten as

$$\frac{d}{dt}\left(\sum_{k=1}^{3} \frac{I_k \omega_k^2}{2}\right) = 0.$$

Thus, the function $(t, \omega_1, \omega_2, \omega_3) \to \sum_{k=1}^{3} \frac{1}{2} I_k \omega_k^2$ is a first integral of system (6). This is what is known as the *energy integral*, for this expression is the kinetic energy of the rotational motion of the rotator.

Now let us multiply the first equation of system (6) by $I_1\omega_1$, the second by $I_2\omega_2$, and the third by $I_3\omega_3$. Addition by sides yields

$$\frac{1}{2} \cdot \frac{d}{dt}\left(\sum_{k=1}^{3} I_k^2 \omega_k^2\right) = 0.$$

We thus infer that the function $(t, \omega_1, \omega_2, \omega_3) \to \sum_{k=1}^{3} I_k^2 \omega_k^2$ is a first integral of system (6). This is the *angular momentum integral*. We thus have

$$I_1\omega_1^2 + I_2\omega_2^2 + I_3\omega_3^2 = c_1,$$
$$I_1^2\omega_1^2 + I_2^2\omega_2^2 + I_3^2\omega_3^2 = c_2.$$

Hence

$$(7) \qquad \begin{aligned} \omega_1^2 &= a_1\omega_3^2 + b_1, \\ \omega_2^2 &= a_2\omega_3^2 + b_2, \end{aligned}$$

344

where a_1, a_2, b_1, b_2 are constants associated with the constants I_1, I_2, I_3, c_1, c_2 by algebraic equations. When formulae (7) are inserted into system (6), this system reduces to a single equation

$$I_3 \frac{d\omega_3}{dt} = (I_1 - I_2) \sqrt{a_1 \omega_3^2 + b_1} \sqrt{a_2 \omega_3^2 + b_2}.$$

The solution is obtained by means of the elliptic integral

$$\int \frac{d\omega_3}{\sqrt{(a_1 \omega_3^2 + b_1)(a_2 \omega_3^2 + b_2)}} = \frac{I_1 - I_2}{I_3} t + C.$$

Remark. Elliptic integrals are tabulated; they are related to doubly periodic functions of a complex variable which are called *elliptic functions*. The fundamentals of elliptic function theory are given in Chapter XXIII and further details can be found in any extensive textbook on functions of a complex variable.

17. DYNAMIC SYSTEMS

Consider the differential equation

(1) $\qquad \dfrac{dx}{dt} = f(x)$

with the initial condition $x(t_0) = x_0$. The right-hand side of equation (1) does not depend on t. Such an equation (system of equations) will be called a *dynamic system*. The solution of equation (1) satisfying the initial condition $x(t_0) = x_0$ will be written as

$$x(t) = R(t, t_0)(x_0),$$

where $R(t, t_0)$ is a mapping of a Banach space X into itself. This mapping is not in general linear since the equation is not linear. It is readily noted, however, that

$$R(t_0, t_0) = I, \qquad R(t, s) \cdot R(s, t_0) = R(t, t_0),$$
$$R(t_0, t_0) = (R(t_0, t))^{-1}.$$

The mapping $R(t, t_0)$ is the *resolvent* of equation (1).

Remark. In the discussion above we must, of course, assume that equation (1) satisfies the hypotheses of the theorem on the uniqueness of the solvability of the initial-value problem.

THEOREM IX.17.1. Let $R(\cdot, \cdot)$ be the resolvent of the dynamic system

$$\frac{dx}{dt} = f(x),$$

satisfying the hypotheses of the theorem on the uniqueness of the solvability of the initial-value problem. Then, for every $a \in R^1$

$$R(t+a, t_0+a) = R(t, t_0).$$

PROOF. Let the function $t \to x(t) = u(t, t_0, x_0)$ be a solution of equation (1) satisfying the initial condition $x(t_0) = x_0$. Let

$$v(t) := u(t+a, t_0+a, x_0),$$

$$\frac{dv(t)}{dt} = \frac{du(t+a, t_0+a, x_0)}{d(t+a)} \frac{d(t+a)}{dt}$$

$$= f\big(u(t+a, t_0+a, x_0)\big).$$

Hence the function $t \to v(t)$ satisfies equation (1):

$$v(t_0) = u(t_0+a, t_0+a, x_0) = x_0.$$

The function $v(\cdot)$ thus satisfies the same equation as function $u(\cdot)$ does and satisfies the same initial condition. The uniqueness of the solution implies that

$$v(t) \underset{t}{\equiv} u(t, t_0, x_0)$$

whence

$$u(t+a, t_0+a, x_0) \underset{t}{\equiv} u(t, t_0, x_0).$$

This equation, written in terms of the resolvent, assumes the form

$$R(t+a, t_0+a)(x_0) = R(t, t_0)(x_0).$$

It holds for every $x_0 \in X$. Therefore

$$R(t+a, t_0+a) = R(t, t_0). \quad \square$$

On setting $a = -t_0$ in Theorem IX.17.1, we have

$$R(t-t_0, 0) = R(t, t_0).$$

Next we define a one-parameter family of mappings of the space X into itself:

$$T_t := R(t, 0), \quad t \in R^1.$$

The family $(T_t)_{t \in R^1}$ has the following properties:

(i) $T_0 = R(0, 0) = I$,

(ii) $T_{-t} = R(-t, 0) = R(0, t) = \big(R(t, 0)\big)^{-1} = (T_t)^{-1}$,

346

(iii) $T_{t+s} = R(t+s, 0) = R(t, -s) = R(t, 0) \cdot R(0, -s) = R(t, 0) \cdot R(s, 0)$
$= T_t \circ T_s$.

DEFINITION. A family $(T_t)_{t \in R^1}$ of mappings of a Banach space into itself, with properties (i), (ii), and (iii), is called a *one-parameter group of mappings of the space X.*

Thus we have proved the following theorem:

THEOREM IX.17.2. A dynamic system defines a one-parameter group of mappings of a Banach space X.

DEFINITION. *Infinitesimal generator* of a one-parameter group $(T_t)_{t \in R^1}$ of mappings of a Banach space is the name given to the mapping

$$X \ni x \rightarrow \frac{d}{dt} \left((T_t(x)) |_{t=0} \in X. \right.$$

When a one-parameter group of mappings is determined by a dynamic system $dx/dt = f(x)$, the function $f(\cdot)$ is the infinitesimal generator of the group. For we have

$$\frac{d}{dt} (T_t(x_0)) = \frac{d}{dt} R(t, 0) (x_0)|_{t=0} = \frac{dx(t)}{dt} \bigg|_{t=0}$$

$$= f(x(0)) = f(x_0).$$

On the other hand, if a group $(T_t)_{t \in R^1}$ of mappings of a Banach space is prescribed *a priori*, it is difficult to find its infinitesimal generator. Even the relatively "easy" case of the group of continuous linear mappings $T_t \in L(X, X)$ has a rich and elegant theory. In this case, the infinitesimal generator is a linear operator which in general is not continuous and not defined on the entire space X. When X is a complex Hilbert space, and (T_t) is a group of unitary operators (that is, $(T_t x | T_t y) = (x|y)$, $x, y \in X$), then Stone's theorem states that the infinitesimal generator of that group is the operator $A = -iH$, where H is a self-adjoint operator (in general, unbounded). If $x(t) = T_t(x)$, then

$$\frac{dx(t)}{dt} = \frac{d}{ds} x(s+t)|_{s=0} = \frac{d}{ds} (T_s \circ T_t(x))|_{s=0} = -iHT_t(x)$$

$$= -iHx(t).$$

Thus we have obtained the equation

(2) $\qquad i\dfrac{dx}{dt} = Hx.$

This is a case of fundamental importance in quantum mechanics. Equation (2) is there called the *Schrödinger equation* and the operator *H*, the *Hamiltonian (energy) operator*.

Examples of Infinitesimal Generators. 1. *Linear equation with constant coefficients*:

$$\frac{dx}{dt} = A \cdot x, \quad R(t, 0) = e^{tA}, \quad \frac{dT_t}{dt}\bigg|_{t=0} = A.$$

Here, the operator A is the infinitesimal generator

2. *System of Hamilton's equations.* Let

$$X = R^{2n} = R^n \times R^n, \quad x = (q_1, \ldots, q_n, p_1, \ldots, p_n),$$

where q_i are generalized coordinates and p_i are generalized momenta. We have

$$\frac{dq_i}{dt} = \frac{\partial H(p, q)}{\partial p_i},$$
$$\frac{dp_i}{dt} = -\frac{\partial H(p, q)}{\partial q_i}, \qquad i = 1, 2, \ldots, n.$$

This is a dynamic system which generates a one-parameter group of transformations of the space R^{2n}, known as *canonical transformations*. In Chapter XXII we shall show that the mappings of this group preserve volume in the space R^{2n}.

18. FIRST-ORDER PARTIAL DIFFERENTIAL EQUATION. THE CHARACTERICTICS METHOD

Let a function

$$R^{2n+1} = R^n \times R^1 \times R^n \ni (x, z, p) \to F(x, z, p) \in R^1$$

of class C^3 be given. We consider a *first-order partial differential equation* for scalar-valued functions $z(\cdot)$ of n variables

(1) $\qquad F(x, z, z') = 0.$

As in the case of ordinary differential equations, (1) may have many (different) solutions. Usually we look for a solution assuming prescribed values on some given $(n-1)$-dimensional surface in R^n; such a problem is called the *Cauchy probtem* for the equation (1).

DEFINITION. If a mapping

$$U \ni u \to \Phi(u) \in \mathbf{R}^s$$

is a C^k-injection, where $U \subset \mathbf{R}^r$ is open, $r \leqslant s$, $k > 0$, and at every point $u \in U$ the derivative $\Phi'(u)$ has the maximal rank, i.e.

(2) $\qquad \operatorname{rank} \Phi' = r$,

then we say that there is given an *r-dimensional* (*parametrized*) *surface of class C^k in the space $\mathbf{R}^s$.*

Suppose that we are given an $(n-1)$-dimensional surface of class C^3 in $\mathbf{R}^n$, whose parametrizing mapping Φ has the form

(3) $\qquad \mathbf{R}^{n-1} \supset U \ni u$
$$:= (u^1, \ldots, u^{n-1}) \to \left(\bar{\xi}^1(u^1, \ldots, u^{n-1}), \ldots, \bar{\xi}^n(u^1, \ldots, u^{n-1}) \right) \in \mathbf{R}^n.$$

Denote $M := \Phi(U) \subset \mathbf{R}^n$. The condition (2) reads now

(4) $\qquad \operatorname{rank} \left(\dfrac{\partial \bar{\xi}^i}{\partial u^j} \right) = n-1.$

It ensures that M is indeed $(n-1)$-dimensional. By *Cauchy data* (on the surface M) we mean a function g on M. In the sequel we assume that all occurring Cauchy data depend C^3-differentiably on the parameters u^k. This means that the following function of the parameters

(5) $\qquad U \ni u \to \zeta(u) := g\left(\bar{\xi}(u) \right) \in \mathbf{R}^1, \qquad \bar{\xi} = (\bar{\xi}^1, \ldots, \bar{\xi}^n)$

is of class C^3. Often, we use for M the term *Cauchy surface*, and by Cauchy data we mean also the pair (M, g).

Now we look for a solution $z(\cdot)$ of (1), which coincides with g on M. Such a solution can be obtained by the so-called *characteristics method* due to Monge and Cauchy. This method leads to the construction of an n-dimensional surface in $\mathbf{R}^{2n+1}$, which is the graph of the solution $z(\cdot)$. Following Cauchy, it is useful to consider the n-dimensional surface $\mathcal{M}$ in $\mathbf{R}^{2n+1}$ given by

(6) $\qquad \mathbf{R}^n \supset V \ni x \to \left(x, z(x), p(x) \right) \in \mathbf{R}^{2n+1},$

where V is open, $z(\cdot)$ is the solution and $p(x) := z'(x)$, i.e.

(7) $\qquad p_j = \dfrac{\partial z}{\partial x^j}; \quad x = (x^1, \ldots, x^n), \ p = (p_1, \ldots, p_n).$

The surface $\mathcal{M}$ is the graph of $z(\cdot)$ and its derivative, however, we will refer to it as to the graph of $z(\cdot)$.

Since $z(\cdot)$ is a solution of (1),

$$(8) \qquad F\big(x, z(x), p(x)\big) \equiv 0$$

which means that $\mathcal{M}$ lies in the 0-level surface of the function F, i.e.

$$(9) \qquad F|\mathcal{M} \equiv 0.$$

Now, let us consider the inverse problem: under what condition a surface $\mathcal{M}$ in R^{2n+1} given by

$$(10) \qquad (x, z, p) = \big(\xi(s), \zeta(s), \pi(s)\big) \in R^{2n+1}$$

($s = (s^1, ..., s^n)$ runs over an open set $W \subset R^n$) is the graph of a solution $z(\cdot)$ of (1) ?

A necessary condition for $\mathcal{M}$ to be the graph of some function $z(\cdot)$ is the possibility of parametrizing it by $x^1, ..., x^n$. This means that the mapping $\xi \colon W \to R^n$ is a diffeomorphism onto its image; in other words, the parameters s^i can be calculated as functions of x^i, i.e. $s = \sigma(x)$. Substituting this to (10) we obtain the parametrization (6) of $\mathcal{M}$, where

$$(11) \qquad z(x) := \zeta\big(\sigma(x)\big), \quad p(x) := \pi\big(\sigma(x)\big).$$

Such surfaces in R^{2n+1} which admit a (unique) parametrization of type (6) will be called *transversal* to the canonical projection $\mathrm{pr}_x \colon R^{2n+1} \to R^n$, $\mathrm{pr}_x(x, z, p) = x$.

It is easy to note that if the surface $\mathcal{M}$ is transversal to pr_x (i.e. functions $z(\cdot)$ and $p(\cdot)$ can be defined by (11)), then the equations (7) and (9) give a necessary and sufficient condition for $\mathcal{M}$ to be the graph of a solution of (1).

Now we shall describe the characteristics method. Firstly we look for an $(n-1)$-dimensional subsurface $\mathcal{M}_0$ of the graph $\mathcal{M}$, consisting of the points which are mapped by pr_x into the Cauchy surface M. The surface $\mathcal{M}_0$ will be parametrized by the same parameters as is M:

$$(12) \qquad \begin{aligned} x &= \bar{\xi}(u^1, ..., u^{n-1}), \\ z &= \bar{\zeta}(u^1, ..., u^{n-1}), \\ p &= \bar{\pi}(u^1, ..., u^{n-1}), \end{aligned}$$

where functions $\bar{\xi}$ are as in (3) and $\bar{\zeta}$ is given by Cauchy data as in (5). It remains to define the functions $\bar{\pi}(\cdot)$. We have to do it consistently with equations (7) and (8). But, at the moment, the function $z(\cdot)$ is defined only on the surface M. Therefore we know derivatives of $z(\cdot)$ only in the directions tangent to M. But vectors tangent to M form the space spanned by the following basis:

$$(13) \qquad e_k := \frac{\partial \bar{\xi}}{\partial u^k} = \left(\frac{\partial \bar{\xi}^1}{\partial u^k}, \dots, \frac{\partial \bar{\xi}^n}{\partial u^k} \right);$$

this implies

$$(14) \qquad \nabla_{e_k} z = \sum_{i=1}^{n} \frac{\partial z}{\partial x^i} \frac{\partial \bar{\xi}^i}{\partial u^k}.$$

Hence

$$(15) \qquad \nabla_{e_k} z = \frac{\partial \bar{\zeta}}{\partial u^k}.$$

This and (7) give the equations

$$(16) \qquad \frac{\partial \bar{\zeta}}{\partial u^k} = \sum_{i=1}^{n} p_i \frac{\partial \bar{\xi}^i}{\partial u^k}, \qquad k = 1, \dots, n-1.$$

For a given point $x \in M$ (16) is a system of $n-1$ linear equations for n unknown p_i. By virtue of (4) the set of solutions is 1-dimensional, i.e. it is a line. But we want the points of $\mathcal{M}_0$ to satisfy also equation (8), which now reads:

$$(17) \qquad F(\bar{\xi}(u), \bar{\zeta}(u), p) = 0.$$

DEFINITION. Cauchy data (M, g) for equation (1) are called *regular* if for every $u \in U$ (i.e., for every point of the surface M) the system of n equations (16), (17) has a unique solution $p \in R^n$, which is a function of class C^2 in the parameters u^k (the coefficients in the equation (16) are of class C^2 as the derivatives of $\bar{\zeta}$ and $\bar{\xi}$, which are of class C^3).

If (M, g) are regular, then for each $u \in U$ we take $\bar{\pi}(u)$ as the solution of (16), (17). This way the surface $\mathcal{M}_0$ with the parametrization (12) is completely defined.

In what follows we shall prove that the graph $\mathcal{M}$ consists of the integral curves (solutions) of the following system of ordinary differential equations in R^{2n+1}:

$$\frac{dx^i}{dt} = \frac{\partial F}{\partial p_i}(x, z, p),$$

$$(18) \qquad \frac{dz}{dt} = \sum_{j=1}^{n} p_j \cdot \frac{\partial F}{\partial p_j}(x, z, p),$$

$$\frac{dp_i}{dt} = -\frac{\partial F}{\partial x^i}(x, z, p) - p_i \frac{\partial F}{\partial z}(x, z, p_,.$$

This system is called the *characteristic system* for the equation (1) Respectively, its solutions are called *characteristic curves*, or simply *characteristics*, of (1).

Now, the characteristics passing through $\mathscr{M}_0$ will be of interest for us. We introduce the following notation. For fixed values of parameters u^k the curve given parametrically by the equations

$$(19) \qquad \begin{aligned} x &= \xi(u, t), \\ z &= \zeta(u, t), \\ p &= \pi(u, t) \end{aligned}$$

will denote the characteristic (solution of (18)) which fulfils the initial conditions

$$(20) \qquad \begin{aligned} \xi(u, 0) &= \bar{\xi}(u), \\ \zeta(u, 0) &= \bar{\zeta}(u), \\ \pi(u, 0) &= \bar{\pi}(u). \end{aligned}$$

The curve (19) is parametrized by the variable t.

Now we consider the whole family of all curves (19) corresponding to all values of parameters $u^1, \ldots, u^{n-1}$. In other words, we take the family of all characteristics passing through the points of $\mathscr{M}_0$. The most interesting case is when this family forms an n-dimensional surface transversal to pr_x. In this case the Cauchy data (M, g) are said to be noncharacteristic.

By virtue of Theorem IX.4.2 (on differentiable dependence of solutions of differential equation on initial conditions) the mapping

$$(21) \qquad (u, t) \rightarrow (\xi(u, t), \zeta(u, t), \pi(u, t))$$

is of class C^2. The transversality condition means that ξ is a diffeomorphism. Using the Inverse Mapping Theorem we can write the transversality condition in the form:

$$(22) \qquad \det\left(\left(\frac{\partial \xi^i}{\partial u^k}\right), \frac{\partial \xi^i}{\partial t}\right) \neq 0.$$

For a point lying on $\mathcal{M}_0$ (i.e. for $t = 0$) the above condition reads

$$(23) \qquad \det\left(\left(\frac{\partial \bar{\xi}^i}{\partial u^k}\right), \frac{\partial F}{\partial p_i}\right) \neq 0,$$

because $\dfrac{\partial \xi^i}{\partial t} = \dfrac{\partial F}{\partial p_i}$ (cf. (18)).

DEFINITION. The Cauchy data (M, g) are called *non-characteristic* for (1) if they are regular and if (23) is satisfied on $\mathcal{M}_0$.

THEOREM IX.18.1. If (M, g) are non-characteristic Cauchy data, then the surface $\mathcal{M}$

$$x = \xi(u, t),$$
$$z = \zeta(u, t),$$
$$p = \pi(u, t),$$

where ξ, ζ, and π are constructed above, is the graph of the unique solution $z(\cdot)$ of (1), defined in a neighbourhood of M and equal to g on M.

PROOF. The condition (23) implies that M is transversal to pr_x at least in a neighbourhood V of M. This means that the inverse mapping to ξ:

$$V \ni x \to \big(u(x), t(x)\big)$$

can be defined. We define $z(\cdot)$ and $p(\cdot)$ by the substitution:

$$(24) \qquad \begin{aligned} z(x) &:= \zeta\big(u(x), t(x)\big), \\ p(x) &:= \pi\big(u(x), t(x)\big). \end{aligned}$$

If $x \in M$, then $t(x) = 0$ and $z(x) = \zeta(u(x), 0) = \bar{\zeta}(u(x)) = g(x)$ which proves that our Cauchy condition is satisfied.

In order to prove that (9) is satisfied, consider the function

$$\varphi(u, t) = F\big(\xi(u, t), \zeta(u, t), \pi(u, t)\big).$$

Since functions $\xi(u, \cdot)$, $\zeta(u, \cdot)$, $\pi(u, \cdot)$ satisfy equations (18), we get

$$\frac{\partial \varphi}{\partial t} = \sum_i \frac{\partial F}{\partial x^i} \frac{\partial \xi^i}{\partial t} + \frac{\partial F}{\partial z} \frac{\partial \zeta}{\partial t} + \sum_i \frac{\partial F}{\partial p_i} \frac{\partial \pi_i}{\partial t}$$

$$= \sum_i \frac{\partial F}{\partial x^i} \frac{\partial F}{\partial p_i} + \frac{\partial F}{\partial z} \sum_i \pi_i \frac{\partial F}{\partial p_i} -$$

$$- \sum_i \frac{\partial F}{\partial p_i} \left(\frac{\partial F}{\partial x^i} + \frac{\partial F}{\partial z} \pi_i \right) = 0.$$

But $\varphi(u, 0) \equiv 0$ because $p = \bar{\pi}$ is a solution of the equation (17). This implies $\varphi \equiv 0$, which proves (9).

In order to prove (7) we shall show that the following equations hold:

$$(25) \qquad v_j = \frac{\partial \zeta}{\partial u^j} - \sum_i \pi_i \frac{\partial \xi^i}{\partial u^j} \equiv 0,$$

$$(26) \qquad v = \frac{\partial \zeta}{\partial t} - \sum_i \pi_i \frac{\partial \xi^i}{\partial t} \equiv 0.$$

Using equations (18) for functions $\xi(u, \cdot)$, $\zeta(u, \cdot)$, $\pi(\cdot, \cdot)$ we get

$$\frac{\partial v_j}{\partial t} = \frac{\partial^2 \zeta}{\partial t \partial u^j} - \sum_i \frac{\partial \pi_i}{\partial t} \frac{\partial \xi^i}{\partial u^j} - \sum_i \pi_i \frac{\partial^2 \xi^i}{\partial t \partial u^j}$$

$$= \frac{\partial}{\partial u^j} \left(\sum_i \pi_i \frac{\partial F}{\partial p_i} \right) + \sum_i \left(\frac{\partial F}{\partial x^i} + \frac{\partial F}{\partial z} \pi_i \right) \frac{\partial \xi^i}{\partial u^j} -$$

$$- \sum_i \pi_i \frac{\partial}{\partial u^j} \left(\frac{\partial F}{\partial p_i} \right)$$

$$= \sum_i \frac{\partial \pi_i}{\partial u^j} \frac{\partial F}{\partial p_i} + \sum_i \left(\frac{\partial F}{\partial x^i} + \frac{\partial F}{\partial z} \pi_i \right) \frac{\partial \xi^i}{\partial u^j}$$

$$= \sum_i \frac{\partial F}{\partial p_i} \frac{\partial \pi_i}{\partial u^j} + \sum_i \frac{\partial F}{\partial x^i} \frac{\partial \xi^i}{\partial u^j} + \frac{\partial F}{\partial z} \left(\frac{\partial \zeta}{\partial u^j} - v_j \right)$$

$$= - \frac{\partial F}{\partial z} \cdot v_j + \frac{\partial \varphi}{\partial u^j} = - \frac{\partial F}{\partial z} \cdot v_j.$$

The last equality follows from the fact that $\varphi \equiv 0$. But for $t = 0$ equation (25) is reduced to (16), which is satisfied. Consequently $v_j(u, 0) = 0$. The unique solution of the equation

$$\frac{\partial v_j}{\partial t} = -\frac{\partial F}{\partial z} v_j$$

with the vanishing initial condition is $v_j \equiv 0$, i.e. (25) is satisfied.

The equation (26) is satisfied by virtue of the second equation of the system (18).

Since $\zeta = z \circ \xi$, we obtain

(27)
$$\frac{\partial \zeta}{\partial u^j} = \sum_i \frac{\partial z}{\partial x^i} \frac{\partial \xi^i}{\partial u^j},$$

$$\frac{\partial \zeta}{\partial t} = \sum_i \frac{\partial z}{\partial x^i} \frac{\partial \xi^i}{\partial t}.$$

Comparing (27) with (25) and (26) and using the condition (22) we see that the derivatives $\dfrac{\partial z}{\partial x^i}$ are equal to $\pi_i = p_i$ on $\mathcal{M}$. Hence (7) is also satisfied.

We have thus proved that (21) is the graph of the solution looked for.

Now we shall prove the uniqueness. Suppose $\tilde{z}(\cdot)$ is any other solution. Let us denote

(28)
$$\tilde{p}(x) = \tilde{z}'(x), \quad \text{i.e.}$$

$$\tilde{p}_i(x) = \frac{\partial \tilde{z}}{\partial x^i}(x).$$

Let $t \to x(t)$ be a solution of the ordinary differential equation in R^n

(29)
$$\frac{dx^i}{dt} = \frac{\partial F}{\partial p_i}(x, \tilde{z}(x), \tilde{p}(x)),$$

with the initial condition $x(0) = x_0 \in M$. We shall prove that the curve

$$t \to \left(x(t), \tilde{z}(x(t)), \tilde{p}(x(t))\right) \in R^{2n+1}$$

is a characteristic. Indeed, (29) shows that this curve fulfils the first equation of (18); the second equation is satisfied because of

$$\frac{d\tilde{z}}{dt} = \sum_i \frac{\partial \tilde{z}}{\partial x^i} \cdot \frac{dx^i}{dt} = \sum_i \tilde{p}_i \frac{\partial F}{\partial p_i}.$$

Write

$$\psi(x) := F(x, \tilde{z}(x), \tilde{p}(x));$$

clearly, $\psi \equiv 0$, and so

$$0 = \frac{\partial \psi}{\partial x^i} = \frac{\partial F}{\partial x^i} + \frac{\partial F}{\partial z} \frac{\partial \tilde{z}}{\partial x^i} + \sum_j \frac{\partial F}{\partial p_j} \frac{\partial \tilde{p}_j}{\partial x^i}$$

$$= \frac{\partial F}{\partial x^i} + \frac{\partial F}{\partial z} \cdot \tilde{p}_i + \sum_j \frac{\partial F}{\partial p_j} \frac{\partial \tilde{p}_i}{\partial x^j}$$

because of the symmetry of second-order derivatives

$$\frac{\partial \tilde{p}_j}{\partial x^i} = \frac{\partial \tilde{p}_i}{\partial x^j}.$$

Hence

$$\frac{d\tilde{p}_i}{dt} = \sum_j \frac{\partial \tilde{p}_i}{\partial x^j} \cdot \frac{d\tilde{x}^j}{dt} = \sum_j \frac{\partial \tilde{p}_i}{\partial x^j} \frac{\partial F}{\partial p_j}$$

$$= -\left(\frac{\partial F}{\partial x^i} + \frac{\partial F}{\partial z} \tilde{p}_i \right)$$

and the third equation of (18) holds. But for $x_0 \in M$ we have $z(x_0) = \tilde{z}(x_0)$. Moreover, $p(x_0) = \tilde{p}(x_0)$, because equations (16), (17) have only one solution. Thus the curve (29) is the characteristic curve passing through the point $(x_0, z(x_0), p(x_0)) \in \mathcal{M}_0$, i.e. it lies completely in $\mathcal{M}$. Since this holds for any point $x_0 \in M$, the graphs of $\tilde{z}(\cdot)$ and $z(\cdot)$ are composed of the same curves, i.e. $z(\cdot) = \tilde{z}(\cdot)$. $\square$

In theoretical mechanics one considers the so-called *Hamilton–Jacobi equation*

$$\frac{\partial S}{\partial \tau} + H\left(x^1, \ldots, x^k, \tau, \frac{\partial S}{\partial x^1}, \ldots, \frac{\partial S}{\partial x^k} \right) = 0,$$

where H is the Hamiltonian of the mechanical system in question. To pass to the previous notation let us denote $n = k+1$, $x^n = \tau$, $z = S$, $p_i = \dfrac{\partial S}{\partial x^i}$ for $i \leqslant k = n-1$, $p_n = \dfrac{\partial S}{\partial \tau}$,

$$F(x, z, p) = p_n + H(x^1, \ldots, x^n, p_1, \ldots, p_{n-1}).$$

The characteristic equations (18) have the following form:

$$\frac{dx^i}{dt} = \frac{\partial F}{\partial p_i} = \frac{\partial H}{\partial x^i} \quad \text{for } i < n,$$

$$\frac{dx^n}{dt} = \frac{\partial F}{\partial p_n} = 1,$$

$$\frac{dp_i}{dt} = -\frac{\partial F}{\partial x^i} = -\frac{\partial H}{\partial x^i} \quad \text{for } i < n,$$

$$\frac{dp_n}{dt} = -\frac{\partial F}{\partial x^n} = -\frac{\partial H}{\partial \tau}.$$

If we take

$$M = \{(x^1, ..., x^k, t): t = 0\}$$

as the Cauchy surface, then the second equation yields

$$\tau = x^n = t.$$

It follows that the characteristic equations coincide with the Hamilton equations of motion:

$$\frac{dx^i}{dt} = \frac{\partial H}{\partial p_i},$$

$$\frac{dp_i}{dt} = -\frac{\partial H}{\partial x^i}.$$

The variable p_n has the physical interpretation as minus energy: $p_n = -E$. Hence

$$\frac{dE}{dt} = \frac{\partial H}{\partial t}.$$

The characteristics of the Hamilton–Jacobi equation are thus trajectories of the mechanical system in the phase-space (the space of variables x, t, p, E).

The similar phenomenon occurs in geometrical optics, where pr_x, the projections of characteristic curves of the so-called *eikonal equation*, are light-rays.

19. THE FROBENIUS–DIEUDONNÉ THEOREM

The present chapter will end with a theorem which is important for differential geometry.

Let U and V be open sets in Banach spaces X and Y. Let F be a mapping of the set $U \times V$ into $L(X, Y)$ (the space of continuous linear mappings from X to Y). Consider the differential equation

(1) $\qquad y' = F(x, y).$

We seek the solution of this equation, that is, the mapping $X \ni x \to y(x) \in Y$ satisfying (1) and the initial condition $y(x_0) = y_0$, $(x_0, y_0) \in U \times V$. Our equation is not an ordinary one, and hence it cannot be expected always to have a solution under suitable smoothness conditions imposed on F. Equation (1) will be said to be *integrable on* $U \times V$ if for every $(x_0, y_0) \in U \times V$ equation (1) has exactly one solution $x \to y(x)$ defined in some neighbourhood of the point x_0 and satisfying the condition $y(x_0) = y_0$.

THEOREM IX.19.1. Let X and Y be Banach spaces, let U and V be open sets in X and Y, respectively, and let F be a mapping of the product $U \times V$ into $L(X, Y)$, this mapping being continuously differentiable on $U \times V$. The equation (1) is integrable on $U \times V$ if and only if for all $(h_1, h_2) \in X \times X$, $(x, y) \in U \times V$, the following Frobenius condition is satisfied:

$$(2) \qquad \big(F_X'(x, y) \cdot h_1\big) \cdot h_2 + \big(F_Y'(x, y) \cdot \big(F(x, y) \cdot h_1\big)\big) h_2$$
$$= \big(F_X'(x, y) \cdot h_2\big) \cdot h_1 + \big(F_Y'(x, y) \cdot \big(F(x, y) \cdot h_2\big)\big) \cdot h_1.$$

PROOF. *Necessity.* Let $(x, y) \in U \times V$. Let the mapping $x \to y(x)$ satisfy the equation

$$(3) \qquad y'(x) = F(x, y(x)).$$

Then for every $h_2 \in X$ we have

$$(4) \qquad y'(x) \cdot h_2 = F(x, y(x)) \cdot h_2.$$

Differentiating (4) with respect to x, we have for every $h_1 \in X$,

$$(5) \qquad \big(y''(x) \cdot h_1\big) \cdot h_2$$
$$= \big(F_X'(x, y(x)) \cdot h_1\big) \cdot h_2 + \big(F_Y'(x, y(x)) \, (y'(x) \cdot h_1)\big) \cdot h_2.$$

Use of (3) yields

$$(6) \qquad \big(y''(x) \cdot h_1\big) \cdot h_2$$
$$= \big(F_X'(x, y(x)) \cdot h_1\big) \cdot h_2 + \big(F_Y'(x, y(x)) \cdot \big(F(x, y(x)) \cdot h_1\big)\big) \cdot h_2.$$

The symmetry of the second derivative implies that

$$(7) \qquad \big(y''(x_0) \cdot h_1\big) \cdot h_2 = \big(y''(x_0) \cdot h_2\big) \cdot h_1,$$

which implies (2).

Sufficiency. If the mapping $K(x_0, r) \ni x \to y(x) \in X$ satisfies the equation $y'(x) = F(x, y(x))$, $y(x_0) = y_0$, then the mapping

(8) $\qquad R^1 \ni \lambda \to u(\lambda, x-x_0) = y(x_0 + \lambda(x-x_0)) \in Y$

satisfies the ordinary differential equation

(9) $\qquad \dfrac{du(\lambda, x-x_0)}{d\lambda} = F(x_0 + \lambda(x-x_0), u(\lambda, x-x_0)) \cdot (x-x_0)$

for $|\lambda| \leqslant 2$, $x \in K\left(x_0, \frac{1}{2}r\right)$. Moreover

(10) $\qquad u(0, x-x_0) = y_0,$

(11) $\qquad u(1, x-x_0) = y(x).$

We introduce the notation

(12) $\qquad z = x-x_0,$

(13) $\qquad f(\lambda, u, z) := F(x_0 + \lambda z, u) \cdot z \in Y.$

Thus we see that if a solution of our problem exists, the equation

(14) $\qquad \dfrac{du(\lambda, z)}{d\lambda} = f(\lambda, u(\lambda, z), z)$

has, for all $z \in K\left(O_X, \frac{1}{2}r\right)$, the solution $[-2, 2] \ni \lambda \to u(\lambda, z) \in Y$ satisfying the condition

(15) $\qquad u(0, z) = y_0.$

We shall now show that the existence of a solution for the problem (14)–(15) implies the solution of our problem.

From the continuity of the mappings

(16) $\qquad (x, y) \to F(x, y), \quad (x, y) \to F'_Y(x, y)$

it follows that there exists a $0 < \varrho < 1$, $1 \leqslant N$, such that

(17) $\qquad \left. \begin{array}{l} \|F'_Y(x, y)\| \leqslant N \\ \|F(x, y)\| \leqslant N \end{array} \right\} \quad \text{for } x \in K(x_0, \varrho), \ y \in K(y_0, \varrho).$

By formula (13) we have

$$\|f'_Y(\lambda, u, z)\| \leqslant \|F'_Y(x_0 + \lambda z, u)\| \cdot \|z\|,$$
$$\|f(\lambda, u, z)\| \leqslant \|F(x_0 + \lambda z, u)\| \cdot \|z\|.$$

Thus

$$\|f'_Y(\lambda, u, z)\| \leqslant \tfrac{1}{2}\varrho := L,$$
$$\|f(\lambda, u, z)\| \leqslant \tfrac{1}{2}\varrho := M$$

for $\lambda \in [-2, 2]$, $u \in K(y_0, \varrho)$, $z \in K(O_x, \varrho/2N)$. Then, as is known, it follows from the proof of Theorem IX.3.2, p. 276, that for every $z \in K(O_x, \varrho/2N)$ equation (14) has exactly one solution defined on the interval $]-\tau, \tau[$, where $\tau = \min(\varrho/M, 1/L) = \min(2, 2/\varrho) = 2$, satisfying the initial condition $u(0, z) = y_0$.

We shall demonstrate that the mapping $K(x_0, \varrho/2N) \ni x \rightarrow y(x)$ $= u(1, x-x_0) \in Y$ is a solution of our problem.

For $x = x_0$ equation (14) reduces to the form

$$\frac{du(\lambda, 0)}{d\lambda} = 0.$$

The constant mapping

$$u(\lambda, 0) \underset{\lambda \in]-2, 2[}{\equiv} y_0$$

is the only solution satisfying the condition $u(0, 0) = y_0$ and therefore $y(x_0) = u(1, 0) = y_0$. Initial condition is thus satisfied. It will now be shown that $y(\cdot)$ satisfies equation (1).

By Theorem IX.3.5 we know that the mapping $\lambda \rightarrow u'_X(\lambda, z) \in L(X, Y)$ satisfies the differential equation

$$(18) \qquad \frac{d}{d\lambda} u'_X(\lambda, z) = f'_Y(\lambda, u(\lambda, z), z) \circ u'_X(\lambda, z) + f'_X(\lambda, u(\lambda, z), z)$$

with the initial condition

$$(19) \qquad u'_X(0, z) = 0.$$

Let $h \in X$, and let

$$(20) \qquad g(\lambda) := u'_X(\lambda, z) \cdot h.$$

From (18) and (20) we have

$$(21) \qquad g'(\lambda) = f'_Y(\lambda, u(\lambda, z), z) \cdot g(\lambda) + f'_X(\lambda, u(\lambda, z), z) \cdot h,$$

$$(22) \qquad g(0) \overset{\triangle}{=} 0.$$

The definition of f can be used so as to rewrite the system (21), (22) in the form

$$(23) \qquad g'(\lambda) = \left(F'_Y(x_0 + \lambda z, u(\lambda, z)) \cdot g(\lambda) \right) \cdot z +$$
$$+ F(x_0 + \lambda z, u(\lambda, z)) \cdot h + \left(\lambda F'_X(x_0 + \lambda z, u(\lambda, z)) \cdot h \right) \cdot z,$$

$$(24) \qquad g(0) = 0.$$

360

By direct evaluation we shall now demonstrate that the solution of the linear equation (23) with the initial condition (24) is of the form

$$(25) \qquad g(\lambda) = \lambda F\big(x_0 + \lambda z, u(\lambda, z)\big) \cdot h.$$

For we have

$$(26) \qquad \frac{d}{d\lambda}\Big(\lambda F\big(x_0 + \lambda z, u(\lambda, z)\big) \cdot h\Big) = F\big(x_0 + \lambda z, u(\lambda, z)\big) \cdot h +$$

$$+ \Big(\lambda F'_X\big(x_0 + \lambda z, u(\lambda, z)\big) \cdot z\Big) \cdot h +$$

$$+ \Big(\lambda F'_Y\big(x_0 + \lambda z, u(\lambda, z)\big) \frac{d}{d\lambda} u(\lambda, z)\Big) \cdot h.$$

But by (13) and (14)

$$(27) \qquad \frac{d}{d\lambda} u(\lambda, z) = F\big(x_0 + \lambda z, u(\lambda, z)\big) \cdot z.$$

When we substitute (27) into (26) and next change the order of z and h by virtue of Frobenius condition (2) we obtain

$$(28) \qquad \frac{d}{d\lambda}\Big(\lambda F\big(x_0 + \lambda z, u(\lambda, z)\big) \cdot h\Big)$$

$$= F\big(x_0 + \lambda z, u(\lambda, z) \cdot h\big) + \Big(\lambda F'_X\big(x_0 + \lambda z, u(\lambda, z)\big) \cdot h\Big) \cdot z +$$

$$+ \Big(\lambda F'_Y\big(x_0 + \lambda z, u(\lambda, z)\big)\Big) \cdot F\big(x_0 + \lambda z, u(\lambda, z) \cdot h\big) \cdot z.$$

Now we recognize that (28) is equivalent to (23) if the substitution (25) is done. The function (25) satisfies also initial condition (24). It means that equality (25) holds everywhere. Setting $z = x - x_0$, $y(x) = u(1, x - x_0)$, we have

$$y'(x) \cdot h = u'_X(1, x - x_0) \cdot h = g(1) \cdot h$$
$$= F\big(x_0 + x - x_0, y(x)\big) \cdot h = F\big(x, y(x)\big) \cdot h.$$

Consequently

$$y'(x) = F\big(x, y(x)\big). \quad \square$$

If $X = R^n$, $Y = R^n$, the equation $y'(x) = F\big(x, y(x)\big)$ reduces to a system of m differential equations with partial derivatives

$$\frac{\partial y_j}{\partial x_i} = F^i_j\big(x_1, \ldots, x_n, y_1(x), \ldots, y_n(x)\big),$$

$$i = 1, \ldots, n, \quad j = 1, \ldots, m.$$

361

The integrability conditions

$$\frac{\partial^2 y_j}{\partial x_i \partial x_k} = \frac{\partial^2 y_j}{\partial x_k \partial x_i}, \quad j = 1, \ldots, m, \quad i, k = 1, \ldots, n,$$

reduce to a system of $\frac{1}{2}n(n-1)$ equations

$$\frac{\partial F_j^i}{\partial x^k}(x, y) + \sum_{s=1}^{m} \frac{\partial F_j^i}{\partial y_s}(x, y) F_s^k(x, y)$$

$$= \frac{\partial F_j^k}{\partial x^i}(x, y) + \sum_{s=1}^{m} \frac{\partial F_j^k}{\partial y_s}(x, y) F_s^i(x, y),$$

which must hold for every $(x, y) \in U \times V$.

We can now present a theorem on the continuous dependence of the solution of the equation $y'(x) = F(x, y(x))$ on the initial conditions.

THEOREM IX.19.2 (of Frobenius and Dieudonné). Suppose that a mapping F is continuously differentiable on $U \times V$ and satisfies the Frobenius condition. Then, for every point $(\tilde{x}, \tilde{y}) \in U \times V$ there are $r_1 > 0$, $r_2 > 0$ such that for every $(x_0, y_0) \in K_1(\tilde{x}, r_1) \times K_2(\tilde{y}, r_2)$ the equation $y' = F(x, y(x))$ has exactly one solution satisfying the initial condition $y(x_0) = y_0$. This solution is defined in $K_1(\tilde{x}_1, r_1)$ and takes values in $K_2(\tilde{y}, r_2)$, and the mapping

$$K_1(\tilde{x}, r_1) \times K_1(\tilde{x}, r_1) \times K_2(\tilde{y}, r_2) \ni (x, x_0, y_0)$$
$$\rightarrow y(x, x_0, y_0) \in K_2(\tilde{y}, r_2)$$

is continuously differentiable on $K_1 \times K_1 \times K_2$.

If F is p times differentiable on $U \times V$, then the mapping $(x, x_0, y_0) \rightarrow y(x, x_0, y_0)$ is p times continuously differentiable on $K_1 \times K_1 \times K_2$.

PROOF. Consider the ordinary differential equation dependent on a parameter:

$$(29) \qquad \frac{du(\lambda, z, x_0, y_0)}{d\lambda} = F(x_0 + \lambda z, y_0 + u(\lambda)) \cdot z$$

$$= f(\lambda, u(\lambda), z, x_0, y_0).$$

From the continuity of the mapping F and its derivative, it follows that there exist a $1 > \sigma > 0$ and an $N \geq 1$ such that

$$\|F(x, y)\| \leq N, \quad \|F_Y'(x, y)\| \leq N$$

$$\text{for } (x, y) \in K(\tilde{x}, \sigma) \cdot K(\tilde{y}, \sigma).$$

Therefore

$$\|f(\lambda, u, z, x_0, y_0)\| \leqslant \tfrac{1}{4}\sigma, \quad \|f_Y'(\lambda, u, x_0, y_0)\| \leqslant \tfrac{1}{4}\sigma$$

for $-2 < \lambda < 2$, $u \in K\left(\tilde{y}, \tfrac{1}{2}\sigma\right)$, $z \in K(O_x, \sigma/4N)$, $x_0 \in K\left(\tilde{x}, \tfrac{1}{2}\sigma\right)$, and $y_0 \in K\left(\tilde{y}, \tfrac{1}{2}\sigma\right)$. Now $z = x - x_0$, and thus if $\|x_0 - \tilde{x}\| \leqslant \sigma/8N$, then for every $x_0 \in K(\tilde{x}, \sigma/8N)$ and every $y_0 \in K\left(\tilde{y}, \tfrac{1}{2}\sigma\right)$ equation (29) has exactly one solution $u(\lambda, x - x_0, x_0, y_0)$ which is defined for $x \in K(\tilde{x}, \sigma/8N)$, $\lambda \in \,]-2, 2[$, $x_0 \in K(\tilde{x}, \sigma/8N)$, $y_0 \in K\left(\tilde{y}, \tfrac{1}{2}\sigma\right)$ and

$$u(0, x - x_0, x_0, y_0) = 0.$$

Then, as follows from the preceding theorem, the mapping $x \to y_0 + u(1, x - x_0, x_0, y_0)$ is the solution of our problem, satisfying the initial condition $y(x_0) = y_0$. Moreover, Theorem IX.3.5 implies that the mapping $(x, x_0, y_0) \to y(x, x_0, y_0)$ is continuously differentiable on $K(\tilde{x}, \sigma/8N) \times K(\tilde{x}, \sigma/8N) \times K(\tilde{y}, \sigma/2)$.

If the mapping F is p times continuously differentiable on $U \times V$, then on invoking Theorem IX.3.5 we obtain the C^p-differentiability of the mapping $(x, x_0, y_0) \to y(x, x_0, y_0)$. $\square$

X. THE THEORY OF CURVES IN E^n

1. THE CURVE AND LENGTH OF ARC.
THE NATURAL DESCRIPTION (PARAMETRIZATION)

DEFINITION. The *parametric description of a curve* in a Banach space X is a continuously differentiable mapping $x(\cdot)$ of the interval $[a, b]$ into X:

$$[a, b] \ni t \to x(t) \in X.$$

Example. Let $X = R^2$. The parametric description of a circle with radius r and centre at the point $(0, 0)$ is of the form

$$x_1(t) = r\cos t, \quad x_2(t) = r\sin t, \qquad t \in [0, 2\pi].$$

It is seen, however, that the parametrization

$$x_1(t) = r\cos 3t, \quad x_2(t) = r\sin 3t, \qquad t \in \left[0, \frac{2}{3}\pi\right],$$

is a description of the same circle. Thus, an equivalence relation needs to be introduced into the set of all parametrizations.

DEFINITION. The parametrization of a curve

$$[a_1, b_1] \ni t \to x_1(t) \in X$$

is said to be *equivalent* to the parametrization

$$[a_2, b_2] \ni t \to x_2(t) \in X$$

if there exists a differentiable mapping f_{12} of the interval $[a_1, b_1]$ into the interval $[a_2, b_2]$, satisfying the conditions:

(i) $f'_{12} > 0$;

(ii) $x_2 \circ f_{12} = x_1$.

We verify that this is an equivalence relation:

(i) The reflexivity is obvious.

(ii) Since $f'_{12} > 0$, the mapping f is invertible; defining $f_{21} := (f_{12})^{-1}$ we demonstrate that the relation is reflexive.

(iii) Transitivity: let

$$([a_1, b_1]; x_1) \sim ([a_2, b_2]; x_2),$$

$$([a_2, b_2]; x_2) \sim ([a_3, b_3]; x_3);$$

let

$$f_{12}: [a_1, b_1] \to [a_2, b_2],$$
$$f_{23}: [a_2, b_2] \to [a_3, b_3];$$
$$f_{13} := f_{23} \circ f_{12}, \quad f_{13}: [a_1, b_1] \to [a_3, b_3],$$
$$f'_{13} = f'_{23} \circ f'_{12} > 0.$$

Then

$$x_3 \circ f_{13} = x_3 \circ f_{23} \circ f_{12} = x_2 \circ f_{12} = x_1.$$

The definition has thus been shown to be correct.

DEFINITION. A class of equivalent parametrizations is called an *oriented curve*.

We speak of an oriented curve since we allow functions f with positive derivatives.

We shall examine certain features of curves which do not depend on the parametrization.

LEMMA X.1.1. Let F be a continuous function on the product $X \times X$,

$$X \times X \ni (x, y) \to F(x, y) \in R.$$

Let F be positive homogeneous in the second variable, i.e.

$$F(x, cy) = c \cdot F(x, y) \quad \text{for } c > 0.$$

If $([a, b]; x)$ is a parametric description of some curve, the expression

$$L(x) = \int_a^b F\left(x(t), \frac{dx(t)}{dt}\right) dt$$

does not depend on the description, i.e. is the same for equivalent descriptions.

PROOF. Let $([a_1, b_1]; x_1) \sim ([a, b]; x)$ and

$$f: [a_1, b_1] \to [a, b], \quad t = f(t_1),$$
$$x(f(t_1)) = x_1(t_1).$$

Then

$$L(x_1) = \int_{a_1}^{b_1} F\left(x_1(t_1), \frac{dx_1(t_1)}{dt_1}\right) dt_1$$

$$= \int_{a_1}^{b_1} F\left((x \circ f)(t_1), \frac{dx(f(t_1))}{d(f(t_1))} f'(t_1)\right) dt_1$$

$$= \int_{a_1}^{b_1} F\left((x \circ f)\,(t_1), \frac{dx(f(t_1))}{d(f(t_1))}\right) f'(t_1)\, dt_1\,.$$

Changing variables, we obtain

$$L(x_1) = \int_a^b F\left(x(t), \frac{dx}{dt}\,(t)\right) dt = L(x).$$

Therefore $L(x_1) = L(x)$. $\square$

In what follows we shall deal with a function F of the particular form

$$F\left(x(t), \frac{dx}{dt}\,(t)\right) = \left\|\frac{dx}{dt}\,(t)\right\|.$$

The number

$$s(x, t) := \int_a^t \left\|\frac{dx}{du}\,(u)\right\| du$$

is called the *length of the arc* from the point $x(a)$ to the point $x(t)$. Lemma X.1.1 shows that the length of the arc does not depend on the choice of description. The arc length thus is a property of the curve.

The function $[a, b] \ni t \to s(t) \in R^1$ is continuously differentiable and

$$s'(t) = \left\|\frac{dx}{dt}\,(t)\right\|.$$

As we see, the derivative of the function s is nonnegative and hence s is nondecreasing. If $s(t_1) = s(t_2)$ for $t_2 > t_1$, then

$$\int_{t_1}^{t_2} \left\|\frac{dx}{du}\,(u)\right\| du = 0,$$

and therefore $\dfrac{dx}{du}\,(u) = 0$ for $u \in [t_1, t_2]$.

Let us eliminate this "pathological" case and let us suppose that $\dfrac{dx}{dt}\,(t)$ does not vanish at any point of the interval $[a, b]$. Then the function $t \to s(t)$ is strictly increasing. Hence, there is an inverse function $t = g(s)$. We can, therefore, replace parameter t by parameter s, expressing t in terms of s. We then have the description

$$[0, s(b)] \ni s \to x\bigl(g(s)\bigr) \in X.$$

A description of this kind is called *a natural description* and s is called a *natural parameter*.

DEFINITION. A *natural description of a curve* is a parametric description in which the arc length of that curve is the parameter.

Example. A *helix* in three-dimensional space:

$$[0, 2\pi[\ni t \to x(t) = (a\cos t, a\sin t, h \cdot t).$$

We have

$$s(t) = \int_0^t \sqrt{a^2 \sin^2 u + a^2 \cos^2 u + h^2} \, du,$$

$$s(t) = \int_0^t \sqrt{a^2 + h^2} \, du.$$

Hence $s(t) = \sqrt{a^2 + h^2} \cdot t$, and consequently

$$t = \frac{s}{(a^2 + h^2)^{1/2}}.$$

The natural description is of the form

$$x(s) = \left(a \cos \frac{s}{(a^2 + h^2)^{1/2}}, a \sin \frac{s}{(a^2 + h^2)^{1/2}}, \frac{h \cdot s}{(a^2 + h^2)^{1/2}} \right).$$

2. THE SCHMIDT ORTHONORMALIZATION

Henceforth we shall concern ourselves exclusively with curves in E^n.

DEFINITION. An n-dimensional real Hilbert space will be called (n-dimensional) *Euclidean space* and denoted by E^n.

We recall (cf. Section VII.4) that a complete vector space with scalar product is called a Hilbert space H. The scalar product $(\cdot \mid \cdot)$ is the linear mapping of the product $H \times H$ into R^1 satisfying the conditions:

(i) $(x|x) \geqslant 0$,

(ii) $(x|y) = (y|x)$,

(iii) $(\alpha x_1 + \beta x_2 | y) = \alpha(x_1|y) + \beta(x_2|y)$, $\alpha, \beta \in R^1$,

(iv) $(x|x) = 0 \Leftrightarrow x = 0$.

Conditions (i), (ii), and (iii) imply the Schwarz inequality

$$|(x|y)| \leqslant (x|x)^{1/2}(y|y)^{1/2}.$$

In the space H we have the norm $\|x\| := (x|x)^{1/2}$. Since E^n is an n-dimensional Hilbert space, we can choose in it a basis. The most interesting case is that of the orthonormal basis, i.e. a basis $e_1, \ldots, e_n$ such that $(e_i|e_j) = \delta_{ij}$. The question is whether an orthonormal basis exists in every finite-dimensional space? We answer in the affirmative to this question by constructing an orthonormal basis of the space E^n out of a given set of n linearly independent vectors. Given n linearly independent vectors $x_1, \ldots, x_n$. We define the system of vectors $e_1, \ldots, e_n$:

$$e_1 := \frac{x_1}{\|x_1\|},$$

$$e_2 := \frac{x_2 - (x_2|e_1)e_1}{\|x_2 - (x_2|e_1)e_1\|},$$

$$\cdot \quad \cdot \quad \cdot \quad \cdot \quad \cdot \quad \cdot \quad \cdot \quad \cdot \quad \cdot \quad \cdot \quad \cdot$$

$$e_k := \frac{x_k - \sum_{i=1}^{k-1} (x_k|e_i)e_i}{\left\|x_k - \sum_{i=1}^{k-1} (x_k|e_i)e_i\right\|},$$

$$\cdot \quad \cdot \quad \cdot \quad \cdot \quad \cdot \quad \cdot \quad \cdot \quad \cdot \quad \cdot \quad \cdot \quad \cdot$$

$$e_n := \frac{x_n - \sum_{i=1}^{n-1} (x_n|e_i) \cdot e_i}{\left\|x_n - \sum_{i=1}^{n-1} (x_n|e_i) \cdot e_i\right\|}.$$

The linear independence of the vectors $x_1, \ldots, x_n$ ensures that for every $1 \leqslant k \leqslant n$

$$x_k - \sum_{i=1}^{k-1} (x_k|e_i)e_i \neq 0.$$

The reader may easily verify that the set $e_1, \ldots, e_n$ is orthonormal. It implies that

$$(1) \qquad 1 = (e_n|e_n) = \frac{1}{\left\|x_n - \sum_{i=1}^{n-1} (x_n|e_i)e_i\right\|} \cdot (x_n|e_n),$$

i.e.

$$(x_n|e_n) > 0.$$

The geometrical meaning of our definition is following: vector e_n is the unique vector satisfying the conditions:

(a) e_n belongs to the subspace spanned by vectors $x_1, x_2, ..., x_n$,

(b) e_n is orthogonal to the subspace spanned by vectors $x_1, x_2, ...$ ..., x_{n-1},

(c) $\|e_n\| = 1$,

(d) $(x_n|e_n) > 0$.

The foregoing procedure is known as the *Gram–Schmidt orthonormalization*. It lends itself to generalization to any Hilbert space.

If we have an orthonormal basis $e_1, ..., e_n$ in E^n, then any vector x can be expressed as a linear combination of basis vectors

$$x = \sum_{i=1}^{n} x_i \cdot e_i.$$

Scalar multiplication of both sides of this equation by the vector e_j gives us

$$(x|e_j) = x_j,$$

whereby

$$x = \sum_{i=1}^{n} (x|e_i) \cdot e_i.$$

Thus

$$\|x\|^2 = (x|x) = \left(\sum_{j=1}^{n} (x|e_j) \cdot e_j \middle| \sum_{i=1}^{n} (x|e_i) \cdot e_i \right) = \sum_{i=1}^{n} (x|e_i)^2.$$

Hence $\|x\| = \sqrt{\sum_{i=1}^{n} (x|e_i)^2}$.

In the space E^n a set of n scalar functions

$$x_1(t), ..., x_n(t), \quad \text{where} \quad x_i(t) = (x(t)|e_i)$$

in the orthonormal basis $e_1, ..., e_n$ corresponds to the parametric description of the curve $[a, b] \ni t \rightarrow x(t) \in E^n$. The formula for the arc length takes the form

$$s(x; t) = \int_a^t \left\| \frac{dx}{du}(u) \right\| du = \int_a^t \left(\frac{dx}{du}(u) \middle| \frac{dx}{du}(u) \right)^{1/2} du.$$

Therefore

$$s(x; t) = \int_a^t \left(\sum_{i=1}^{n} \left(\frac{dx_i}{du}(u) \right)^2 \right)^{1/2} du.$$

3. THE FRENET FORMULAE

Given the parametric description of a curve of class $i+1$ in E^n (that is, $x \in C^{i+1}([a, b]; E^n)$). If $\frac{dx}{dt}(t) \neq 0$ for $t \in [a, b]$, then the corresponding natural description $s \to x(s)$ is also of class $i+1$. The curve will further be assumed not degenerate, i.e. for every s the vectors

$$\frac{dx}{ds}(s), \quad \frac{d^2x}{ds^2}(s), \quad \ldots, \quad \frac{d^nx}{ds^n}(s)$$

are linearly independent. Geometrically this means that the curve does not lie entirely in any plane of dimension less than n. We introduce the notation

$$x^{(1)}(s) = \frac{dx}{ds}(s),$$

$$x^{(2)}(s) = \frac{d^2x}{ds^2}(s),$$

$$\cdots \cdots \cdots \cdots$$

$$x^{(n)}(s) = \frac{d^nx}{ds^n}(s).$$

The vectors $x^{(1)}(s), x^{(2)}(s), \ldots, x^{(n)}(s)$ are linearly independent and hence an orthonormal set $e_1(s), e_2(s), \ldots, e_n(s)$ can be formed out of them by the Gram–Schmidt method. If follows from the construction that

(1) $$e_k(s) = \sum_{j=1}^{k} b_{kj}(s) x^{(j)}(s),$$

(2) $$x^{(j)}(s) = \sum_{k=1}^{j} c_{jk}(s) e_k(s).$$

Because e_k is orthogonal to the vectors $x^{(1)}, \ldots, x^{(k-1)}$ the formula (1) gives

$$1 - (e_k(s)|e_k(s)) = \sum_{j=1}^{k} b_{kj}(x^{(j)}(s)|e_k(s)) = b_{kk}(x^{(k)}(s)|e_k(s))$$

which implies

$$b_{kk} = (x^{(k)}(s)|e_k(s))^{-1} > 0$$

373

by virtue of inequality (1) from the preceding section.

Now we shall prove that

(3) $\qquad \|x^{(1)}(s)\| = 1, \quad$ that is $x^{(1)}(s) = e_1(s)$,

(4) $\qquad (x^{(1)}(s)|x^{(2)}(s)) = 0, \quad$ that is $e_2(s) = \dfrac{x^{(2)}(s)}{\|x^{(2)}(s)\|}$.

For we have

$$\|x^1(s)\| = \left\|\frac{dx}{ds}(s)\right\| = \left\|\frac{dx}{dt}\frac{dt}{ds}\right\| = \left\|\frac{dx}{dt}\right\| \cdot \left|\frac{dt}{ds}\right| = \left\|\frac{dx}{dt}\right\| \cdot \left(\left|\frac{ds}{dt}\right|\right)^{-1}.$$

But $\dfrac{ds}{dt} = \left\|\dfrac{dx}{dt}\right\|$, and hence $\|x^{(1)}(s)\| = 1$.

Now

$$\frac{d}{ds}\left(x^1(s)|x^1(s)\right) = 2\left(\frac{dx^1}{ds}(s)|x^1(s)\right) = 2(x^2(s)|x^1(s))$$

which implies $\left(x^2(s), x^1(s)\right) = 0$ because $\left(x^1(s)|x^1(s)\right) \equiv 1$.

Let us decompose the vector $\dfrac{de_j}{ds}(s)$ into basis vectors:

(5) $\qquad \dfrac{de_j}{ds}(s) = \displaystyle\sum_{k=1}^{n} a_{jk}(s)e_k(s), \quad j = 1, \dots, n.$

From the preceding section it follows that

(6) $\qquad a_{jk}(s) = \left(\dfrac{de_j}{ds}(s)\middle|e_k(s)\right) = \left(e_k(s)\middle|\dfrac{de_j}{ds}(s)\right).$

Differentiating both sides of the equation

$$\left(e_j(s)|e_k(s)\right) \underset{s}{\equiv} \delta_{jk},$$

we have

(7) $\qquad \left(\dfrac{de_j}{ds}(s)\middle|e_k(s)\right) + \left(e_j(s)\middle|\dfrac{de_k}{ds}(s)\right) = 0.$

Substitution of (6) into (7) results in

$$a_{jk}(s) + a_{kj}(s) = 0,$$

whence

(8) $\qquad a_{jk}(s) = -a_{kj}(s).$

Formulae (1) and (2) imply that $\dfrac{de_j}{ds}(s)$ is a linear combination of the vectors $e_1(s), e_2(s), \ldots, e_j(s), e_{j+1}(s)$, and thus

(9) $\qquad a_{jk}(s) = 0 \quad$ for $k > j+1$.

From formulae (8) and (9) it follows that the matrix $a_{jk}(s)$ is of the form

$$[a_{jk}(s)] = \begin{bmatrix} 0 & q_1(s) & 0 & 0 & \cdots & 0 \\ -q_1(s) & 0 & q_2(s) & 0 & \cdots & 0 \\ 0 & -q_2(s) & 0 & q_3(s) & \cdots & 0 \\ \cdots & \cdots & \cdots & \cdots & \cdots & \cdots \\ & \cdots & & & 0 & -q_{n-1}(s) \\ & \cdots & & & -q_{n-1}(s) & 0 \end{bmatrix},$$

inasmuch as only $a_{k,k+1}(s)$ and $a_{k+1,k}(s)$ are nonzero. We have here introduced the notation $q_k(s) := a_{k,k+1}(s)$. The functions $q_1(s), \ldots$ $\ldots, q_{n-1}(s)$ are called, respectively, the *first, second, ..., (n−1)-th curvature of the curve*. Formula (5) assumes the form

(10) $\qquad \dfrac{de_j}{ds}(s) = q_j(s)e_{j+1}(s) - q_{j-1}(s)e_{j-1}(s), \quad j = 1, \ldots, n.$

These are *Frenet's formulae*.

DEFINITION. The orthonormal basis $\{e_1(s), \ldots, e_n(s)\}$ is called the *Frenet reper* of the curve.

Remark. In the case $n = 3$, the first curvature is called simply the *curvature*, and the second curvature is called the *torsion*.

Now we shall prove that all curvatures are positive: $q_k(s) > 0$. To do this we differentiate the formula (1):

$$\frac{d}{ds} e_k(s) = \sum_{j=1}^{k} \frac{db_{kj}(s)}{ds} x^{(j)}(s) + \sum_{j=1}^{k} b_{kj}(s) x^{(j+1)}(s).$$

But we know that $e_{k+1}(s)$ is orthogonal to the subspace spanned on vectors $x^{(1)}(s), x^{(2)}(s), \ldots, x^k(s)$. It means that

$$q_k(s) = \left(\frac{de_k}{ds}(s) | e_{k+1}(s) \right) = b_{kk}(s) \big(x^{(j+1)}(s) | e_{k+1}(s) \big) > 0.$$

375

4. DEGENERATE CURVES

In deriving the Frenet formulae, we assumed that the vectors

$$\frac{dx}{ds}(s), \quad \frac{d^2x}{ds^2}(s), \quad \ldots, \quad \frac{d^nx}{ds^n}(s)$$

are linearly independent for every s. If these vectors are linearly independent at some point s_0, then the continuity of all the derivatives implies that they are linearly independent in a neighbourhood of the point s_0. We can then confine ourselves to this neighbourhood and derive the Frenet formulae on it.

Let us now consider the case when the vectors $x^{(1)}(s), x^{(2)}(s), \ldots$ $\ldots, x^{(n)}(s)$ are linearly dependent in a neighbourhood of the point s. The curve is then said to be *degenerate*. Suppose in particular that

$$(1) \qquad \frac{d^{k+1}x}{ds^{k+1}}(s) = \sum_{j=1}^{k} c_j(s)\frac{d^jx}{ds^j}(s) \quad \text{for } s \in [s_0-r, s_0+r];$$

the functions $c_j(\cdot)$ are continuous. It is readily seen that the higher-order derivatives are also expressed by linear combinations of the vectors $\frac{dx}{ds}(s), \ldots, \frac{d^kx}{ds^k}(s)$. Of course, we assume that the vectors $x^{(1)}(s), \ldots, x^{(k)}(s)$ are linearly independent for $s \in [s_0-r, s_0+r]$.

Let us span the subspace $E^{(k)}(s_0)$ on the vectors $x^{(1)}(s_0), \ldots, x^{(k)}(s_0)$. Let $(E^k(s_0))^\perp$ be the orthogonal complement of the subspace $E^k(s_0)$

$$(E^k(s_0))^\perp := \{v \in E^n: (v|u) = 0, \ u \in E^k(s_0)\}.$$

We take an arbitrary vector $v \in (E^k(s_0))^\perp$ and we define the set of functions

$$(2) \qquad g_m(s) := \left(\frac{d^mx}{ds^m}(s)\middle|v\right), \quad m = 1, \ldots, k.$$

It is easily seen that

$$(3) \qquad g_m(s_0) = 0, \quad m = 1, \ldots, k.$$

On differentiating formula (2), we have

$$\frac{dg_m}{ds}(s) = \left(\frac{d^{m+1}x}{ds^{m+1}}(s)\middle|v\right) = g_{m+1}(s), \quad m = 1, \ldots, k-1,$$

$$\frac{dg_k}{ds}(s) = \left(\frac{d^{k+1}x}{ds^{k+1}}(s)\bigg|v\right) = \left(\sum_{j=1}^{k} c_j(s)\frac{d^j x}{ds^j}(s)\bigg|v\right)$$

$$= \sum_{j=1}^{k} c_j(s)g_j(s).$$

The functions $g_j(\cdot)$ thus satisfy the system of equations:

$$\frac{dg_m}{ds} = g_{m+1}, \quad m = 1, \ldots, k-1,$$

(4)

$$\frac{dg_k}{ds} = \sum_{j=1}^{k} c_j(s)g_j,$$

with the initial condition $g_m(s_0) = 0$, $m = 1, 2, \ldots, k$. This is a system of linear equations. It is readily verified that the functions

(5) $$g_m(s) \underset{s}{\equiv} 0, \quad m = 1, 2, \ldots, k,$$

satisfy system (4) and the initial condition. Since the initial-value problem for a linear function is uniquely solvable, the functions (5) are unique solutions. In view of the fact that $g_1(s) \underset{s}{\equiv} 0$, we have

$$\int_{s_0}^{s} g_1(u)\,du = 0.$$

But $g_1(u) = \left(\dfrac{dx}{ds}(u)\bigg|v\right)$, and therefore

$$\int_{s_0}^{s}\left(\frac{dx}{ds}(u)\bigg|v\right) du = 0,$$

whereby

$$\left(\int_{s_0}^{s}\frac{dx}{du}(u)\,du\bigg|v\right) = 0.$$

Integration yields

$$(x(s) - x(s_0)|v) = 0 \quad \text{for } s \in K(s_0, r).$$

The vector $v \in (E^k(s_0))^\perp$ was chosen arbitrarily, whereby the vector $x(s) - x(s_0)$ is perpendicular to the space $(E^k(s_0))^\perp$ and thus lies in the subspace $E^k(s_0)$. We can write

$$x(s) \in \{x(s_0) + E^k(s_0)\}.$$

If the curve is translated by a distance equal to the vector $x(s_0)$, then locally it lies in $E^k(s_0)$; we shall consider it in a k-dimensional space.

If for every s

$$\frac{d^{k+1}x}{ds^{k+1}}(s) = \sum_{j=1}^{k} c_j(s)\frac{d^j x}{ds^j}(s),$$

then after translation by a distance equal to the vector $x(s_0)$ the curve lies entirely in the subspace $E^k(s_0)$, where it is now a nondegenerate curve.

5. THE FUNDAMENTAL THEOREM OF THE THEORY OF CURVES

In Section 3 we demonstrated that to every nondegenerate curve of class $n+1$ in the space E^n there correspond $n-1$ functions of the variable s (length of arc) $q_1(s), ..., q_{n-1}(s)$ called *curvatures*. Now we shall show that giving the curvatures determines a curve uniquely up to Euclidean motions (i.e. compositions of translation and orthogonal mapping (rotation)).

THEOREM X.5.1. Given $n-1$ positive functions of class C^{n-1}, $q_1(s), ..., q_{n-1}(s)$, defined on the interval $[0, a]$. There then is a curve of class C^{n+1} in E^n for which:

(i) s is the natural parameter,

(ii) the functions $q_1(s), ..., q_{n-1}(s)$ are curvatures.

This curve is determined up to Euclidean motions (i.e. composition of translation and rotation).

PROOF. The proof is by construction. Let $\mathring{e}_1, ..., \mathring{e}_n$ be an arbitrary orthonormal basis in the space E^n. Take the system of equations

(1) $$\frac{de_j}{ds}(s) = \sum_{k=1}^{n} a_{jk}(s)e_k(s), \quad j = 1, 2, ..., n,$$

with the initial condition

(2) $$e_j(0) = \mathring{e}_j, \quad j = 1, 2, ..., n.$$

The functions $a_{jk}(s)$ form a skew-symmetric matrix and are related to the functions $q_1(s), ..., q_{n-1}(s)$ in the following manner:

$$a_{J,J+1}(s) := q_j(s),$$

(3) $$a_{J+1,J}(s) := -q_j(s), \quad j = 1, \ldots, n-1,$$

$$a_{jk}(s) = 0 \quad \text{dla } k \neq j+1 \text{ and } k \neq j-1.$$

The solutions of system (1) are functions $e_1(s), \ldots, e_n(s)$ with values in E^n. Denote by $A(s) \in L(E^n, E^n)$ the linear, antisymmetric operator given (with respect to the basis $\mathring{e}_1, \ldots, \mathring{e}_n$) by the matrix $(a_{ij}(s))$. Now take a matrix $(v_{kj}(s))$ composed of coordinates of vectors $e_k(s)$ with respect to the same basis:

(4) $$e_k(s) = \sum_{j=1}^{n} v_{kj}(s)\mathring{e}_j.$$

Vector $e_k(s)$ is thus the k-th column of $(v_{kj}(s))$. By $V(s) \in L(E^n, E^n)$ denote the operator corresponding to the matrix $(v_{kj}(s))$. Now equation (1) reads

(5) $$\frac{dV(s)}{ds} = A(s) \circ V(s)$$

with the initial condition $V(0) = I$.

We know (see Theorem IX.9.2 and next Lemma IX.11.1) that the operators $V(s)$ are orthogonal. It means that vectors $e_1(s), \ldots, e_n(s)$ form an orthonormal basis at each point of the interval $[0, a]$.

Let

(6) $$\frac{dx}{ds}(s) := e_1(s).$$

Integration of equation (6) yields

(7) $$x(s) = x_0 + \int_0^s e_1(u)\,du.$$

Thus we have constructed a curve with origin at the point x_0. We calculate its arc length $\sigma(s)$:

$$\sigma(s) - \sigma(0) = \int_0^s \left\| \frac{dx}{du}(u) \right\| du = \int_0^s \|e_1(u)\|\,du = \int_0^s du = s.$$

We thus have

(8) $$\sigma(s) = s + \sigma(0).$$

The parameter s hence is the arc length of the obtained curve.

Now what remains to prove is that the Frenet reper $\tilde{e}_1(s), \tilde{e}_2(s), ..., \tilde{e}_n(s)$ obtained from $x^{(1)}(s), x^{(2)}(s), ..., x^{(n)}(s)$ by orthonormalization (where $x^{(i)}(s) = \dfrac{d^i x}{ds^i}(s)$) coincides with our reper $e_1(s), e_2(s), ..., e_n(s)$. We shall prove it by induction:

From equation (6) we see that $\left\| \dfrac{dx}{ds}(s) \right\| = \|e_1(s)\| = 1$, thus

(9) $\dot{e}_1(s) = e_1(s).$

Now assume that $\tilde{e}_j(s) = e_j(s)$ for $j \le k$. It means that

(10) $e_j(s) = \displaystyle\sum_{i=1}^{j} b_{ji}(s) x^{(i)}(s), \quad b_{jj} > 0$

for $j \le k$. Differentiating both sides of (10) and using (1) and (3) we get for $j = k$

(11) $q_k(s) e_{k+1}(s) - q_{k-1}(s) e_{k-1}(s) = \dfrac{de_k}{ds}(s)$

$= \displaystyle\sum_{i=1}^{k} \dfrac{db_{ki}}{ds}(s) \cdot x^{(i)}(s) + \sum_{i=1}^{k} b_{ki}(s) x^{(i+1)}(s)$

which implies

(12) $e_{k+1}(s) = \dfrac{1}{q_k} \left\{ q_{k-1}(s) e_{k-1}(s) + \displaystyle\sum_{i=1}^{k} \dfrac{db_{ki}}{ds}(s) \cdot x^{(i)}(s) + \right.$

$\left. + \displaystyle\sum_{i=1}^{k-1} b_{ki}(s) x^{(i+1)}(s) \right\} + \dfrac{1}{q_k} b_{kk} x^{(k+1)}(s).$

By virtue of our assumption the first summand belongs to the space spanned by vectors $x^{(1)}(s), ..., x^{(k)}(s)$, i.e. by vectors $e_1(s), ..., e_k(s)$. It means that

(13) $0 \le \big(e_{k+1}(s) | e_{k+1}(s)\big) = \dfrac{1}{q_k} \left(b_{kk}(e_{k+1}(s) | x^{(k+1)}(s)) \right),$

thus

(14) $\big(e_{k+1}(s) | x^{(k+1)}(s)\big) > 0.$

The vector $e_{k+1}(s)$ lies moreover in the subspace spanned by $x^{(1)}(s), ..., x^{(k+1)}(s)$ and is orthogonal to the space spanned by $x^{(1)}(s), ..., x^{(k)}(s)$.

But the above three conditions uniquely define the orthonormalized vector $\tilde{e}_{k+1}(s)$. It means that

(15) $\tilde{e}_{k+1}(s) = e_{k+1}(s),$

which implies that vectors obtained from $x^{(1)}(s), \ldots, x^{(n)}(s)$ by orthonormalization procedure satisfy the equation (1). It means that functions $q_i(s)$ are true curvatures of our curve.

Now we shall show that the curve $x(s)$ is determined uniquely up to Euclidean motions with reflections. Suppose that in addition to the curve

$$x(s) = x_0 + \int_0^s e_1(u)\,du$$

there exists a curve $x_1(s)$ with the same curvatures. The orthonormal set of vectors associated with the curve $x_1(s)$ will be denoted by $'e_1(s), \ldots$ $\ldots, 'e_n(s)$. Now, let us perform an orthogonal transformation of the space E^n so that

$$('e_1(0), \ldots, 'e_n(0)) \to (''e_1(0), \ldots, ''e_n(0))$$
$$= (e_1(0), \ldots, e_n(0)) = (\mathring{e}_1, \ldots, \mathring{e}_n).$$

The curve obtained from $x_1(s)$ by this transformation will be called $x_2(s)$. Since an orthogonal transformation does not change scalar products and norms, the parameter s is also the arc length of the curve $x_2(s)$, and the set of vectors $''e_1(s), \ldots, ''e_n(s)$, being an image of the set $'e_1(s), \ldots, 'e_n(s)$, is an orthonormal system associated with the curve. The curvatures of $x_2(s)$ are equal to the curvatures of $x(s)$ for curvatures are scalar products of certain vectors. The sets of vectors $''e_1(s), \ldots$ $\ldots, ''e_n(s)$ and $e_1(s), \ldots, e_n(s)$ thus satisfy the system of equations (1) with the initial condition (2). From the uniqueness of the solution of the initial-value problem it follows that these sets of vectors are equal to each other. In particular, $''e_1(s) = e_1(s)$. By formula (3), Section 3, we have

$$\frac{dx_2}{ds}(s) = ''e_1(s),$$

whereby

$$x_2(s) = ''x_0 + \int_0^s ''e_1(u)\,du = ''x_0 + \int_0^s e_1(u)\,du.$$

381

On the other hand

$$x(s) = x_0 + \int_0^s e_1(u)\, du.$$

Thus we have

$$x_2(s) - x(s) \underset{s \in [0,a]}{\equiv} {''}x_0 - x_0.$$

The curve $x_2(s)$ thus coincides with $x(s)$ after displacement by the vector ${''}x_0 - x_0$. It means that $x_1(s)$ and $x(s)$ coincide after orthogonal transformation of the space E^n. $\square$

Examples. 1. The *Frenet formulae for a helix in R^3*. The natural description of the helix is of the form (cf. Section 1)

$$x(s) = \left(a\cos \frac{s}{(a^2+h^2)^{1/2}}\,,\; a\sin \frac{s}{(a^2+h^2)^{1/2}}\,,\; \frac{h\cdot s}{(a^2+h^2)^{1/2}} \right),$$

$$x^{(1)}(s) = \frac{1}{(a^2+h^2)^{1/2}} \left(-a\sin\frac{s}{(a^2+h^2)^{1/2}}\,,\; a\cos\frac{s}{(a^2+h^2)^{1/2}}\,, h \right),$$

$$x^{(2)}(s) = \frac{a}{(a^2+h^2)} \left(-\cos\frac{s}{(a^2+h^2)^{1/2}}\,,\; -\sin\frac{s}{(a^2+h^2)^{1/2}}\,, 0 \right),$$

$$x^{(3)}(s) = \frac{a}{(a^2+h^2)^{3/2}} \left(\sin\frac{s}{(a^2+h^2)^{1/2}}\,,\; -\cos\frac{s}{(a^2+h^2)^{1/2}}\,, 0 \right).$$

Orthonormalization results in

$$x^{(1)}(s) = e_1(s)$$

$$= \frac{1}{\sqrt{a^2+h^2}} \left(-a\sin \frac{s}{(a^2+h^2)^{1/2}}\,,\; a\cos \frac{s}{(a^2+h^2)^{1/2}}\,, h \right),$$

$$e_2(s) = \frac{x^{(2)}(s) - \big(x^{(2)}(s)|e_1(s)\big)e_2(s)}{||x^{(2)}(s) - \big(x^{(2)}(s)|e_1(s)\big)e_1(s)||} = \frac{x^{(2)}(s)}{||x^{(2)}(s)||}$$

$$= \left(-\cos \frac{s}{(a^2+h^2)^{1/2}}\,,\; -\sin \frac{s}{(a^2+h^2)^{1/2}}\,, 0 \right),$$

$$e_3(s) = \frac{x^{(3)}(s) - \big(x^{(3)}(s)|e_1(s)\big)e_1(s) - \big(x^3(s)|e_2(s)\big)e_2(s)}{||x^{(3)}(s) - \big(x^{(3)}(s)|e_1(s)\big)e_1(s) - \big(x^{(3)}(s)|e_2(s)\big)e_2(s)||}$$

But

$$\big(x^{(3)}(s)|e_1(s)\big) = -\frac{a^2}{(a^2+h^2)^2}\,, \qquad \big(x^{(3)}(s)|e_2(s)\big) = 0,$$

382

and therefore

$$e_3(s) = \frac{1}{\sqrt{a^2+h^2}}\left(h\sin\frac{s}{\sqrt{a^2+h^2}}, \ -h\cos\frac{s}{\sqrt{a^2+h^2}}, \ a\right).$$

We evaluate the curvatures

$$q_1(s) = \left(\frac{de_1}{ds}(s)\Big| e_2(s)\right) = \frac{a}{a^2+h^2},$$

$$q_2(s) = \left(\frac{de_2}{ds}(s)\Big| e_3(s)\right) = \frac{h}{a^2+h^2}.$$

2. Find the curve in R^2, knowing that $q(s) = 1/a \cdot s$. We integrate the system of equations

$$\frac{de_1}{ds}(s) = \frac{1}{a \cdot s}e_2(s),$$

$$\frac{de_2}{ds}(s) = -\frac{1}{a \cdot s}e_1(s).$$

We introduce the new variable $t = \log s$; the equations take the form

$$\frac{de_1}{dt}(t) = \frac{1}{a}e_2(t),$$

$$\frac{de_2}{dt}(t) = -\frac{1}{a}e_1(t).$$

The solution of this equation satisfying the initial conditions $e_1(0) = (1, 0)$, $e_2(0) = (0, 1)$ is of the form

$$e_1(t) = \left(\cos\frac{t}{a}, \ \sin\frac{t}{a}\right), \quad e_2(t) = \left(-\sin\frac{t}{a}, \ \cos\frac{t}{a}\right).$$

We must now solve the equation

$$\frac{dx}{ds} = e_1(s).$$

But

$$\frac{dx}{ds} = \frac{dx}{dt}\frac{dt}{ds} = e_1(s),$$

and hence, on substituting, we obtain

$$\frac{dx_1}{dt} = e^t\cos\frac{t}{a}, \quad \frac{dx_2}{dt} = e^t\sin\frac{t}{a},$$

383

wherefrom, integration yields

$$x_1(t) = \frac{1}{1+a^2} a^2 e^t \left(\cos\frac{t}{a} + \frac{1}{a}\sin\frac{t}{a} \right),$$

$$x_2(t) = \frac{1}{1+a^2} a^2 e^t \left(\sin\frac{t}{a} - \frac{1}{a}\cos\frac{t}{a} \right).$$

When we introduce d such that

$$\sin d = \frac{1}{\sqrt{1+a^2}}, \quad \cos d = \frac{a}{\sqrt{1+a^2}},$$

we obtain

$$x_1(t) = \frac{a}{(1+a^2)^{1/2}} e^t \cos\left(\frac{t}{a} - d \right),$$

$$x_2(t) = \frac{a}{(1+a^2)^{1/2}} e^t \sin\left(\frac{t}{a} - d \right).$$

Setting $t/a - d = \varphi$, we have

$$r(\varphi) = \frac{a}{\sqrt{1+a^2}} e^{ad} e^{a\varphi}, \quad r = (x_1^2 + x_2^2)^{1/2}.$$

The curve obtained above is known as the *logarithmic spiral*. Its length of arc is proportional to the radius r, for we have

$$s = e^t = \frac{\sqrt{1+a^2}}{a} r.$$

XI. FAMILIES OF CONTINUOUS FUNCTIONS ON A PRECOMPACT SPACE

1. PRECOMPACTNESS. THE ASCOLI THEOREM

DEFINITION. A metric space (X, d) is called *precompact* when a Cauchy subsequence can be selected from every sequence of its elements.

The definition of compact metric space immediately implies the equivalence of the foregoing definition with the one which follows:

DEFINITION. A metric space (X, d) is *precompact* if its completion $(\tilde{X}, \tilde{d})$ is a compact space.

(For a discussion of completion of metric space, see Chapter II.)

DEFINITION. A subset $K \subset X$ (where (X, d) is a metric space) is said to be *precompact* if the space $(K, d|_{K \times K})$ is precompact.

We recall that if we have a mapping $T: X \to Y$, the symbol $T|_Z$ denotes a restriction of the mapping T to the set $Z \subset X$.

The definitions given above lead to

COROLLARY XI.1.1. In a complete space (X, d) a set $K \subset X$ is precompact if and only if the set K is compact.

The following theorem gives an interesting characterization of precompact sets:

THEOREM XI.1.2 (Hausdorff). A space (X, d) is precompact if and only if for every number $\varepsilon > 0$ there exists a finite set of n points $x_1(\varepsilon), \ldots$
$\ldots, x_n(\varepsilon) \in X$ such that $\bigcup\limits_{i=1}^{n(\varepsilon)} K(x_i, \varepsilon) = X$.

The set of points $\{x_1, \ldots, x_n\}$ is often called an "ε-net" since an arbitrary point $x \in X$ is less than ε away from some "element of the net".

PROOF. $\Rightarrow$: Since X is a dense set in the completed space $\tilde{X}$,

$$\bigwedge_{\varepsilon > 0} \bigcup_{x \in X} \tilde{K}(x, \varepsilon) = \tilde{X}.$$

But $\tilde{X}$ is compact, and thus a finite covering can be selected from each of its coverings with open sets. Hence, there is a finite number of points $x_1, \ldots, x_n$ such that

$$\bigcup_{i=1}^{n} \tilde{K}(x_i, \varepsilon) = \tilde{X}.$$

Finally,

$$X = X \cap \tilde{X} = \bigcup_{i=1}^{n} \{X \cap \tilde{K}(x_i, \varepsilon)\} = \bigcup_{i=1}^{n} K(x_i, \varepsilon). \quad \square$$

We give yet another proof, based on the first definition:

$\Rightarrow$: Let $x_1 \in X$. If $K(x_1, \varepsilon) = X$, the point x_1 constitutes the desired ε-net. If this is not the case, let us take $x_2 \notin K(x_1, \varepsilon)$.

If $K(x_1, \varepsilon) \cup K(x_2, \varepsilon) = X$, then x_1 and x_2 constitute the ε-net. Otherwise, a point x_3 can be taken exterior to the balls $K(x_1, \varepsilon)$ and $K(x_2, \varepsilon)$.

This process must end after n steps (i.e. we obtain $\bigcup_{i=1}^{n} K(x_i, \varepsilon) = X$) since otherwise the sequence $(x_i)_1^\infty$ would possess the property $d(x_i, x_j) > \varepsilon$ for $i \neq j$, and it would thus not be possible to select a Cauchy subsequence from it.

Remark. In the case of a subset $K \subset X$, the ε-net may be taken from the space X.

$\Leftarrow$: Given a sequence $(a_n)_1^\infty$, $a_n \in X$.

Let $\varepsilon = 1/k$ and let $\{x_1^{(k)}, \ldots, x_{n(k)}^{(k)}\}$ be a $(1/k)$-net.

From the equation

$$\bigcup_{i=1}^{n(1)} K\left(x_i^{(1)}, \frac{1}{1}\right) = X$$

it follows that one of these balls (e.g. labelled i_1) contains an infinite number of elements of the sequence (a_n). Let us take one of them, e.g. labelled n_1. We denote

$$K(x_{i_1}^{(1)}, 1) = Z^1.$$

But

$$Z^1 = Z^1 \cap \left(\bigcup_{i=1}^{n(2)} K\left(x_i^{(2)}, \frac{1}{2}\right)\right) = \bigcup_{i=1}^{n(2)} \left(Z^1 \cap K\left(x_i^{(2)}, \frac{1}{2}\right)\right).$$

Since the set Z^1 contained infinitely many elements of the sequence (a_n), one of the sets $Z^1 \cap K\left(x_i^{(2)}, \frac{1}{2}\right)$ (e.g. labelled i_2) also contains an inifinite number of these elements.

Let us take one of them, e.g. with label $n_2 > n_1$. We introduce the notation

$$Z^1 \cap K\left(x_{i_2}^{(2)}, \frac{1}{2}\right) = Z^2 \subset Z^1.$$

Proceeding further in the same way, we obtain a sequence of sets $Z^1 \supset Z^2 \supset Z^3 \supset \ldots$, where $Z^k = Z^{k-1} \cap K(x_{i_k}^{(k)}, 1/k)$, and a sequence of points $\{a_{n_1}, a_{n_2}, \ldots\}$, $a_{n_k} \in Z^k$, $n_j > n_{j-1}$. We shall demonstrate that the sequence $(a_{n_k})_1^\infty$ is a Cauchy sequence.

Given a number $\delta > 0$. Take a natural number $N(\delta)$ such that $1/N < \frac{1}{2}\delta$. Then for $k, j > N(\delta)$ we have

$$a_{n_j} \in Z^j \subset Z^N \subset K\left(x_{i_N}^{(N)}, \frac{1}{N}\right), \quad a_{n_k} \in Z^k \subset Z^N \subset K\left(x_{i_N}^{(N)}, \frac{1}{N}\right),$$

and thus

$$d(a_{n_j}, a_{n_k}) \leqslant d(a_{n_j}, x_{i_N}^{(N)}) + d(a_{n_k}, x_{i_N}^{(N)}) \leqslant \frac{1}{N} + \frac{1}{N} < \delta.$$

Therefore the sequence (a_{n_k}) is the desired Cauchy subsequence of the sequence (a_n). □

In turns out that the following lemma holds:

LEMMA XI.1.3. A precompact set is bounded.

PROOF (a.a.). Let (X, d) be a metric space and let $P \subset X$ be a precompact set. Suppose that P is not bounded:

$$\bigwedge_{x \in X} \bigwedge_{M > 0} P \nsubseteq K(x, M).$$

Let us take $x_1 \in P$, $M_1 = 1$, $P \ni x_2 \notin K(x_1, M_1)$, $M_2 = d(x_1, x_2) + 1$, $P \ni x_3 \notin K(x_1, M_2)$, $M_3 = d(x_3, x_1) + 1$, $P \ni x_4 \notin K(x_1, M_3)$, etc.

Let $i > j$. Then

$$d(x_i, x_j) \geqslant d(x_i, x_1) - d(x_j, x_1)$$
$$\geqslant d(x_{i-1}, x_1) + 1 - d(x_j, x_1) \geqslant 1.$$

We have thus constructed a sequence (x_n) from which a Cauchy subsequence cannot be selected. □

Remark 1. Possession of a finite ε-net for any $\varepsilon > 0$ is often called *total boundedness*. Accordingly, Theorem XI.1.2 could be reformulated as follows:

$$((X, d)\text{—precompact}) \Leftrightarrow ((X, d)\text{—totally bounded}).$$

Lemma XI.1.3 justifies the term "total boundedness".

Remark 2. A subset Z of a Hausdorff space is *conditionally compact* if it is a subset of an compact set.

The reader should prove that:

(i) In a metric space a relatively (or conditionally) compact set is precompact.

(ii) Not every precompact set is conditionally compact (e.g. when (X, d) is precompact but not compact).

(iii) The two concepts coincide in complete spaces.

Examples of Precompact Sets. 1. A compact set is a precompact set.

2. Any subset of a precompact set is precompact since $\bar{P} \subset \bar{K}$ ($\bar{K}$ is compact), and therefore $\bar{P}$, being a closed subset of a compact set, is compact.

3. A bounded set in a finite-dimensional space R^n is precompact as a subset of a ball $\overline{K(0, M)}$ which is a compact set in R^n.

4. The ball $K(0, 1) \subset B(R^1, R^1)$ is not precompact, even though it is bounded.

For instance, suppose that

$$f(x) := \begin{cases} 1-x & \text{for } 0 \leqslant x \leqslant 1, \\ 1+x & \text{for } -1 \leqslant x \leqslant 0, \\ 0 & \text{for } |x| \geqslant 1; \end{cases}$$

we denote $f_n(x) := f(x-2n), f_n \in B(R^1, R^1)$.

Then for $n \neq k$

$$d(f_n, f_k) = \sup_{x \in R} |f(x-2n) - f(x-2k)| = 1.$$

This means that a Cauchy subsequence cannot be selected from the sequence (f_n).

5. A finite sum of precompact sets is a precompact set, since the finite sum of finite ε-nets is a finite ε-net.

We shall now give the Ascoli theorem. This theorem makes it possible easily to identify an extensive class of precompact sets in spaces of continuous functions and is a basic tool in the proofs of compactness of sets in other function spaces.

DEFINITION. Let (X, d), (Y, ϱ) be metric spaces. A family $F = \{f_\alpha\}$ of mappings $f_\alpha: X \to Y$ is said to be *equicontinuous at the point* $x \in X$ when

$$\bigwedge_{\varepsilon > 0} \bigvee_{\delta(x,\varepsilon) > 0} \bigwedge_{x' \in X} \bigwedge_{f \in F} \left(d(x', x) < \delta(x, \varepsilon) \right) \Rightarrow \left(\varrho(f(x), f(x')) < \varepsilon \right).$$

The family F is *uniformly equicontinuous* when

$$\bigwedge_{\varepsilon > 0} \bigvee_{\delta(\varepsilon) > 0} \bigwedge_{x, x' \in X} \bigwedge_{f \in F} \left(d(x', x) < \delta(\varepsilon) \right) \Rightarrow \left(\varrho(f(x'), f(x)) < \varepsilon \right).$$

Thus, the "equi"-concepts are obtained from concepts concerning one function by adjoining the quantifier $\bigwedge\limits_{f \in F}$ at the end.

THEOREM XI.1.4 (Ascoli).

Assumptions: (X, d) is a precompact space and (Y, ϱ) is a metric space. $F \subset C(X, Y)$ is a family of mappings with the following properties:

(i) For every $x \in X$, $F(x) := \{f(x) : f \in F\}$ is a precompact subset of the space Y.

(ii) F is a uniformly equicontinuous family.

Thesis: (i) $F \subset B(X, Y)$ (the mappings $f \in F$ are bounded).

(ii) F is a precompact set in $B(X, Y)$.

PROOF. Given a number $\varepsilon > 0$. Let $\{x_1, \ldots, x_n\}$ be a $\delta\left(\frac{1}{4}\varepsilon\right)$-net of the space X, that is,

$$X = \bigcup_{i=1}^{n} K\left(x_i, \delta\left(\tfrac{1}{4}\varepsilon\right)\right),$$

where $\delta\left(\frac{1}{4}\varepsilon\right)$ is taken from the definition of uniform equicontinuity. We introduce the following notations:

$$Z^1 = K(x_1, \delta), \quad Z^2 = K(x_2, \delta) - K(x_2, \delta) \cap Z^1, \quad \ldots$$

$$\ldots, \quad Z^k = K(x_k, \delta) - K(x_k, \delta) \cap \left(\bigcup_{i=1}^{k-1} Z^i\right).$$

We now have

$$X = \bigcup_{i=1}^{n} Z^i, \quad Z^i \cap Z^j = \emptyset \quad \text{for } i \neq j \text{ and } Z^i \subset K(x_i, \delta).$$

Let $\{Z^{i_p}\}$ be the set of all non-empty Z^i. Now, out of each set Z^{i_p} we select one element

$$\overset{p}{x} \in Z^{i_p}, \quad p = 1, 2, \ldots, h \leqslant n.$$

The set $P = \bigcup\limits_{p=1}^{h} F(\overset{p}{x})$ is precompact as the union of a finite number of precompact sets. Let $\{y_1, \ldots, y_m\}$ be the $\frac{1}{4}\varepsilon$-net of the set P. Let T be the (finite) set of all mappings of the set of numbers $\{1, 2, \ldots, h\}$ into the set of numbers $\{1, 2, \ldots, m\}$. The set T has m^h elements $t_s \in T$ ($s = 1, 2, \ldots, m^h$).

We define m^h bounded mappings of the space X into Y:

$$r_s(Z^{i_p}) := y_{t_s(p)} \quad (t_s(p) \in \{1, 2, \ldots, m\}), \quad r_s \in B(X, Y).$$

Next we show that

$$(1) \qquad \bigcup_{s=1}^{m^h} K\left(r_s, \tfrac{1}{2}\varepsilon\right) \supset F, \quad \text{where} \quad K\left(r_s, \tfrac{1}{2}\varepsilon\right) \subset B(X, Y).$$

Let $f \in F$ and take $t_s \in T$ such that

$$f(\overset{p}{x}) \in K\left(y_{t_s(p)}, \tfrac{1}{4}\varepsilon\right).$$

Clearly, $f(\overset{p}{x}) \in P$, and hence lies in one of the balls $K\left(y_i, \tfrac{1}{4}\varepsilon\right)$.

$$\tilde{d}(f, r_s) = \sup_{x \in X} \varrho(f(x), r_s(x)) = \max_{1 \leqslant p \leqslant h}\left(\sup_{x \in Z^{ip}} \varrho(f(x), r_s(x))\right)$$

$$= \max_p\left(\sup_{x \in Z^{ip}} \varrho(f(x), y_{t_s(p)})\right) = \max_p\left(\sup_{x \in Z^i}\left(\varrho(f(x), f(\overset{p}{x})\right) + \right.$$

$$\left. + \varrho(f(\overset{p}{x}), y_{t_s(p)})\right)\right) \leqslant \max_p\left(\tfrac{1}{4}\varepsilon + \tfrac{1}{4}\varepsilon\right) = \tfrac{1}{2}\varepsilon.$$

We have made use here of the fact that $d(x, x') < \delta\left(\tfrac{1}{4}\varepsilon\right)$.

Thus we have proved that for every function $f \in F$, there exists among m^h functions r_s a function such that $f \in K\left(r_s, \tfrac{1}{2}\varepsilon\right)$, which proves the validity of formula (1).

The functions r_s do not yet constitute the ε-net of the set F since they are not elements of F. However, on selecting one element $f_s \in F$ from each ball $K\left(r_s, \tfrac{1}{2}\varepsilon\right)$, for each function lying in $K\left(r_s, \tfrac{1}{2}\varepsilon\right)$ we have

$$d(f, f_s) \leqslant \tilde{d}(f, r_s) + \tilde{d}(r_s, f_s) \leqslant \tfrac{1}{2}\varepsilon + \tfrac{1}{2}\varepsilon = \varepsilon.$$

Consequently, $K\left(r_s, \tfrac{1}{2}\varepsilon\right) \subset K(f_s, \varepsilon)$, wherefrom it immediately follows that

$$\bigcup_{s=1}^{m^h} K(f_s, \varepsilon) \supset F$$

($\{f_s\}$ in an ε-net of the set F). $\square$

When X is a compact space, the equicontinuity of the family F at every point $x \in X$ implies its uniform equicontinuity:

LEMMA XI.1.5. In a compact space the concept of equicontinuity (at all points) and the concept of uniform equicontinuity coincide.

PROOF. (i) The fact that equicontinuity at all points follows from uniform equicontinuity is trivial.

(ii) $X = \bigcup_{x \in X} K\left(x, \frac{1}{2}\delta\left(x, \frac{1}{2}\varepsilon\right)\right)$ (the number $\delta(x, \varepsilon)$—see the definition of equicontinuity). From this covering we select the finite covering

$$X = \bigcup_{i=1}^{n} K\left(x_i, \frac{1}{2}\delta\left(x_i, \frac{1}{2}\varepsilon\right)\right).$$

We take

$$\delta(\varepsilon) := \frac{1}{2} \min_{1 \leqslant i \leqslant n} \left(\delta\left(x_i, \frac{1}{2}\varepsilon\right)\right).$$

Now let $d(x, x') < \delta(\varepsilon)$ and let $f \in F$. There exists a number $i \leqslant n$, such that $x \in K\left(x_i, \frac{1}{2}\delta\left(x_i, \frac{1}{2}\varepsilon\right)\right)$. We have

$$d(x', x_i) \leqslant d(x', x) + d(x, x_i)$$
$$\leqslant \frac{1}{2}\delta\left(x_i, \frac{1}{2}\varepsilon\right) + \frac{1}{2}\delta\left(x_i, \frac{1}{2}\varepsilon\right) = \delta\left(x_i, \frac{1}{2}\varepsilon\right),$$
$$d\big(f(x), f(x')\big) \leqslant d\big(f(x), f(x_i)\big) + d\big(f(x_i), f(x')\big)$$
$$\leqslant \frac{1}{2}\varepsilon + \frac{1}{2}\varepsilon = \varepsilon. \quad \square$$

Hence, in the case of compact spaces for the Ascoli theorem to be valid it is sufficient to require the family F to be equicontinuous at every point of the space X.

Note, moreover, that in the case of a compact space, the converse of the Ascoli theorem holds:

THEOREM XI.1.6.

Assumptions: (X, d) is a compact space, (Y, ϱ) is a metric space, and the set $F \subset C(X, Y)$ is precompact in $B(X, Y)$ (that is, in the sense of uniform convergence).

Thesis: (i) For every $x \in X$ the set $F(x) = \{f(x) : f \in F\}$ is precompact in Y.

(ii) F is a uniformly equicontinuous family.

PROOF. (i) Let $\{f_1, ..., f_n\}$ be an ε-net for F. Let $y \in F(x)$, that is, $y = f(x), f \in F$. Take a function f_i for which $\tilde{d}(f_i, f) = \sup_x \varrho\big(f(x), f_i(x)\big) < \varepsilon$.

Then also $\varrho\big(f(x), f_i(x)\big) < \varepsilon$. Thus the set

$$\{f_1(x), ..., f_n(x)\}$$

in an ε-net for $F(x)$.

(ii) Each of the functions constituting a $\frac{1}{3}\varepsilon$-net $\{f_1, \ldots, f_m\}$ for f is uniformly continuous (being continuous on a compact set)

$$\bigwedge_i \bigwedge_{\varepsilon > 0} \bigvee_{\delta_i(\varepsilon) > 0} \bigwedge_{x, x'} \left(d(x, x') < \delta_i(\varepsilon)\right) \Rightarrow \left(\varrho\left(f_i(x), f_i(x')\right) < \varepsilon\right).$$

We take $\delta(\varepsilon) := \min \delta_i\left(\frac{1}{3}\varepsilon\right)$. Let $f \in F$ and let f_i be that function from the $\frac{1}{3}\varepsilon$-net for which $\tilde{d}(f, f_i) \leqslant \frac{1}{3}\varepsilon$. If $d(x, x') < \delta$, then

$$\varrho\left(f(x), f(x')\right) \leqslant \varrho\left(f(x), f_i(x)\right) + \varrho\left(f_i(x), f_i(x')\right) +$$
$$+ \varrho\left(f_i(x'), f(x')\right) \leqslant \frac{1}{3}\varepsilon + \frac{1}{3}\varepsilon + \frac{1}{3}\varepsilon = \varepsilon. \quad \square$$

Remark. This theorem is valid for any compact (not necessarily metric) space X. How this theorem should be constructed will be shown in Chapter XIV. It turns out that when X is a compact space and Y is a finite-dimensional space, then condition (i) of the Ascoli theorem (or of the thesis of Theorem XI.1.6) can be replaced by another, equivalent one:

LEMMA XI.1.7. If (X, d) is a compact space and the family $F \subset C(X, R^n)$ is equicontinuous, the following conditions are equivalent:

(i) The set $F(x)$ is precompact for every $x \in X$.

(ii) The family F is uniformly bounded, i.e.:

$$\bigvee_M \bigwedge_{x \in X} \bigwedge_{f \in F} \|f(x)\| < M.$$

Remark. The symbol $\| \ \|$ denotes a norm in the space R^n.

PROOF. If condition (ii) holds, all the sets $F(x)$ are bounded in R^n, and hence are precompact.

Next, let condition (i) hold. Since precompact sets are bounded, then for every x there is a number M_x such that for every $f \in F$ we have

$$\|f(x)\| \leqslant M_x.$$

Equicontinuity implies the existence of a constant $\delta(x)$ for which

$$\bigwedge_{f \in F} \bigwedge_{x' \in K(x, \delta(x))} \varrho\left(f(x), f(x')\right) < 1.$$

Therefore

$$\|f(x')\| = \varrho\left(f(x'), 0\right) \leqslant \varrho\left(f(x'), f(x)\right) + \varrho\left(f(x), 0\right) \leqslant 1 + M_x.$$

But $X = \bigcup_{x \in X} K(x, \delta(x))$, and a finite covering can be selected from this covering, since X is compact

$$X = \bigcup_{i=1}^{n} K(x_i, \delta(x_i)).$$

Now let us take $M := \{\max_i M_{x_i}\} + 1$. For any $x \in X$ there exists a $j \in \{1, \ldots, n\}$, such that $x \in K(x_j, \delta(x_j))$ and hence for every $f \in F$ $\|f(x)\| \leqslant M_{x_j} + 1 \leqslant M$. $\square$

Thus, by invoking Theorem XI.1.6, and Lemmas XI.1.5 and XI.1.7, we can give the following formulation of the Ascoli theorem in the case when X is a compact space and Y is a finite-dimensional space:

THEOREM XI.1.8. Let (X, d) be a compact space and $F \subset C(X, \mathbf{R}^n)$. The following conditions are then equivalent:

(i) F is a precompact set in $B(X, \mathbf{R}^n)$.

(ii) F is an equicontinuous and bounded family.

Remark. A different proof of the Ascoli theorem will be given in Chapter XIV.

2. THE STONE–WEIERSTRASS THEOREM. UNIFORM APPROXIMATION OF CONTINUOUS FUNCTIONS ON COMPACT SETS

We introduce the following notation:

If f and g are functions on the metric space (X, d) (that is, f: $X \to \mathbf{R}^1$), then

$$(f \cup g)(x) := \max(f(x), g(x)),$$

$$(f \cap g)(x) := \min(f(x), g(x)).$$

These definitions can be rewritten if use is made of the fact that $|a-b| = a-b$ for $a > b$ and $|a-b| = b-a$ for $a < b$:

$$f \cup g = \tfrac{1}{2}(f+g+|f-g|),$$

$$f \cap g = \tfrac{1}{2}(f+g-|f-g|).$$

If $\mathscr{H} \subset C(X, \mathbf{R}^1) \cap B(X, \mathbf{R}^1)$, the symbol $A(\mathscr{H})$ will be used to denote the closed algebra spanned by elements of the set $\mathscr{H}$, or the closure (in $B(X, \mathbf{R}^1)$, that is, in the topology of uniform convergence) of the

395

set of all polynomials of functions from $\mathscr{H}$. We recall that a polynomial of a function from $\mathscr{H}$ is a linear combination of functions in the form $f_1^{\alpha_1} f_2^{\alpha_2} \ldots f_n^{\alpha_n}$, where $f_i \in \mathscr{H}$, and α_i is a nonnegative integral exponent.

Remark. A constant function (e.g. $f(x) = a$) always belongs to $A(\mathscr{H})$ as a zeroth power of an arbitrary function from $\mathscr{H}$, multiplied by a.

DEFINITION. A family of functions $\mathscr{H}$ *separates the points of the set* X if the proposition

$$\bigwedge_{x, y \in X} \bigvee_{f \in \mathscr{H}} (x \neq y) \Rightarrow \left(f(x) \neq f(y)\right)$$

is true. These definitions can be used to formulate

THEOREM XI.2.1 (M.H. Stone). Let (X, d) be a compact space. Let the family $\mathscr{H} \subset C(X, R^1)$ separate the points of the set X.

Then $A(\mathscr{H}) = C(X, R^1)$.

The thesis of this theorem can be formulated in a different way:

Every continuous function on a compact space X can, with arbitrary accuracy, be uniformly approximated by polynomials of functions from the family $\mathscr{H}$.

Examples. 1. If $X = [a, b] \subset R^1$, and we take one function $f(x) = x$ for the family $\mathscr{H}$, then that function separates the points of the compact set X, whereas its polynomials are ordinary polynomials of the variable x, familiar from algebra, for example. In this case, therefore, the Stone Theorem can be formulated as follows:

Every continuous function on the interval $[a, b]$ can be uniformly approximated by polynomials.

This is the form in which Weierstrass gave this theorem. Hence, the Stone Theorem is a generalization of the classical Weierstrass Theorem.

2. Let $X = R^p$. We take $\mathscr{H} := \{f_1, \ldots, f_p\}$, where f_i is the i-th co-ordinate, that is, $f_i \colon (x_1, \ldots, x_p) \to x_i \in R^1$. Let $\alpha = (\alpha_1, \ldots, \alpha_p)$ be a multiple label. We denote $x^\alpha := x_1^{\alpha_1}, \ldots, x_p^{\alpha_p}$. If $x \neq y$, they must differ by at least one coordinate, that is, one of the functions f_i satisfies the condition

$$f_i(x) \neq f_i(y).$$

Thus, if the set X is compact, every continuous function on X can be uniformly approximated by polynomials of the variables $\{x_1, \ldots \ldots, x_p\}$. Accordingly

$$\bigwedge_f \bigwedge_{\varepsilon>0} \bigvee_{w_\varepsilon} w_\varepsilon(x) = \sum_{|\alpha|\leqslant n(\varepsilon)} a_\alpha^\varepsilon x^\alpha, \quad \tilde{d}(f, w_\varepsilon) \leqslant \varepsilon.$$

By $|\alpha|$ we have denoted the number $\alpha_1 + \alpha_2 + \ldots + \alpha_n$.

Remark. The coefficients of the polynomials w_ε in general depend on ε. They are independent of ε if and only if f is an analytic function. The proof of the Stone Theorem will be broken up into five items:

(i) $(f \in \mathscr{H}) \Rightarrow (|f| \in A(\mathscr{H}))$.

(ii) $(f, g \in A(\mathscr{H})) \Rightarrow (f \cup g, f \cap g \in A(\mathscr{H}))$.

(iii) $(x, y \in X, x \neq y, \alpha, \beta \in R^1) \Rightarrow (\bigvee_{f \in A(\mathscr{H})} f(x) = \alpha, f(y) = \beta)$.

(iv) $\bigwedge_{\varepsilon>0} \bigwedge_{f \in C(X, R^1)} \bigwedge_{x \in X} \bigvee_{g_x \in A(\mathscr{H})} g_x(x) = f(x), \quad g_x \leqslant f + \varepsilon$.

(v) $\bigwedge_{\varepsilon>0} \bigwedge_{f \in C(X, R^1)} \bigvee_{\varphi \in A(\mathscr{H})} f - \varepsilon \leqslant \varphi \leqslant f + \varepsilon$.

Obviously, the final item, stating that

$$|||f - \varphi||| = \tilde{d}(f, \varphi) = \sup_{x \in X} |f(x) - \varphi(x)| \leqslant \varepsilon,$$

is the thesis of the Stone Theorem.

Proof of the Stone Theorem. (i) We prove that on the interval $[0, 1]$ the function $f(t) = \sqrt{t}$ can be uniformly approximated by polynomials of the variable t.

If this were true, the function

$$\frac{|f(x)|}{|||f|||} = \sqrt{\frac{f^2(x)}{|||f|||^2}}$$

could be uniformly approximated by polynomials of the function $f^2/|||f|||^2$. But

$$\sqrt{t} = \sqrt{1 - (1 - t)} = \sum_{n=0}^{\infty} c_k \cdot (1 - t)^k,$$

where $c_k = \dfrac{(2k-3)!!}{(2k)!!}$. If now we assume that $a_0 = 1$, and a_k

$= \dfrac{(2k-1)!!}{(2k)!!}$, we obtain

$$a_{k-1} - a_k = \frac{(2k-3)!!}{(2(k-1))!!} \cdot \frac{2k}{2k} - \frac{(2k-1)!!}{(2k)!!}$$

$$= \frac{(2k-3)!! \cdot (2k \cdot (2k-1))}{(2k)!!} = c_k.$$

Consequently $c_1 + c_2 + \ldots + c_k = a_0 - a_k \leqslant 1$, and thus

$$\sum_{i=1}^{\infty} |c_k| < \infty.$$

Since $|c_k(1-t)| \leqslant |c_k|$ for $t \in [0, 1]$, the series

$$\sum_{i=1}^{\infty} c_k(1-t)^k$$

is also absolutely convergent.

(ii) Clearly, $A\big(A(\mathcal{H})\big) = A(\mathcal{H})$ (the closure of a closed set is equal to the set itself and the closure of an algebra is an algebra). Now $f, g \in A(\mathcal{H})$, and so $(f-g) \in A(\mathcal{H})$, and consequently $|f-g| \in A(\mathcal{H})$ and $f \cup g = \frac{1}{2}(f+g+|f-g|) \in A(\mathcal{H})$. Identically, $f \cap g \in A(\mathcal{H})$.

(iii) Let us take an arbitrary function $g \in \mathcal{H}$ separating the points x and y (that is, $g(x) \neq g(y)$). Taking

$$f(t) := \alpha + \frac{\beta - \alpha}{g(y) - g(x)} \big(g(t) - g(x)\big),$$

we obtain the desired function.

(iv) Let $z \in X$, $f \in C(X, \mathbf{R}^1)$. Let $\alpha = f(x)$, $\beta = f(z)$.

From the previous item of the proof it follows that there exists a function $h_z \in A(\mathcal{H})$ such that $h_z(x) = \alpha = f(x)$, $h_z(z) = \beta = f(z)$. The continuity of the function h_z implies the existence of a number $r_z > 0$ such that

$$\big(y \in K(z, r_z)\big) \Rightarrow \big(h_z(y) \leqslant f(y) + \varepsilon\big).$$

But

$$X = \bigcup_{z \in X} K(z, r_z).$$

Since X is compact, a finite covering can be selected:

$$X = \bigcup_{i=1}^{n} K(z_i, r_{z_i}).$$

Now, taking $A(\mathcal{H}) \ni g_x := h_{z_1} \cap h_{z_2} \cap \ldots \cap h_{z_n}$ we have $g_x(x) = f(x)$, since $h_z(x) = f(x)$ for any z.

For an arbitrary $y \in X$, that is, lying in one of the balls $K(z_i, r_{z_i})$, at least one of the functions h_{z_i} (e.g. labelled j) satisfies the condition $h_{z_j}(y) \leqslant f(y) + \varepsilon$.

Therefore $g_x = \min(h_{z_i}) \leqslant f + \varepsilon$.

(v) Since the function g_x constructed in item (iv) is continuous, there exists a number R_x such that

$$(y \in K(x, R_x)) \Rightarrow (g_x(y) \geqslant f(y) - \varepsilon).$$

But

$$X = \bigcup_{x \in X} K(x, R_x).$$

The compactness of the space X can be used so as to select a finite covering

$$X = \bigcup_{i=1}^{m} K(x_i, R_{x_i}).$$

Taking now $A(\mathscr{H}) \ni \varphi = g_{x_1} \cup g_{x_2} \cup \ldots \cup g_{x_m}$, we conclude that for any $y \in X$ (that is, lying in one of the balls $K(x_i, R_{x_i})$) at least one of the functions g_x (e.g. labelled j) satisfies the condition $g_{x_j}(y) \geqslant f(y) - \varepsilon$, and therefore $\max_i (g_{x_i}(y)) \geqslant f(y) - \varepsilon$. Clearly $g_x \leqslant f + \varepsilon$, and hence

$$f(y) - \varepsilon \leqslant \varphi(y) \leqslant f(y) + \varepsilon. \quad \square$$

Now we give the Stone Theorem for complex functions. For this purpose, we introduce the following notation:

Given a family $\mathscr{H} \subset C(K, C)$,

$$\tilde{\mathscr{H}} := \{h \in C(K, C) \colon \bar{h} \in \mathscr{H}\}, \quad \text{where } \bar{h}(x) := \overline{h(x)}.$$

THEOREM XI.2.2. Let (X, d) be a compact space. Let a family $\mathscr{H} \subset C(X, C)$ separate the points of the set X and let $\mathscr{H} = \tilde{\mathscr{H}}$. Then $A(\mathscr{H}) = C(X, C)$.

PROOF. Since $\mathscr{H} = \tilde{\mathscr{H}}$, the formulae

(1) $\qquad \operatorname{Re} h = \frac{1}{2}(h + \bar{h}) \in A(\mathscr{H})$,

(2) $\qquad \operatorname{Im} h = \frac{1}{2}(h - \bar{h}) \in A(\mathscr{H})$

hold for every function $h \in \mathscr{H}$.

If now we define the families

$$C(X, R^1) \supset \operatorname{Re}\mathscr{H} := \{f \colon \bigvee_{h \in \mathscr{H}} f = \operatorname{Re} h\},$$

$$C(X, R^1) \supset \operatorname{Im}\mathscr{H} := \{f \colon \bigvee_{h \in \mathscr{H}} = \operatorname{Im} h\},$$

then the family $\text{Re}\mathscr{H} \cup \text{Im}\mathscr{H}$ separates the points of the set X (if $h(x) \neq h(y)$, then $\text{Re}h(x) \neq \text{Re}h(y)$ or $\text{Im}h(x) \neq \text{Im}h(y)$).

Now, let $g \in C(X, C)$. Then, by virtue of Stone's Theorem both $\text{Re}g$ and $\text{Im}g$ can be approximated uniformly by polynomials of functions from the family $\text{Re}\mathscr{H} \cup \text{Im}\mathscr{H} \subset \mathscr{H}$. Therefore $\text{Re}g \in A(\mathscr{H})$ and $\text{Im}g \in A(\mathscr{H})$, that is $g = \text{Re}g + i\text{Im}g \in A(\mathscr{H})$.

3. PERIODIC AND ALMOST PERIODIC FUNCTIONS

DEFINITION. A mapping T of a vector space X into a set Y is said to be *periodic with period* $\omega \in X$ if $T(x+\omega) = T(x)$ for every $x \in X$.

By $C_\omega(R, C)$ let us denote a set of continuous periodic functions with period ω, defined on R, with complex values.

By S^1 we denote the unit circle in C, that is, the set of solutions of the equation $|z| = 1$.

The following lemma holds:

LEMMA XI.3.1. There exists an isometric isomorphism of Banach spaces:

$$C_\omega(R, C) \subsetneq B(R, C) \quad \text{and} \quad C(S^1, C) \subset B(S^1, C).$$

PROOF. Let

$$z = e^{\frac{2\pi it}{\omega}}, \quad t = \frac{\omega}{2\pi i}\log z, \quad \text{where } i = \sqrt{-1}.$$

If $t \in R^1$, then $|z| = 1$ and conversely.

We define the isomorphism J as

$$(J\varphi)(t) := \varphi(e^{\frac{2\pi it}{\omega}}) \in C_\omega(R, C), \quad \text{when } \varphi \in C(S^1, C).$$

It is readily seen that

$$(J^{-1}f)(z) = f\left(\frac{\omega}{2\pi i}\log z\right) \in C(S^1, C), \quad \text{when } f \in C_\omega(R, C).$$

Since f is periodic, the function $(J^{-1}f)$ is well-defined on S^1, even though $\log z$ is defined up to $2\pi i$ (see Chapter XIV).

The mapping J is an isometry since

$$|||(J\varphi)||| = \sup_{t \in R^1}|\varphi(e^{\frac{2\pi it}{\omega}})| = \sup_{z \in S^1}|\varphi(z)| = |||\varphi|||. \quad \square$$

Note that the two-element family of functions

$$C(S^1, C) \supset \mathscr{H} = \{f_1(z) = z, f_2(z) = \bar{z}\}$$

satisfies the hypotheses of Theorem XI.2.2. Therefore $C(S^1, C) = A(\mathscr{H})$. Invoking Lemma XI.3.1, we have immediately

THEOREM XI.3.2. The formula

$$C_\omega(R, C) = A(\{e^{\frac{2\pi i t}{\omega}}, e^{\frac{-2\pi i t}{\omega}}\})$$

holds.

PROOF. Let $f \in C_\omega(R, C)$. Then there exists a function $\varphi \in C(S^1, C)$ such that $J\varphi = f$. But φ can be uniformly approximated by polynomials of the functions f_1 and f_2. Therefore, by virtue of Lemma XI.3.1, f can be uniformly approximated by polynomials of the functions Jf_1 and Jf_2. But

$$(Jf_1)(t) = e^{\frac{2\pi i t}{\omega}}, \quad (Jf_2)(t) = e^{\frac{-2\pi i t}{\omega}},$$

and this completes the proof. $\square$

This theorem can be formulated as follows:

Every continuous periodic function with period ω on R can be uniformly approximated by functions of the form

$$f_\varepsilon(t) = \sum_{k=-n(\varepsilon)}^{n(\varepsilon)} a_k^\varepsilon \cdot e^{\frac{2\pi i k t}{\omega}}, \quad \text{where } |||f - f_\varepsilon||| < \varepsilon.$$

We now prove an important property of periodic functions. For this purpose we introduce the following notations: Let $f \in C(R, C)$, $h \in R$; then

$$(\tau_h f)(x) := f(x - h).$$

LEMMA XI.3.3. If a function $f \in C(R, C)$ is periodic on R (with period ω), the family of functions

$$F := \{\tau_h f : h \in R\}$$

is a (pre)compact set in $B(R, C)$.

PROOF. The set $F \subset B(R, C)$, being a continuous image of an open interval $[0, \omega]$ in the metric space $B(R, C)$, is compact. $\square$

DEFINITION. A function $f \in C(R, C)$ is said to be *almost periodic* (a.p.) if the set $\{\tau_h f : h \in R\}$ is precompact in the space $B(R, C)$.

401

COROLLARY XI.3.4. A periodic function is a.p.

As is known, periodic functions do not form an algebra: the sum or product of periodic functions cannot be a periodic function if their periods ω_1 and ω_2 are not commensurable.

It turns out that the set of periodic functions is made into an algebra by the adjunction of all other a.p. functions. This is stated by the following theorem:

THEOREM XI.3.5. Let $f, g \in C(R, C)$ be a.p. functions. Let $a \in R$. Then the functions

   (i) af,

   (ii) $f+g$,

   (iii) $f \cdot g$,

are a.p. functions.

PROOF. (i) The straightforward proof is left to the reader.

(ii) Let $\{\tau_{h_1}f, \tau_{h_2}f, ..., \tau_{h_n}f\}$ be a $\frac{1}{4}\varepsilon$-net of the set $\{\tau_h f\}$, and $(\tau_{k_1}g, \tau_{k_2}g, ..., \tau_{k_m}g)$ a $\frac{1}{4}\varepsilon$-net of the set $\{\tau_k g\}$.

Take any function of the family $\{\tau_r(f+g)\}$,

$$\tau_r(f+g) = \tau_r f + \tau_r g.$$

There exist indices (i, j) such that

$$|||\tau_r f - \tau_{h_i}f||| \leqslant \frac{1}{4}\varepsilon, \quad |||\tau_r g - \tau_{k_j}g||| \leqslant \frac{1}{4}\varepsilon.$$

Hence $|||\tau_r(f+g) - (\tau_{h_i}f + \tau_{k_j}g)||| < \frac{1}{2}\varepsilon$.

Denoting $\tau_{h_i}f + \tau_{k_j}g := u_{ij}$, we can write what has been proved in the following form:

$$\{\tau_r(f+g): r \in R\} \subset \bigcup_{i=1}^{n} \bigcup_{j=1}^{m} K\left(u_{ij}, \tfrac{1}{2}\varepsilon\right).$$

Now, if one function $w_{ij} = \tau_{r_{ij}}(f+g) \in K\left(u_{ij}, \tfrac{1}{2}\varepsilon\right)$, is selected from each ball $K\left(u_{ij}, \tfrac{1}{2}\varepsilon\right)$, then for every $r \in R$ there exist indices (i, j) such that

$$\tau_r(f+g) \in K\left(u_{ij}, \tfrac{1}{2}\varepsilon\right) \subset K(w_{ij}, \varepsilon),$$

since

$$|||\tau_r(f+g) - w_{ij}||| \leqslant |||\tau_r(f+g) - u_{ij}||| + |||u_{ij} - w_{ij}||| \leqslant \varepsilon.$$

Thus, $\{w_{ij}\}$ is the desired ε-net.

(iii) The functions f and g are bounded. Take $M = \max(|||f|||, |||g|||)$.

Let $\{\tau_{h_i}f, ..., \tau_{h_n}f\}$ be an $\dfrac{\varepsilon}{4M}$-net of the set $\{\tau_h f\}$, and $\{\tau_{k_1}g, ...$

$..., \tau_{k_m}g\}$ an $\dfrac{\varepsilon}{4M}$-net of the set $\{\tau_k g\}$.

Take an arbitrary function from the set $\{\tau_r(f \cdot g)\}$. There exist indices (i, j) such that

$$|||\tau_r f - \tau_{k_i}f||| \leqslant \varepsilon/4M, \qquad |||\tau_r g - \tau_{k_j}g||| \leqslant \varepsilon/4M.$$

But $\tau_r(f \cdot g) = (\tau_r f) \cdot (\tau_r g)$, and therefore

$$|||(\tau_r f)(\tau_r g) - (\tau_{h_i} f)(\tau_{k_j}g)|||$$
$$\leqslant |||(\tau_r f)(\tau_r g) - (\tau_{h_i} f)(\tau_r g)||| +$$
$$+ |||(\tau_{h_i} f)(\tau_r g) - (\tau_{h_i} f)(\tau_{k_j}g)|||$$
$$\leqslant |||\tau_r g||| \cdot |||\tau_r f - \tau_{h_i} f||| + |||\tau_{h_i} f||| \cdot |||\tau_r g - \tau_{k_j}g|||$$
$$\leqslant M\frac{\varepsilon}{4M} + M\frac{\varepsilon}{4M} = \frac{\varepsilon}{2}.$$

Setting $u_{ij} := (\tau_{h_i} f)(\tau_{k_j}g)$ we have

$$\{\tau_r(f \cdot g): r \in R\} \subset \bigcup_{i=1}^{n} \bigcup_{j=1}^{m} K(u_{ij}, \varepsilon/2).$$

The rest of the proof is as in (ii). $\square$

The following theorem holds:

THEOREM XI.3.6 (Fundamental Theorem on A. P. Functions). The set of a.p. functions is the smallest algebra spanned by the set of periodic functions.

The proof of this theorem is not given here (cf. K. Maurin, *Methods of Hilbert Spaces*, Warszawa 1967).

APPENDIX. INTEGRATION OF RATIONAL FUNCTIONS

1. INTEGRATION OF RATIONAL FUNCTIONS

For many years the conviction was current that every integral of elementary functions could be expressed by elementary functions. Liouville was the first to show that this is so only in "exceptional" cases. Rational functions are among them. The theorems given below on rational functions (on the existence of roots of a polynomial and on the decomposition of a rational function into simple fractions) will be proved in Part II of this book, for that is where they actually belong. At this point, we give these theorems and show how they are used in evaluating the primitive functions.

DEFINITION. A *rational function* is a function of the form $f(x) = \dfrac{P_n(x)}{R_m(x)}$, where $P_n(x)$ is a polynomial of degree n, and $R_m(x)$ is a polynomial of degree m. Clearly, $f(x)$ is a continuous function, apart from points for which $R_m(x) = 0$.

Several theorems concerning the general properties of polynomials will be given without proof.

THEOREM A.1 (The Fundamental Theorem of Algebra). Every polynomial of degree n with complex coefficients,

$$P_n(z) = \sum_{k=0}^{n} a_k z^k, \quad a_k \in C^1,$$

has n complex roots.

This should be understood as follows:

There exist complex numbers $\alpha_1, \alpha_2, \ldots, \alpha_n$ (not necessarily different) such that

$$P_n(z) = a_n(z - \alpha_1)(z - \alpha_2) \ldots (z - \alpha_n).$$

Suppose that $P_n(z)$ is a polynomial with real coefficients (that is, $\bar{a}_i = a_i$). Suppose further that α is a root of this polynomial. Then $\bar{\alpha}$ is also a root of this polynomial since

$$P_n(\bar{\alpha}) = a_0 + a_1 \bar{\alpha} + a_2 \bar{\alpha}^2 + \ldots + a_n \bar{\alpha}^n$$
$$= \overline{\alpha_0} + \overline{a_1 \alpha} + \overline{a_2 \alpha^2} + \ldots + \overline{a_n \alpha^n} = \overline{P_n(\alpha)} = \bar{0} = 0.$$

Thus, if a factor $z - \alpha$ appears in the factorization of a polynomial with real coefficients, then $z - \bar{\alpha}$ also appears. Note that $(z - \alpha)(z - \bar{\alpha})$

$= z^2 - z(\alpha + \overline{\alpha}) + \alpha\overline{\alpha}$. But $\alpha + \overline{\alpha} = 2\,\mathrm{Re}\,\alpha \in R$; $\alpha\overline{\alpha} = |\alpha|^2 \in R$. Hence a polynomial with real coefficients decomposes into the product of factors $z - \alpha$ (where α is real) and quadratic trinomials $z^2 + bz + c$ (where b and c are real).

THEOREM A.2 (On Decomposition into Simple Fractions).

Hypothesis: Let $f(z) = \dfrac{P_n(z)}{R_m(z)}$ be a rational function of a complex variable; let

$$R_m(z) = (z - \alpha_1)^{n_1}(z - \alpha_2)^{n_2} \ldots (z - \alpha_i)^{n_i},$$

where $\displaystyle\sum_{j=1}^{i} n_j = m$.

Thesis: There exist complex numbers A_i $(i = 1, 2, \ldots)$ such that

$$f(z) = Q_{n-m}(z) + \frac{A_1}{z - \alpha_1} + \frac{A_2}{(z - \alpha_1)^2} + \ldots + \frac{A_{n_1}}{(z - \alpha_1)^{n_1}} +$$

$$+ \frac{A_{n_1+1}}{(z - \alpha_2)} + \frac{A_{n_1+2}}{(z - \alpha_2)^2} + \ldots + \frac{A_{n_1+n_2}}{(z - \alpha_2)^{n_2}} +$$

$$+ \ldots + \frac{A_m}{(z - \alpha_i)^{n_i}},$$

where $Q_{n-m}(z)$ is a polynomial of degree $n - m$ (if $n < m$, then $Q_{n-m} = 0$).

As it turns out, for polynomials with real coefficients the terms corresponding to the roots α and $\overline{\alpha}$ can be collected and rearranged in the simple form:

$$\frac{A_\alpha}{(z - \alpha)^n} + \frac{A_{\overline{\alpha}}}{(z - \overline{\alpha})^n} = \frac{Bz + D}{(z^2 - z(\alpha + \overline{\alpha}) + \alpha\overline{\alpha})^n}.$$

Hence, real-valued rational functions

$$f(x) = \frac{P_n(x)}{R_m(x)},$$

$$R_m(x) = (x - \alpha_1)^{n_1} \ldots (x - \alpha_i)^{n_i}(x^2 + b_1 x + c_1)^{k_1}$$
$$\ldots (x^2 + b_s x + c_s)^{k_s},$$

where $\displaystyle\sum_{j=1}^{i} n_j + 2\sum_{j=1}^{s} k_j = m$ $(\alpha_i, b_i, c_i$ are real), are decomposable as follows:

$$f(x) = Q_{n-m}(x) + \frac{A_1}{(x - \alpha_1)} + \ldots + \frac{A_n}{(x - \alpha_1)^{n_1}} +$$

$$+ \ldots + \frac{B_1 x + C_1}{(x^2 + b_1 x + c_1)} + \frac{B_2 x + C_2}{(x^2 + b_1 x + c_1)^2} +$$

$$+ \ldots + \frac{B_{k_1} x + C_{k_1}}{(x^2 + b_1 x + c_1)^{k_1}} + \ldots + \frac{B_{k_s} x + C_{k_s}}{(x^2 + b_s x + c_s)^{k_s}}.$$

To evaluate the integral of a rational function, we decompose it into simple fractions. We must know how to integrate the following types of integrals in this form:

I.

$$\int \frac{dx}{(x-a)^k} = \begin{cases} \log|x-a| & \text{for } k = 1, \\ \dfrac{1}{-(k-1)(x-a)^{k-1}} & \text{for } k > 1. \end{cases}$$

II. Integrals of the expressions $\dfrac{Bx + C}{(x^2 + bx + c)^{k-1}}$ are evaluated in the following manner:

$$x^2 + bx + c = x^2 + bx + \left(\frac{b}{2}\right)^2 - \left(\frac{b}{2}\right)^2 + c$$

$$= \left(x + \frac{b}{2}\right)^2 + \frac{4c - b^2}{4} = \left[\left(\frac{2x+b}{\sqrt{-\Delta}}\right)^2 + 1\right] \cdot \frac{(-\Delta)}{4}$$

(where $4c - b^2 = -\Delta > 0$). Setting $(2x+b)/\sqrt{-\Delta} = t$, we reduce the problem to one of calculating integrals of the type

$$\int \frac{Dt + E}{(t^2 + 1)^n} \, dt.$$

Examples.

1. $\displaystyle\int \frac{t \, dt}{(t^2 + 1)^n} = \begin{cases} \dfrac{1}{-2(n-1)} \dfrac{1}{(t^2+1)^{n-1}} & \text{for } n > 1, \\ \frac{1}{2} \log(t^2 + 1) & \text{for } n = 1, \end{cases}$

2. $\displaystyle\int \frac{dt}{(1 + t^2)^n} = \begin{cases} \arctan t & \text{for } n = 1, \\ \displaystyle\int \frac{dt}{(1+t^2)^{n-1}} - \int \frac{t^2 \, dt}{(1+t^2)^n} & \text{for } n > 1. \end{cases}$

The second integral above is evaluated by parts:

$$\int \frac{t^2 \, dt}{(1 + t^2)^n} = \int \frac{t}{(1+t^2)^n} \, t \, dt$$

$$= -\frac{t}{2(n-1)(1+t^2)^{n-1}} + \frac{1}{2(n-1)} \int \frac{dt}{(1+t^2)^{n-1}}.$$

Thus we have obtained a recursion formula which, after $n-1$ steps, allows the problem to be brought down to that of evaluating the integral

$$\int \frac{dt}{(1+t^2)} = \arctan t.$$

2. IMPORTANT SUBSTITUTIONS, INTEGRALS, FUNCTIONS, AND SERIES

Table of Important Substitutions

By $R(\cdot)$ we denote a rational function of the expressions written in the parentheses.

Integral	Substitution
(1) $\int R(\sin x, \cos x, \tan x)\,dx$	$t = \tan \frac{1}{2} x$
(2) $\int R(\sinh x, \cosh x, \tanh x)\,dx$	$t = \tanh \frac{1}{2} x$
(3) $\int R\left(x, \sqrt{\dfrac{ax+b}{cx+d}}\right) dx$	$t = \sqrt{\dfrac{ax+b}{cx+d}}$
(4) $\int R(x, \sqrt{ax^2+bx+c})\,dx$	
bringing the trinomial $ax^2 + +bx+c$ to canonical form, we obtain one of three possibilities:	
(a) $\Delta < 0,\ a > 0, \Delta = b^2 - 4ac$	
$\int R(x, \sqrt{x^2+\alpha^2})\,dx$	$x = \alpha \sinh t$
(b) $\Delta > 0,\ a > 0$	
$\int R(x, \sqrt{x^2-\alpha^2})\,dx$	$x = \alpha \cosh t$
(c) $\Delta > 0,\ a < 0$	
$\int R(x, \sqrt{\alpha^2-x^2})\,dx$	$x = \alpha \sin t$ or $x = \alpha \cos t$.

Table of Primitive Functions and Important Integrals

$f(x)$	$F(x)+$const, where $F' = f$

$$\frac{1}{x^2+2bx+c}$$

$$\begin{cases} \dfrac{1}{\sqrt{c-b^2}}\arctan\dfrac{x+b}{\sqrt{c-b^2}} & \text{for } c > b^2, \\[3ex] \dfrac{1}{2\sqrt{b^2-c}}\log\dfrac{x+b-\sqrt{b^2-c}}{x+b+\sqrt{b^2-c}} & \text{for } c < b^2 \end{cases}$$

$$\frac{Ax+B}{(x^2+2bx+c)^n}$$

$$\frac{-A}{2(n-1)(x^2+2bx+c)^{n-1}} -$$

$$-\frac{Ab-B}{(c-b^2)^{n-1,2}}\int\frac{dt}{(1+t)^{n-1}},$$

$$t := (x+b)/\sqrt{c-b^2}, \quad c > b^2$$

$$\frac{1}{(1+x^2)^n}$$

$$\frac{x}{2(n-1)(1+x^2)^{n-1}} +$$

$$+\frac{2n-3}{2n-2}\int\frac{dx}{(1+x^2)^{n-1}} \quad \text{for } n \neq 1$$

$$\frac{1}{\sqrt{ax^2+2bx+c}}$$

$$\frac{1}{\sqrt{a}}\log(b+ax+\sqrt{a}\,\sqrt{ax^2+2bx+c}) \quad \text{for } a > 0,$$

$$b^2-ac \neq 0$$

$$\frac{1}{\sqrt{-ax^2+2bx+c}}$$

$$\frac{1}{\sqrt{a}}\arcsin\frac{ax-b}{\sqrt{b^2+ac}} \quad \text{for } a > 0,\ b^2+ac \neq 0$$

$$\int(\sin x)^m(\cos x)^n = -\frac{(\sin x)^{m-1}(\cos x)^{n+1}}{m+n} +$$

$$+\frac{m-1}{m+n}\int(\sin x)^{m-2}(\cos x)^n dx = \frac{(\sin x)^{m+1}(\cos x)^{n-1}}{m+n} +$$

$$+\frac{n-1}{m+n}\int(\sin x)^m(\cos x)^{n-2}dx,\ m+n \neq 0$$

Functions Defined as Integrals with a Parameter

$$\Gamma(x+1) = \Pi(x) := \int_0^\infty e^{-t}t^x dt;$$

function equations:

$$\Gamma(z+1) = z\Gamma(z),$$

$$\Pi(z+1) = (z+1)\Pi(z), \quad \Gamma(z)\Gamma(1-z) = \frac{\pi}{\sin \pi z}.$$

$$B(x,y) := \int_0^1 t^x(1-t)^y dt = 2\int_0^1 t^{2x+1}(1-t^2)^y dt$$

$$= 2\int_0^{\pi/2} (\sin t)^{2x+1}(\cos t)^{2y+1} dt = \frac{\Pi(x)\Pi(y)}{\Pi(x+y+1)},$$

$$\frac{\pi}{\sin(\pi x)} = \int_0^\infty \frac{t^{x-1}}{1+t} dt, \quad 0 < x < 1.$$

Important Integrals

$$\int_0^{\pi/2} (\sin x)^n (\cos x)^m dx$$

$$= \begin{cases} \left(\dfrac{(n-1)!!(m-1)!!}{(n+m)!!}\right) \cdot \dfrac{\pi}{2} & \text{even } n, m, \\[3mm] \dfrac{(n-1)!!(m-1)!!}{(n+m)!!} & \text{all other cases.} \end{cases}$$

$$\int_0^\infty e^{-x^2} dx = \tfrac{1}{2}\sqrt{\pi};$$

this formula is generalized to R^n as follows:

$$\int_{R^n} e^{-\frac{1}{2}(Cx|x)} dx = (2\pi)^{n/2}(\det C)^{-1/2},$$

where $C \in L(R^n)$ is positive;

$$\left(\frac{\det C}{(2\pi)^n}\right)^{1/2} \int_{R^n} (Ax|x) e^{-\frac{1}{2}(Cx|x)} dx = \operatorname{tr}(AC^{-1}),$$

$$A \in L(R^n), \quad (Cx|x) \geqslant a\|x\|^2, \quad a > 0.$$

Power Series

Binomial coefficients:

$$\binom{x}{m} := \frac{x(x-1)\ldots(x-m+1)}{1\cdot 2\cdot \ldots \cdot m},$$

$$\binom{n}{k} = \frac{n!}{k!(n-k)!} = \binom{n}{n-k}, \quad n, k-\text{natural.}$$

412

Bernoulli numbers B_k are defined as the coefficients of the series

$$\frac{x}{e^x-1} =: \sum_{k=0}^{\infty} B_k \frac{x^k}{k!},$$

and hence $B_0 = 1$, $B_1 = -\frac{1}{2}$, $B_2 = \frac{1}{6}$, $B_{2k+1} = 0$ for $k = 1, 2, \ldots$
$B_4 = -\frac{1}{30}$, $B_6 = \frac{1}{42}$, etc.

$$\sum_{n=0}^{\infty} \frac{1}{n^{2k}} = \frac{(-1)^n (2\pi)^{2k} B_{2k}}{2(2k)!}, \qquad k = 1, 2, \ldots,$$

$$\frac{1}{1-x} = \sum_{k=0}^{\infty} x^k, \qquad |x| < 1,$$

$$(1+x)^a = \sum_{k=0}^{\infty} \binom{a}{k} x^k, \qquad |x| < 1.$$

$$e^z := \sum_{n=0}^{\infty} \frac{z^n}{n!}, \qquad z \in C,$$

$$a^x = \exp(x \log a) = \sum_{k=0}^{\infty} \frac{x^n \log a}{k!},$$

$$\log(1+x) = x - \frac{x^2}{2} + \frac{x^3}{3} - \frac{x^4}{4} + \cdots,$$
$$|x| \leqslant 1, \ x \neq -1,$$

$$\frac{1}{2} \log\left(\frac{1+x}{1-x}\right) = \sum_{n=0}^{\infty} \frac{x^{2n+1}}{2n+1}, \qquad |x| \leqslant 1, \ x \neq \pm 1,$$

$$\tan x = \sum_{n=0}^{\infty} \frac{2^{2n}(2^{2n}-1) B_{2n}}{(2n)!} x^{2n-1}, \qquad |x| < \frac{1}{2}\pi,$$

$$\arctan x = x - \frac{x^3}{3} + \frac{x^5}{5} - \cdots, \qquad |x| \leqslant 1, \ x \neq \pm i,$$

$$\arcsin x = x + \sum_{n=1}^{\infty} \frac{(2n-1)!!}{(2n)!!} \cdot \frac{x^{2n+1}}{(2n+1)}, \qquad |x| < 1.$$

INDEX OF SYMBOLS

$\{x: \mathscr{P}\}$ — the set of all x which possess the property $\mathscr{P}$ 3

$\left.\begin{array}{l} \vee, \wedge, \bigvee\limits_{x} , \bigwedge\limits_{x} \\ \neg, \Rightarrow, \Leftrightarrow \end{array}\right\}$ — logical symbols 3

$a := b$ — a is equal to b by definition 3

$a \in A \ (a \notin A)$, $A \ni a \ (A \not\ni a)$ — element a belongs (does not belong) to the set A 4

$A \cap B$ — the intersection (common part) of sets A and B 4

$A \cup B$ — the union of sets A and B 4

$\bigcap\limits_{i \in I} A_i$ — the intersection of sets A_i 5

$\bigcup\limits_{i \in I} A_i$ — the union of sets A_i 5

$\varnothing$ — the null set 5

$A \subset B, B \supset A$ — A is included in B 5

$\complement A$ — the complement of the set A 6

(a, b) — the ordered pair of elements a and b 6

$\{a, b\}$ — the two-element set 6

$A \times B$ — the Cartesian product of sets A and B 6

$\bigtimes\limits_{i=1}^{n} A_i$ — the Cartesian product of sets A_i 6

$\mathscr{R}, \mathscr{R}^{-1}$ — a relation and its inverse relation 7

$\mathscr{R} \circ \mathscr{S}$ — the composition of relations $\mathscr{R}$ and $\mathscr{S}$ 7

$P_X \mathscr{R}$ — set of elements $x \in X$ which are in the relation $\mathscr{R}$ with an element 7

$\mathrm{id}_X, \mathrm{id}$ — the identity relation (equality) 7

$T: X \to Y$ — the mapping of the set X into the set Y 8

$T(x) = Tx$ — 8

$T(A)$ — the image of the set A 9

$T^{-1}(A)$ — the inverse image of the set A 9

$T_1 = T|X_1$ — T_1 is the restriction of the map T to the subset X_1 9

$S \circ T$ — the composition of maps S and T 10

$(A_i)_{i \in I}$ — a family of sets A_i 10

I — a set of indices 10

N — the set of natural numbers 11

(x_n) — a sequence of elements 11

$\left.\begin{array}{l} (x_{nm}), \ (x_{nm})_{n \in N, m \in N} \\ (x_{nm}) \ (n, m \in N) \end{array}\right\}$ — a double sequence 11

$X/\mathcal{R}$, X/F — a quotient space 12

A_x, $[x]$, $[x]_\mathcal{R}$ — the equivalence classes 12

$x \overset{\mathcal{R}}{\sim} y$, $x \sim y$ — x and y are in equivalence relation 14

$\mathcal{R}_A$ — the relation induced by $\mathcal{R}$ on the set A 15

$\mathcal{F}(X, Y)$, Y^X — the set of all mappings $F: X \to Y$ 15

$\underset{i \in I}{\times} T_i$, $\prod_{i \in I} T_i$ — the product of mappings T_i 17

$(S_1, \ldots, S_n)$ — a family of mappings for a finite set of indices 17

$(X, \leqslant)$ — an order, ordered space, ordered set, the chain 18

R or R^1 — the space of real numbers 20

R^n — the n-dimensional Euclidean space 20

$\mathscr{C}$ — the space of Cauchy sequences 20

Q — the set of rational numbers 20

$\widetilde{\mathscr{P}}$ — the complemented space $\mathscr{P}$ 20

$\mathcal{F}(N, Q)$ — the set of sequences of rational numbers 21

$|a|$ — the absolute values of the number a 21

$[a, b]$, $]a, b[$, $[a, b[$, $]a, b]$, $]-\infty, a]$, $]-\infty, a[$, $[a, \infty[$, $]a, \infty[$ — intervals 25

$\left.\begin{array}{l} \lim\limits_{n \to \infty} x_n = x, \\ x_n \to x,\ n \to \infty \end{array}\right\}$ — the sequence (x_n) has a limit (converges to) x 25

R_+ — the set of nonnegative real numbers 33

d, $d(x, y)$ — the distance between points x and y; a metric 33

(X, d) — a metric space with metric d 33

$K(x_0, r)$ — a ball of radius r and centre x_0 35

$\sup A$ — the least upper bound (supremum) of the set A 35

$\inf A$ — the greatest lower bound (infimum) of the set A 35

$\mathscr{T}$ — the topology of a metric space 37

$\operatorname{int} A$, $\mathring{A}$ — the interior of a set A 38

$\mathcal{O}(x)$ — the neighbourhood of a point x 38

$\overline{K}(x_0, r)$ — a closed ball 39

$\overline{A}$ — the closure of a set A 40

$T = \lim\limits_{x \to x_0} T(x)$ — T is the limit of sequence of mappings $T(x)$ 42

$f'(x_0)$, $f^{(1)}(x_0)$, $\dfrac{df(x_0)}{dx}$, $\dfrac{df}{dx}(x_0)$, $\dfrac{d}{dx}f(x_0)$ — the derivative of the function f at the point x_0 58

$df(x_0, x - x_0)$ — the differential of the function f at the point x_0 58

$\prec$ — the direction of the set 64

$(\Pi, \prec)$ — the directed set 64

$\pi_1 \prec \pi_2$ — the element π_1 is earlier than the element π_2 (π_2 is later than π_1) 64

$(x_\pi)_{\pi \in \Pi}$, (x_π) — a net 65

$\left.\begin{array}{l} \limsup\limits_{\pi \in \Pi} f(\pi) \\ \liminf\limits_{\pi \in \Pi} f(\pi) \end{array}\right\}$ — the upper (lower) limit of the function f with respect to the filtration set Π (limit superior, limit inferior) 66

417

$(f \cup g)(x), f \cup g$ — $\max(f(x), g(x))$ 395
$(f \cap g)(x), f \cap g$ — $\min(f(x), g(x))$ 395
$A(\mathscr{H})$ — the closed algebra spanned by elements of the set $\mathscr{H}$ 395
$\tilde{\mathscr{H}}$ — 399
$C_\omega(R, C)$ — the set of continuous periodic functions with period ω 400
S^1 — the unit circle in C 400

SUBJECT INDEX

Adjunction of union and intersection of sets 4
Associativity of union and intersection of sets 4
Axiom of choice 19
   of maximality 19

Ball 35
   , closed 39
Bijection 9
Boundedness, total 389

Centre of gravity 87
Chain 18
Characteristics 352
Circle of convergence of a power series 119
Closure of a set 40
Common part (intersection) of sets 4
Commutativity of union and intersection of sets 4
Complement, algebraic 247
   of a set 6
Completeness of a space 26
Composition of relations 7
   of maps 10
Condition, Frobenius 358
   , necessary, for the existence of an extremum 105
   , sufficient, for the existence of a local extremum 105
Cone 93
Constant, Euler's 146
   , Lipschitz 267

Contraction 227
Convergence, uniform, of a sequence of mappings 111
Coordinates of a pair 6
Cosine, hyperbolic 81
Covector 159
Covering of a set 48
   , open 48
   , closed 48
Curvature of a curve 375, 378
Curve, characteristic 352
   , degenerate 376
   , oriented 368

Data, Cauchy 349
   , Cauchy, non-characteristic 353
   , Cauchy, regular 351
Derivative, directional 162
   , directional, of a mapping at a point 202, 208
   , Gateaux 208
   , Gateaux, of order p at a point 213
   , Gateaux, of order p on a set 214
   , left-hand, of a mapping at a point 223
   , mixed partial 193
   of a function at a point 58, 124
   of a mapping at a point 161, 222
   of a mapping on a set 161
   of order p, partial 196
   , partial, at a point 178
   , partial, of a mapping 178
   , partial, of a mapping at a point 177
   , p-th, of a mapping 189, 222

421

SUBJECT INDEX

NAME INDEX

Ascoli, G. 296, 391

Jacobi, C. G. J. 165, 356

Bachmann, 109
Banach, S. 150, 228, 237
Bolzano, B. 47
Borel, E. 48
Bourbaki, N. 6, 33, 159

Klein, F. 76
Kuratowski, K. 6, 18, 40

Cantor, G. 20
Cauchy, A. 20, 21, 26, 101, 120, 144, 267,
    294, 348, 349
Coulomb, C. A. 171

Lagrange, J. L. 101, 220, 221, 252, 255
Landau, E. 109
Laplace, P. S. 266
Lebesgue, H. L. 48
Leibniz, G. W. 57, 118
Liouville, J. 316, 407
Lipschitz, R. 267
Lusternik, L. A. 248, 251

Dieudonné, J. 357, 362
Dini, U. 114

Maclaurin, C. 144
Minkowski, H. 85, 92
Euler, L. 127, 143, 146, 171, 220, 221, 243
Monge, G. 349
De Morgan, A. 3

Farkas, 94
Fréchet, R. M. 33, 162
Frenet, F. 375
Frobenius, F. G. 358, 362

Neumann, J. von 232
Newton, I. 57

Peano, G. 296, 297
Gateaux, 163, 208, 213
Pythagoras 128
Gram, J. P. 372
Graves, L.M. 241

Radon, J. 89
Riemann, B. 57, 69
Hadamard, J. 120
Rolle, M. 60
Hamilton, W. 266, 348, 356
Hausdorff, F. 20, 33, 38, 387
Hilbert, D. 85, 169, 170
Schauder, J. 296
L'Hospital, G. F. A. 106
Schlömilch, O. 101

429